Tropentag 2020

International Conference on Research on Food Security, Natural Resource Management and Rural Development

AF592440

Tropentag 2020

International Research on Food Security, Natural Resource Management and Rural Development

Food and nutrition security and its resilience to global crises

Book of abstracts

Editor: Eric Tielkes

Reviewers/scientific committee: Folkard Asch, Jan Banout, Mathias Becker, Michael Brüntrup, Mizeck Chagunda, Falko Feldmann, Ingo Grass, Ulrike Grote, Bettina Haussmann, Christian Hülsebusch, Irmgard Jordan, Jana Juhrbandt, Gudrun B. Keding, Patrick Kolsteren, Bohdan Lojka, Eike Luedeling, Joachim Müller, Thilo Rennert, Antje Risius, Hermann Waibel, Florian Wichern

Editorial assistance: Esther Asare

Impressum

Bibliografische Information der Deutschen Nationalbibliothek
Die Deutsche Nationalbibliothek verzeichnet diese Publikation in der Deutschen Nationalbibliografie; detaillierte bibliografische Daten sind im Internet über http://dnb.ddb.de abrufbar.

Tropentag 2020: Food and nutrition security and its resilience to global crises, Tielkes, E. (ed.) - Witzenhausen, DITSL

© CUVILLIER VERLAG, Göttingen
Nonnenstieg 8, 37075 Göttingen
Telefon: 0551-54724-0
Telefax: 0551-54724-21
http://www.cuvillier.de

Alle Rechte vorbehalten. Ohne ausdrückliche Genehmigung des Verlages ist es nicht gestattet, das Buch oder Teile daraus auf fotomechanischem Weg (Fotokopie, Mikrokopie) zu vervielfältigen.

The authors of the articles are solely responsible for the content of their contribution.
All rights reserved. No part of this publication may be reproduced, stored in a retrieval system, or transmitted in any form or by any means without prior permission of the copyright owners.

ISBN: 978-3-7369-7572-9
eISBN: 978-3-7369-6572-0

Online-Version: https://www.tropentag.de/

Preface

Tropentag is the largest interdisciplinary conference in Europe on development oriented research in the fields of sub-/tropical agriculture, food security, natural resource management and rural development. Taking place annually, *Tropentag* 2020 turned out to be a special challenge. Originally planned to take place in Prague, the Corona pandemic did not allow presence in or travel to Prague for prospective participants. ATSAF took on the challenge to organise a virtual *Tropentag* based on Zoom meetings being streamed on YouTube channels using the Whova as online conference platform from September 7 to 9, 2020.

Since *Tropentag* normally is a whirling pool of people interacting and listening to each other, learning new things, building and refreshing networks, and enjoying science and coffee, the question arises if a virtual *Tropentag* can actually provide a similar ambience. Well, the straight answer is: No! So, what happens if such conferences, such scientific networking activities, such exchanges of knowledge are constrained by global crises? Will people and science suffer? Will project activities have to be abandoned, projects to be terminated and where does it leave the efforts of the last decades in fighting hunger and malnutrition?

We thought this to be an interesting discussion topic for a virtual *Tropentag* and chose *"Food and nutrition security and its resilience to global crises"* as the overarching theme for the *Tropentag* 2020.

We received 460 contributions related to the theme of which 318 were presented in the virtual conference and are now available in these proceedings. The five plenary keynote contributions to this years' *Tropentag* gave us a very good overview of the disciplinary understanding of crises and how resilience or responses to crises are detected and evaluated in the different disciplines. Prof. Dr. Reiner Sauerborn, a paediatrician from Heidelberg University, showed us how children are affected by all kinds of crises ranging from the lean season nutritional crises to exceptional droughts and climate change. Dr. Helen Young, a nutritionist of the Feinstein International Center at Tufts University, was driving home the point that persistent global acute malnutrition is still a major problem for children not only in times of crises, but enforced through any crisis. Dr. Petra Schmitter, an agronomist from IWMI, introduced us to the effects of the global water crisis on food and nutrition security. Dr. Bjoern Ole Sander, a climate change specialist from IRRI, showed us how climate change and rice production influence each other and enhance the crises in global climate change and food security. Finally, Dr. Michael Bruentrup, an economist from the German Development Institute DIE, concluded the plenary keynote by asking: "Which of these crises matters?" and presenting us a clear case of multi-risk management as the way to go forward.

The plenary sessions and the fiat panis award session were the only *Tropentag* 2020 events that were moderated and held physically under strict COVID-19 hygiene regulations at the University of Hohenheim - with an allowance of 66 individually numbered attendees in a lecture hall seating 450 in normal times. The foundation *fiat panis* split its 2020 Hans-H. Ruthenberg award among Sina Bierkamp (Leibniz University Hannover), Franziska Steinhübel (Passau University) and Roberto Villalba (University of Hohenheim) for three outstanding accomplished Masters theses. This years Josef G. Knoll award winners for outstanding PhD achievements in connection with the improvement of the nutritional situation in developing countries were Dr. Addisu Fekadu Andeta (KU Leuven, Belgium), Dr. Thomas Daum (University of Hohenheim) and Dr. Arndt Feuerbacher (Humboldt University Berlin).

The plenary and award session was then followed by 109 virtual oral presentations in 23 sessions from places around the world and in time zones as different as Peru and Vietnam □ as always in parallel sessions. The 209 poster presentations where held as 3-5 min audio/video clips in 28 moderated poster sessions and the discussion was held via zoom meetings with the presenters interfaced with the youtube audience via the whova platform. All these contributions are still available online via the Whova conference tool and accessible to all participants for review.

Tropentag 2020 has clearly shown that the scientific exchange and also the discussions needed to keep the science alive are possible and fruitful even in a virtual conference. A better understanding as to how crises affect food and nutrition security was clearly achieved by the 738 people from 67 countries during this years' *Tropentag*. We have learned some lessons on online conferencing that are particularly suitable to increase *Tropentag*'s outreach and that might spill over also into coming non-virtual *Tropentag* conferences. In the absence of the team of student reporters, Ralph Dickerhof, freelance reporter, ATSAF member and student reporter media trainer of the first hour, fed the *Tropentag* blog with his impressions of the web-based scientific exchange. Likewise, we fed the flickr foto-stream with impressions of the plenary and with behind the scenes images from Eric Tielkes and the IT support team.

Special thanks go to Eric Tielkes and his team for his very valuable support in organising this first ever virtual *Tropentag*, getting the IT setup sorted and distributing the scientific debates to the likewise decentral *Tropentag* family around the globe. Particular thanks are due also to our longstanding donors (listed on the back cover) for their unwaning financial in-kind support, which permitted us to do a 2020 *Tropentag* at all, to go virtual and to keep conference fees at a modest level, especially for junior scientists.

On behalf of ATSAF as the organiser of *Tropentag* 2020:

Folkard Asch and Christian Hülsebusch

Internet, September 2020

Contents

9. Institutions

Crops and systems

Assessing farming systems at different scales

Oral Presentations

Posters

Modelling Land Suitability for Expansion of Irrigated Rice and Dissemination of Alternate Wetting and Drying Water Management in Burkina Faso

Elliott Ronald Dossou-Yovo[1], Komlavi Akpoti[2], Paul Kiepe[1]

[1]*AfricaRice, Ivory Coast*
[2]*University of Energy and Natural Resources, Civil and Environmental Engineering Department, Ghana*

Spatially explicit assessment of land suitability can guide the identification of cropland with the highest potential for irrigated rice development, but for many regions in sub-Saharan Africa, the knowledge is very limited. Besides, reducing water input while maintaining rice yield is important for sustainable rice production in SSA. The objectives of this study were to produce a nation-wide prediction of irrigated rice area and estimate the climatic suitability of the alternative wetting and drying (AWD) technique of irrigation. We applied three environmental niche modelling (ENM) approaches that use machine learning algorithms along with the current distribution of irrigated rice locations in Burkina Faso to determine the extent of the potentially irrigated rice area. We used a simple water balance model to estimate the climatic suitability for AWD for the two main growing seasons: February – June and July – November. The evaluation metrics of the ENMs such as Area Under the Curves (AUC) and Percentage Correctly Classified (PCC) were higher than 90 % and 80 % for both training and testing, respectively. Exchangeable sodium percentage, distance to stream networks, exchangeable potassium, precipitation of the warmest quarter, annual mean temperature, soil depth to bedrock, topographical wetness index, actual evapotranspiration, soil organic carbon stock, and total phosphorous were the top 10 predictors determining a land suitability for irrigated rice development. The modelling predicted that 3 million ha of land are potentially suitable for irrigated rice cultivation in Burkina Faso. Most of these suitable lands are located within the sub-Sahelian and north-Sudanese climatic zones while the Sahelian climatic zone only showed marginal suitability for irrigated rice. About 97 % of the suitable lands for irrigated rice cultivation were found to be appropriate for AWD in the first growing season against 57 % in the second growing season. The results of this study can guide investments in irrigated rice development and large-scale dissemination of AWD in SSA.

Keywords: Alternate wetting and drying, climate change, ecological niche modelling, land suitability, rice

Contact Address: Elliott Ronald Dossou-Yovo, AfricaRice, 01 BP 2551, Bouake, Ivory Coast, e-mail: e.dossou-yovo@cgiar.org

Potentials and Risks of Alternate Wetting and Drying in Rice Production of the Dry Savannah Zone of West Africa

Jean-Martial Johnson[1], J. P. Eric Kabore[2], Samuel A. M. Ottnad[1], Elliott Ronald Dossou-Yovo[3], Mathias Becker[1]

[1]*University of Bonn, Inst. Crop Sci. and Res. Conserv. (INRES), Germany*

[2]*Centre National de la Recherche Scientifique et Technologique, Inst. de l'Environnement et de Recherches Agricoles, Burkina Faso*

[3]*Africa Rice Center, Sustainable Productivity Enhancement Program, Ivory Coast*

Irrigated rice farming plays a vital role in global food security but also requires more water than any other staple crop. Meeting the high demand for rice to feed the growing population under increasing water scarcity is one of the major challenges of the twenty-first century. Alternate wetting and drying (AWD) is one of the most widely advocated water-saving irrigation technologies. The technology was introduced only recently in the dry Savannah zone of West Africa, where it is still largely unknown by farmers. We assessed the effect of AWD on grain yield and irrigation water productivity against the backdrop of possible N losses. Participatory on-farm trials compared AWD to farmers' irrigation practice in four irrigation schemes of Burkina Faso during both the dry and the wet seasons of 2018 and 2019. AWD was compared to farmer' irrigation practice (FP) in 156 pairwise comparisons of AWD and FP plots. In addition, soil nitrate-N dynamics in relation to soil water content was assessed in dry season 2019. Compared to farmers' practice (FP), irrigation water input with AWD technology was reduced by 32 % in the dry and by 25 % in the wet season. With no significant effects on grain yields (mean of 4.9 Mg ha^{-1}) AWD increased the irrigation water productivity by 64 %. However, each AWD cycle resulted in soil N mineralisation of about 3 kg N ha^{-1} and the loss of this nitrate-N upon rewetting. Total N losses increased with soil drying intensity and the number of AWD cycles and reached up to 30 kg ha^{-1}. While AWD appears to be an effective strategy to save irrigation water with no rice yield penalty, the observed nitrate losses point towards possible negative longer-term impacts on soil fertility and productivity in rice irrigation schemes of the dry Savannah zone.

Keywords: AWD, Burkina Faso, *Oryza sativa*, water productivity, water-saving technology

Contact Address: Jean-Martial Johnson, University of Bonn, Inst. Crop Sci. and Res. Conserv. (INRES), Karlrobert-Kreiten-Strasse 13, D-53115 Bonn, Germany, e-mail: j.johnson@uni-bonn.de

Reviving Seed Sharing for Biodiversity Conservation Food Security and Ground Water Recharge

Lorenz Bachmann[1], Tahiratou Alassane[2], Helga Fink[2], Komlan Kpongbegna[2], Günther Rapp[2]

[1]*Agrecol Association, Germany*

[2]*ProSEHA-GIZ, GIRE Water Ressource, Benin*

Ground water protection and recharge is a major concern in West Africa, since water demand is likely to double in next thirty years with the growing population. Climate projections indicate that West Africa will be subjected to increased variability combined with a decline in rainfall. The GIZ Water Program ProSEHA investigated how ground water recharge can be improved alongside improvements of food security.

ProSEHA decided to test the improvement of rice production by increasing rice biodiversity and by encouraging site specific cultivation of old traditional rice varieties. ProSEHA collected a total of 25 varieties with different local names and the farmer-led testing and genetic tests validated 15 well distinguished varieties. Among these varieties were some with very good abilities to grow in deeper water and varieties with very low water needs for upland cultivation. Due to this wider adoption range, farmers were able to cultivate a larger proportion of their traditional watersheds. This increase in production area, helped that less land in the watershed banks remains idle, and thus prone to erosion, and consequently water recharge is enhanced.

167 farmer managed rice testing plots were evaluated. In the first year the plots were very small (<100 sqm) and increased over time as more seeds become available to plot sizes of up to 1 ha. All farmers tested various old varieties against a modern variety (mainly IR841). Farmers were invited to rate the relative performance of cultivars by observation and in addition precise yield measurements were taken. The trials were done without chemical fertilisers or other chemical plant protection measures.

The 4-years results showed local varieties performed at least equal (22 %) or even better (52 %) than modern varieties. Average yield for local rice varieties was 2.35 t ha^{-1} against 1.94 t ha^{-1} for modern varieties (+21 %). Farmer observation revealed that the local varieties offer a broader variation in crop cycle length, flooding and drought and pest resistance. Information and seed were diffused by annual seed sharing fares. The encouraging results led to an increase in participating farmers from below 20 in 2016 to 538 in 2019.

Keywords: Biodiversity, food security, seed exchange, water recharge

Contact Address: Lorenz Bachmann, Agrecol Association, Breiteweg 1, 35415 Pohlheim, Germany, e-mail: l.bachmann@gmx.de

Water Availability and its Interaction with Cropping Intensity Patterns of Rice-Based Systems in Southeast Asia

Christin Heckel, Carlos Angulo

University of Bonn, Inst. Crop Sci. and Res. Conserv. (INRES) - Plant Nutrition, Germany

Agricultural changes in cultivation patterns of rice-based systems in Southeast Asia have been investigated in the frame of the BMBF funded project "RICH-3P" coordinated by the University of Bonn. Six sites were considered, two in each of the listed countries: Cambodia, Myanmar and the Philippines. Alongside other changes, a shift from (rainy season) single rice cultivation to double rice cultivation (rainy and dry season) per year was observed at five out of six locations. For example, in the Central Dry Zone in Meiktila-Myanmar, 108 farmers (out of 160 respondents) indicated that they were producing rice twice a year in the present (2018), whereas about 20 years ago none of them were cultivating rice in both seasons. This development was made possible by the improved access to water in the respective regions. The present study aims to evaluate the external pressures and the (farming system) internal drivers leading to the change of single to double rice cropping. On the one hand, factors such as the improvement or installation of public irrigation infrastructure, cooperative work on municipal irrigation systems, advisory campaigns and assistance in the implementation of pumps, etc. are taken into consideration to explain the increase in water availability for farmers. On the other hand, we are investigating the thresholds of water quantity and availability that would lead farmers to definitely establish a second rice crop. For these purposes, apart from the evaluation of secondary sources, our work focuses on the acquisition and evaluation of optical satellite data (Landsat, Sentinel-2) from 1990 onwards in the corresponding study areas. A field-wise evaluation (polygon- and pixel-wise) of spatial and temporal water cover patterns during the flooding and transplanting status in the study areas will serve as proxy for the assessment of water availability and the determination of thresholds that induce farmers to shift from single to double rice cropping in the regions. The current state of research is taken into account and already implemented algorithms for the recognition of rice fields are applied using the data basis of the project, which also includes georeferenced boundaries of the farmers fields.

Keywords: Agricultural change, cropping patterns, irrigation, remote sensing, rice, water

---

Contact Address: Christin Heckel, University of Bonn, Inst. Crop Sci. and Res. Conserv. (INRES) - Plant Nutrition, Bonn, Germany, e-mail: christin.heckel@uni-bonn.de

Introducing Sustainable Farming Practices in Rice Production to Myanmar's Transitioning Agriculture Sector

Helena Wehmeyer[1], Melanie Connor[2]
[1]*University of Basel, Dept. of Environmental Sciences, Philippines*
[2]*International Rice Research Institute (IRRI), Sustainable Impact Platform, Philippines*

Myanmar has experienced considerable economic and social changes since its political transition in 2011/2012. Its agriculture sector has demonstrated rapid intensification and modernisation. However, rice yield gaps remain an important issue with regard to food security. Reducing rice yield gaps in Myanmar could increase annual production and support efforts to establish food security. Therefore, agricultural best management practices (BMPs) were introduced to rice farmers in the Bago Region from 2012 on to increase sustainable rice production, reduce rice yield gaps, and countervail negative environmental impacts of agricultural intensification. The objective of this study was to determine rice farmers' agronomic development, socioeconomic situation, and livelihood changes due to the adoption of BMPs. Using a digital survey questionnaire application to collect household data, 160 farmers in eight villages were interviewed in 2012 and 2017. Data were analysed using uni- and multivariate statistics. Results showed that farmers who adopted BMPs such as improved rice varieties and optimised fertiliser application demonstrated significantly higher yields, income, and profitability while reducing inputs and labour. Furthermore, after five years significant socioeconomic differences were found between BMP adopters and non-adopters. The study showed that BMP adopters improved their livelihoods due to increased agricultural efficiency. However, yield productivity remains low in Myanmar compared to neighbouring countries. Poor access to inputs, high input prices, and little risk management are factors impeding improved agricultural profitability, and hence rural development. Furthermore, natural conditions as well as economic and social constraints play an important role in the way farmers are able to manage their land. Therefore, further development research and dissemination strategies for the implementation of appropriate sustainable technologies are needed to improve rice farming.

Keywords: Adoption, best management practices, dissemination, impact, Myanmar, rice production, sustainable agriculture

Contact Address: Helena Wehmeyer, University of Basel, Dept. of Environmental Sciences current address: International Rice Research Institute, 4031 Los Baños, Philippines, e-mail: h.wehmeyer@irri.org

Increasing Nutrition Security with Vertical Gardens – Testing Different Systems for Vegetable Production

Sahrah Fischer[1], Saskia Grünwasser[2], Bastian Winkler[1], Thomas Pircher[3], Christine Lambert[4], Thomas Hilger[1], Georg Cadisch[1]

[1]*University of Hohenheim, Inst. of Agric. Sci. in the Tropics (Hans-Ruthenberg-Institute), Germany*

[2]*Nürtingen Geislingen University (HfWU), Germany*

[3]*University of Hohenheim, Research Center for Global Food Security and Ecosystems, Germany*

[4]*University of Hohenheim, Inst. of Biological Chemistry and Nutrition, Germany*

Vertical garden systems have been a largely urban phenomenon, used to cultivate food crops as well as ornamental species in areas that would not normally be suitable for plant growth. Vegetables grown in vertical garden systems can provide an important dietary supplement to households. However not much research has been done concerning systems' or varieties' production efficiency. We aimed to construct vertical garden systems that are low-cost, low-labour, and simple to make with materials available to all. The developed systems were tested for their (i) water holding capacity; (ii) produced biomass and yield; and (iii) which vegetable plant families could be suitable. The three systems designed and constructed were: the Second Wall, Planting Tower, and Bucket System, using three irrigation systems, i.e. cotton cloth, plastic tubes, and drip irrigation. The crops used in the systems represented four vegetable families: field peas (*Pisum sativum* L.; Fabaceae), African spinach (*Beta vulgaris* spp.; Amaranthaceae), black nightshade (*Solanum nigrum* L.; *Solanaceae*) and sukuma (*Brassica oleraceae* L.; Brassicaceae). Soil temperature and moisture were measured through implanted sensors, and yield was recorded. Six systems were constructed at three sites (2 system types per site) in schools located in Kapchorwa, Uganda. The systems and vegetables were compared using a mixed model. The Planting Tower had the highest and most constant water holding capacity, followed by the Second Wall. The soil temperatures of all three systems remained very constant, varying slightly between 18–23°C. The Planting Tower showed the highest yields for all cultivated species, followed by the Second Wall. The Bucket System produced the lowest yield for all vegetables. Both African spinach ($p = 0.020$) and black nightshade ($p = 0.049$) showed significant differences in yield depending on their placement in the system (at the top or the bottom), making them more sensitive to water content than sukuma and field peas. Overall, the systems performed well to produce a mix of nutrient-dense vegetables under different conditions in the field. We consider vertical gardens a promising option to increase surface area to produce a higher amount of diverse vegetables for the household, hence improving their food and nutrition security.

Keywords: Food and nutrition security, home garden, vegetables, vertical garden

Contact Address: Sahrah Fischer, University of Hohenheim, Inst. of Agric. Sci. in the Tropics (Hans-Ruthenberg-Institute), Garbenstr. 13, 70599 Stuttgart, Germany, e-mail: sahrah.fischer@uni-hohenheim.de

Opportunistic Adaptation of Conserved Moisture for Food Sustainability in Arid Zone

Dheeraj Singh[1], M Chaudhary[1], Chandan Kumar[2], M L Meena[2]

[1]*Central Arid Zone Research Institute, Krishi Vigyan Kendra, India*
[2]*CAZRI KVK, India*

Climate change coupled with multiple stressors have compelled subsistence farmers to develop location specific adaptation strategies to sustain their livelihoods in risk-prone ecosystems. A study on the opportunity of using conserved soil moisture for food and livelihood security and its adaptation in rural areas was carried out in Hemawas check dam with its catchment area in Pali district of Rajasthan, India. In arid zone (study zone) farmers are exposed to a set of multiple stressors making their subsistence vulnerable and to sustain their livelihood under these conditions, farmers have adopted specific options to diversify their livelihood options. In recent years as the consequence of climate change the terminal heat and rising temperature in the winter season have diverted farmers in Pali to adapt muskmelon in late winters. This is a short duration crop, cultivated with very least external inputs and moderate vulnerability. As the water dry up in the dam the land is first sown with short duration varieties of wheat, barley, oats, mustard and vegetables. The standing water is utilised for irrigation and as the land becomes devoid of water, muskmelon is grown in the conserved moisture. Normal ploughing is done to open up the soil and then manual sowing is done in the open spaces using local variety of muskmelon. Local plants are used as windbreaks to protect the small plants from cold winds in initial period and from hot winds in later period. Due to surplus moisture on top layer the seed germinates and as the plant grows its long root system draws water from the deeper layers of soil profile. After 15–20 days planking is done on the germinated seed and vines to close the open soil strata and this practice kills insects and parasites hiding in the crevices besides conserving the moisture in the soil layers. The soil is very rich in organic matter and nutrients thus on getting favourable conditions the vines yield ample fruits to sustain farmers livelihood. As no chemicals and fertilisers are used the crop is purely organic and the entire produce is sold at farm gate.

Keywords: Arid zone, moisture conservation, muskmelon, opportunistic adaptation, organic

Contact Address: Dheeraj Singh, Central Arid Zone Research Institute, Krishi Vigyan Kendra, Jodhpur Road, 306401 Pali, India, e-mail: dheerajthakurala@yahoo.com

Smallholder Farms Characterisation and their Use of Productive Resources in the Mt. Elgon Region, Uganda

Christine Arwata Alum[1], Ernst-August Nuppenau[1], Stephanie Domptail[1], Johnny Mugisha[2]

[1]*Justus-Liebig University Giessen, Inst. of Agric. Policy and Market Res., Germany*
[2]*Makerere University, School of Agricultural Sciences, Uganda*

Smallholder farming systems in Uganda are diverse and this diversity relates to the farm structures and production strategies employed by the smallholders. The behaviour of these smallholders regarding resource allocation has implications on their overall farm output and on their contribution to the local food system. They are faced with various challenges and therefore to increase farm production in small rural farming systems through targeting of interventions needs an understanding of the diversity in these farm systems. This study explores the relation between production diversity and the socio-economic characteristics (including resource endowment) of 108 farming systems in the Mt. Elgon region, Uganda by creating a farm typology. Using a multivariate analytic technique of principal components and cluster analysis based on the socio-economic characteristics of the farm, we identified three farm types strongly differing in terms of assets. Farm type 3 had high resource endowment in terms of productive assets such as land, livestock and labour resources compared to households that belonged to farm types 1 and 2. Generally, household production decisions were linked to larger farm sizes, access to a greater number of fields, availability of labour resources, ownership of livestock and off-farm income sources. Farm households that were less endowed on the other hand had low production diversity and were less involved in the sale of crop output. The farm types identified in the study area are a basis for identifying representative farms from which farm models can be constructed. Therefore, identifying homogeneous groups of farms may contribute to target policy recommendations through developing feasible farm strategies and estimation of production potential to realistic farms.

Keywords: Farm household, multivariate analysis, production diversity, typology

Contact Address: Christine Arwata Alum, Justus-Liebig University Giessen, Inst. of Agricultural Policy and Market Research, Senkenbergstrasse 3, 35390 Giessen, Germany, e-mail: tinaalum@yahoo.com

Assessment of Wastewater-Irrigated Urban Vegetable Production and Market Systems in Ethiopia: The Case of Akaki River in Addis Ababa

Solomon Araya[1], Habitamu Alemayehu[2], Zerihun Tadele[3], Ingrid Fromm[1], Alemayehu Masresha[4]

[1]*Bern University of Applied Science, School of Agricultural, Forest and Food Sci., Switzerland*

[2]*Office of Coordination for Research and Community Service Addis Ababa, Ethiopia*

[3]*University of Bern, Inst. of Plant Sciences, Switzerland*

[4]*Ethiopian Environment and Forest Research Institute, Environmental Laboratory Directorate, Ethiopia*

Ethiopia, urban vegetable production using wastewater from Akaki River in the capital city Addis Ababa is a common practice. In a wider sense, the potential of urban agriculture in the area has not been realised due to subsistence agriculture, insufficient land, underdeveloped marketing structure and the state of water used for irrigation. Apart from these, little is known about the production system, opportunities, and challenges. In cognizant of the fact, the study attempts to comprehensively address the issues of production, marketing value chain, challenges and health related implications of urban vegetable production through wastewater irrigation. The study was based on a household survey of 115 respondents (75 producers and 40 consumers) around Akaki River while review of secondary documents supplements the study with evidences. Mixed approaches of quantitative and qualitative data analysis methodologies including descriptive statistics with simple tests coupled with narrative of qualitative findings were employed. In this study, the production system and the perception of the people about the system has been analyzed. The results indicated that this vegetable production system produce mainly cabbage, lettuce, cucumber, and green pepper while the average frequency of production per year is about five cycles. The irrigation practice using wastewater from Akaki River is dominated by furrow irrigation and traditional features which impact the efficiency of water use as manifested in the dry seasons. Moreover, industrial and household wastes released into the river increase health risks as result of contamination added to the low level of safety standards followed by producers. Despite the problems, the wastewater irrigation practice resulted in higher vegetable productivity as compared to the national average. This makes producers profitable although marketing of products directly on the farm could also have a significant impact. Apart

Contact Address: Solomon Araya, Bern University of Applied Science, School of Agricultural, Forest and Food Sciences, Laengasse 85, 3012 Zollikofen, Switzerland, e-mail: Solomongebremichael.araya@students.bfh.ch

from the negative implications, vegetable production using wastewater from Akaki River found to improve plant growth, household vegetable consumption, and create job opportunities. Hence, synergy between local authorities, NGOs, Universities and Research centres for multidimensional interventions is expected to enhance the safe production system with increased benefit from the practice.

Keywords: Akaki River, irrigation, urban agriculture, vegetable, wastewater

Effect of Saffron-Mallow Intercropping Patterns in the Third Year on Possible Cooling of Corms for Climate Change Adaptation

Soroor Khorramdel[1], Vida Varnaseri[2]
[1]*Ferdowsi University of Mashhad, Agrotechnoloogy, Iran*
[2]*University of Zabol, Agrotechnology, Iran*

Intercropping is a traditional agricultural approach which is the growing of multiple plant species at the same time in the same location. Traditionally, intercropping has been used to enhance plant yield and the efficiency of the resource as well decrease risk. Intercropping has been shown to decline the risk of plant failure by increasing the plant yield stability over time. Intercropping creates biodiversity in the agroecosystems, and it is considered to make the ecosystems more resilient against environmental perturbations, thus improving food security. The current study was aimed to investigate the effects of intercropping patterns of mallow as a perennial medicinal plant on stigma yield and quality characteristics of saffron affected as possible cooling of corms for climate change and global warming mitigation. The experiment was carried out at Faculty of Agriculture, Ferdowsi University of Mashhad, Iran. Treatments were 15, 30, 45 and 60-cm row spacings for saffron from mallow planting rows and sole saffron and mallow cultivations. The results revealed that the impact of intercropping patterns with mallow was significant on yield indicators of flower indicators of saffron. In comparison between sole cultivation and intercropped saffron revealed that the highest values for flower number, dried stigma yield and yield of daughter corms were recorded for sole saffron cultivation with 81 flowers m^{-2}, 0.2115 g m^{-2} and 26.51 g m^{-2}, respectively. In comparisons amongst intercropping patterns, the highest value for dried stigma weight was related to 30-cm row spacings from mallow with 13.39 g m^{-2}. However, corcin, picrocrocin and safranal contents were not significantly affected by mallow intercropping patterns. The maximum land equivalent ratio was calculated for 15-cm row spacing with 1.77.

Keywords: Crocin, land equivalent ratio, stigma yield

Contact Address: Soroor Khorramdel, Ferdowsi University of Mashhad, Agrotechnoloogy, Azadi, 0098 Mashhad, Iran, e-mail: khorramdel@um.ac.ir

Ethnobotanical Study of Medicinal Plants Used by Mocho Community in the State of Chiapa, Mexico

Eduardo Alberto Lara Reimers[1], David Jonathan Lara Reimers[2], Uresti Duran Diana[3], Ainura Seitmuratova[4], Cusimamani Fernández Eloy[4]

[1]*Autonomous Agrarian University Antonio Narro, Department of Forestry Engineering, Mexico*

[2]*Chapingo Autonomous University, Dept. of Forestry Engineering, Mexico*

[3]*National Institute of Forest, Agricultural and Livestock Research, CIRNE-CESAL, Mexico*

[4]*Czech University of Life Sciences Prague, Fac. of Tropical AgriSciences, Czech Republic*

Ethnomedicine is still used as primary health care resource by several indigenous communities and people who live in rural areas over the world. In Mexico, most of the indigenous population live in rural areas (61.1 % in communities with less than 2,500 inhabitants). The State of Chiapas is inhabited by 11 ethnic groups with valuable ancestral knowledge in the management, and use of medicinal plants that are transmitted orally from generation to generation. The aim of this study was the documentation of traditional knowledge about medical plants and its usage in traditional medicine by the locals of the Mocho community, located in the state of Chiapas. Ethnobotanical and socio-demographical data were collected using a questionnaire from 43 local informants from Motozintla municipality, the state of Chiapas. In addition, quantitative approaches were used to determine medicinal use value (MUV), use report (UR), frequency of citation (ICF), the relative frequency of citation (RFC), and informant consensus factor (ICF). A total of 83 medicinal plant species belonging to 44 botanical families were documented. Asteraceae was the most dominant family by number of species (6 species) followed by Lamiaceae and Rutaceae (5 species each). The most often used parts are leaves (46%) and the decoction is the most common method of preparation. Diseases of the digestive and gastrointestinal system were dominated with 102 use-reports (27.6%) and diseases of the reproductive system had the highest ICF index (0.76) among other aliment categories. According to RFC and MUV index the most important species were *Verbena litoralis* Kunth., *Matricaria chamomilla* L., *Bursera simaruba* (L.) Sarg., *Dysphania ambrosiodes* (L.) Mosyakin & Clemants, and *Ruta graveolens* L. The collected information represents a base of knowledge for future research in the ethnobotanical field in the state, and it will contribute to the understanding of proper usage of medicinal plants. When knowledge is transformed in goods, culture, income and health it can be promoted through the demand of tourists that visit the place searching better options to treat their ailments.

Keywords: Lamiaceae, traditional medicine, *Verbena litoralis*, Asteraceae, ethnobotany

Contact Address: Eduardo Alberto Lara Reimers, Autonomous Agrarian University Antonio Narro, Department of Forestry Engineering, Calz. Antonio Narro 1923, Buenavista , 25315 Saltillo, Mexico, e-mail: agroforestal33@gmail.com

Agroecology and Organic Coffee: Where Does the Organic Matter Come From? A Resource Accounting Approach

Stephanie Domptail[1], Pardis Aminjavaheri[1], Christopher Sebatta[2], Ernst-August Nuppenau[1], Jeninah Karungi[2]

[1] *Justus-Liebig University Giessen, Inst. of Agricultural Policy and Market Research, Germany*

[2] *Makerere University, Department of Agricultural Production, Uganda*

The adjectives organic and agrocological are often used synonymously. Yet, the current certification for organic products does not consider several aspects that make up agroecology, and especially misses out on systemic dynamics and cycling issues on the farm. In other words, an agroecological agricultural system is potentially more tightly closed than an organic agricultural system. From a metabolic point of view, organic systems may therefore be just as dependent as conventional systems on the input of matter. The matter being not synthetic but rather organic matter, the enrichment of an organic system can be coupled with the depleting of another resource. To the least, the maintenance of the organic system may be dependent on the existence of an external but coupled source of organic matter. The paper analyses farming systems from a metabolic perspective and evaluates their sustainability but also their agroecological character. Our study characterises the metabolism of 5 farms producing coffee in intercrop with banana, chosen among a large sample of farms known from previous research in the Mount Elgon, Uganda. The 5 farms are representative of 3 groups: organic coffee producers, conventional coffee producers and low-input coffee producers. The accounting of material, energy and financial flows through the different production systems provides answers the following questions, all strongly related to agroecological characteristics: How dependent are the farms on material and energy inputs from outside? How related are these with financial flows and funds? How efficient are the farms? And how regenerative are they to their environment? Our findings suggest that through its maintained reliance on external inputs, the organic system shows some metabolic similarities with the conventional system. Much of the organic material needed to obtain good organic coffee yields is imported from distantly accessible savannahs. The reliance on external inputs rather than cycling also has social implications as access to external inputs seem conditional to the wealth of the farm. Our ability to assess agroecological features of farming systems and their relationship to the embedded ecological and institutional landscape, via the metabolic perspective, will increase our capacity to design sustainable farming systems in the future.

Keywords: Comparison, farming systems, material and energy flow analysis

Contact Address: Stephanie Domptail, Justus-Liebig University Giessen, Inst. of Agricultural Policy and Market Research, Semckenbergstrasse 3, 35390 Gießen, Germany, e-mail: stephanie.domptail@agrar.uni-giessen.de

Crop Productivity and Contributing Factors in Organic and Conventional Farming Systems in Kenya - Evidence from a Long-Term Experiment (SysCom)

David Bautze[1], Edward Karanja[2], Martha Musyoka[2], Noah Adamtey[1]

[1]*Research Institute of Organic Agriculture (FiBL), International Cooperation, Switzerland*
[2]*International Centre of Insect Physiology and Ecology (ICIPE), Kenya*

Contributing to the global debate if organic can feed the world, the Research Institute of Organic Argriculture (FiBL) has started two long-term trials in Kenya. The Trial sites at Chuka and Thika are situated in the subhumid zones of the Central Highlands. At each site, conventional farming (Conv) and organic farming (Org) were compared at two input levels: high inputs (High) representing export-oriented, large scale production and low inputs (Low) representing smallholder production mainly for domestic use. The conventional system. received mostly synthetic fertiliser and used synthetic pesticides. The organic system only used organic fertiliser and biopesticides. The differences between input levels were the amount of nutrients supplied and supplementary irrigation. The crop rotation included maize (*Zea mays*), different leafy vegetables (*Brassica oleracea, Beta vulgaris*), leguminous crops (*Phaseolus vulgaris, Glycine max*) and potatoes (*Solanum tuberosum*). After twelve years of continuous cropping, we encountered some trends with regards to productivity in organic and conventional systems: Grain maize, baby corn and common bean were able to achieve similar yield in organic and conventional, whereas cole crops, French beans and potatoes showed significant lower yields in organic. For example, cabbage showed 40 % higher yields in conventional high input system compared to organic high input system. However, our results showed that productivity often depends on the crop and the chosen management practice within a system. Some crops like grain maize were able to achieve similar yields because crop nutrient supply in organic systems was sufficient and pest and disease were of minor importance to the crop development. On the contrary, we experienced that the organic systems were less productive if management practices were guided by conventional mindsets - substitution of synthetic products by biological ones was not sufficient and other more system-based approaches like mixed cropping or pest resistance varieties need to be incorporated to achieve better outputs. Additionally, it must be mentioned that high productivity might not be enough to declare a farming system sustainable or not. Organic farming systems in our trials were also able to show positive effects on soil fertility, human health and biodiversity.

Keywords: Crop productivity, farming system research, organic farming

Contact Address: David Bautze, Research Institute of Organic Agriculture (FiBL), International Cooperation, Ackerstr. 113, 5070 Frick, Switzerland, e-mail: david.bautze@fibl.org

How Do Organic and Conventional Production Systems Perform: Evidence from Long-Term Study in India

Amritbir Riar[1], Bhupendra Singh Sisodia[2], Gurbir Bhullar[1]

[1]*Research Institute of Organic Agriculture (FiBL), Dept. of International Cooperation, Switzerland*

[2]*Biore Association, India*

Organic agriculture has gained a reputation for being ecologically sustainable, however sometimes it faces criticism for productivity and profitability aspects in comparison to conventional farming production. This is specifically a matter of concern when aiming at finding solutions to agricultural challenges of smallholder farmers in the developing countries. Through the 'long-term farming systems comparison (SysCom) program', FiBL together with local partners, runs a network of field experiments in the tropics, which aims at obtaining solid scientific evidence on the performance of organic and conventional production systems.

Here we present the findings from the SysCom long-term experiment located in (Madhya Pradesh State) the Central Indian cotton belt. The climate of the project area is semi-arid and cotton, soybean and wheat are the major crops grown in this region. Four management systems (treatments) namely (a) organic, (b) biodynamic, (c) conventional and (d) conventional farming with genetically modified (GM) cotton, are being tested in the field trial on bioRe research station since 2007. After analysis of twelve years of data, we found that for cotton crop, total production cost was highest in Bt-conventional system followed by conventional system and organic systems (org + bio-dynamic) respectively. On an average, there was a yield gap (Seed yield of Conventional systems – Seed yield of organic systems) of nearly 20 % for cotton and main wheat crop. Average yield of second wheat crop following cotton in rotation was above 50 % for organic systems than conventional systems. However, Soybean crop performed consistently equal in organic and conventional systems. However, the profitability – being the output function of input costs – was not consistent with the productivity outcomes. The detailed analysis of productivity and profitability of different systems, with and without premium prices for cash crop, will be presented.

Keywords: Farming systems, organic vs. conventional, productivity, profitability

Contact Address: Gurbir Bhullar, Research Institute of Organic Agriculture (FiBL), Department of International Cooperation, Ackerstrasse 113, 5070 Frick, Switzerland, e-mail: gurbir.bhullar@fibl.org

Bt Cotton Technology Impacts on Agricultural Land Use Dynamics of Nagpur District of Maharashtra in India

Radhika Channanamchery, Hukumraj L Kharbikar, Nitin G Patil, Udayabhaskar Kethineni

Indian Council of Agricultural Research (ICAR), National Bureau of Soil Survey and Land Use Planning, India

Sustainable land management is now recognised as a major policy instrument due to severe land degradation problem in India. Understanding the temporal dynamics and trends of agricultural land use will help in planning suitable efforts to materialise the long term sustainable land management goals and improvement in life quality standards of the farmers of the region. In this perspective we analyse the temporal dynamics of agricultural land use change in Nagpur district of Maharashtra in the perspective of Bt cotton adoption since the year 2002. This study finds that a 3.84 percent growth in area under cotton cultivation in Nagpur district for the period 2000–20001 to 2017–18 and a 9.27 percent growth in production for the period studied. Cotton area, production and its yield has shown a significant improvement in the state since year 2000–2001. The study finds negative growth in area and production in case of then major crops of the district viz Black gram. Soybean, Green gram and Sorghum. The increase in the area of cotton finds to be at the cost of other competing crops like millets and other pulses, given inelastic supply nature of land. Cotton got an utmost importance in Nagpur district over the study period as revealed by the compound growth rate (CAGR) analysis. Growth performance of the cotton sector of Nagpur district is examined with a break up into two sub-periods viz; 2000–01 to 2008–09 , 2009–10 to 2017–18 and 2000–01 to 2017–18. The significant and positive growth trends in cultivated area for few crops such as cotton, rice and red gram in the studied district clearly show that there is shifting from diversified cultivation to intensive monoculture which adversely affect the sustainable land management in the area.

Keywords: Bt Cotton, compound agricultural growth rate, sustainable land management, temporal land use dynamics

Contact Address: Radhika Channanamchery, Indian Council of Agricultural Research (ICAR), National Bureau of Soil Survey and Land Use Planning, Lup Amravati Road, 440033 Nagpur, India, e-mail: radhika.kerala@gmail.com

Economic Performance and GHG Emissions of Traditional and Organic Cocoa Farms in the Peruvian Amazon

Andrés Charry, Miguel Romero, Matthias Jäger, Marcela Quintero
Alliance of Bioversity International and CIAT, Colombia

Organic certification for agricultural commodities has been promoted as a way to reduce the environmental impacts of food production and improving the welfare of smallholder farmers. Nevertheless, per capita net incomes in both conventional and certified farms appear insufficient to cover the average household basic needs and the share of the profits reaching smallholders from the chocolate industry remains marginal, raising governance questions regarding the effectiveness and fairness of the cocoa value chain and the certification schemes. On the other hand, while the environmental benefits of organic production are well documented, the effect of changing low-input cocoa systems in the Amazon to organic production systems on profitability and GHG emissions is unclear. In this study, we assess the profitability and carbon footprint of cocoa production under traditional (low input) and organic systems in the Amazonian region of Ucayali, Peru. For this, we used the Typical Farm Approach, combining participatory farmer workshops and expert guidance to define farms' typologies and obtaining a detailed quantification of all activities, processes, as well as input and output flows of both production systems. Subsequently, we calculated economic indicators and carbon footprint under different scenarios to assess their performance and trade-offs. Our results show that organic production allows higher yields and farm gate prices compared to traditional production systems, yet current price premiums and yield gains are insufficient to cover the additional costs of engaging in organic certification. Both systems require more than 4 hectares to provide the household with two monthly minimum wages, which is above the average of household cultivated area in the region, and could imply increasing pressure on the forests. Additionally, GHG emission increases with the inclusion of organic fertiliser are not offset by the declared yield gains when compared to traditional farming. Based on our results, we do not discourage organic cocoa production, but raise questions on adequate farm gate prices and fair share of the final product value reaching the producer, as well as on the need for tying production to zero deforestation commitments and compensating the aggregate environmental and social benefits when promoting more sustainable cacao production systems.

Keywords: Carbon footprint, cost-benefit analysis, organic agriculture, typical farm

Contact Address: Andrés Charry, Alliance of Bioversity International and CIAT, Cali, Colombia, e-mail: a.charry@cgiar.org

Land Suitability and Socioeconomic Factors for Pigeonpea Cultivation in Uganda

Angella Kyobutungi[1], Jimmy Obala[2], Ellen Kayendeke[1], Vincent Muwanika[1], Cory Whitney[3], John Robert Stephen Tabuti[1]

[1]*Makerere University, Dept. of Environmental Management, College of Agricultural and Environmental Sciences (CAES), Uganda*

[2]*Kabale University, Dept. of Agricultural Sciences, Fac. of Agriculture and Environmental Sciences, Uganda*

[3]*University of Bonn, Inst. Crop Sci. and Res. Conserv. (INRES), Germany*

Pigeonpea (*Cajanus cajan* (L.) Millsp.) is one of Uganda's many traditional pulses and may have an important role to play in helping the country to achieve food and nutritional security, while at the same time protecting and enhancing natural resources. The sustainable cultivation of pigeonpea will require, among other things, identification of suitable regions for cultivation and a better understanding of the factors that influence its adoption by farmers. To offer guidance on production we performed a suitability analysis by matching land characteristics with crop requirements using a GIS weighted overlay technique and gathered related socio-economic data collected with farmers in regions considered suitable. Our findings indicate that a large area of Uganda is suitable for pigeonpea growing and that farmers are likely to adopt, given the right support. The biophysical requirements for cultivation (i.e. rainfall, temperature, slope, soil drainage) were gathered from the available literature. A digital elevation model of the study sites was downloaded from the United States Geological Survey (USGS, https://earthexplorer.usgs.-gov) to create a data layer for the slope using surface analysis in ArcGIS v10.4 software. We used soil drainage data from the Harmonized World soil database (http://www.fao.org) and temperature and rainfall data from the Global Weat-her Database (https://globalweather.tamu.edu/). Suitability analysis revealed that pigeonpea can be grown on 79 % of land in Uganda. Generally, the highly to moderately suitable areas were found in central and western regions whereas the northern and eastern regions were either marginally or not suitable. To determine the factors that influence adoption of pigeonpea we conducted a household survey with 283 randomly selected farmers from three of the sub counties identified by the suitability analysis. Farmers cited pest and disease, lack of market, lack of extension services and lack of improved varieties as the major factors constraining pigeonpea production. Seed distribution is essentially informal, either self-saved, purchased from neighbours/relatives or from local markets. Farmers' preferred traits included resistance to diseases, early maturing, resistance to drought and short cooking times. We recommend promotion of pigeonpea in suitable areas, farmers are likely to adopt the crop if provided with the right materials and support.

Keywords: Cultivation, land suitability, pigeonpea, socioeconomic factors, Uganda

Contact Address: Cory Whitney, University of Bonn, Inst. Crop Sci. and Res. Conserv. (INRES), Auf Dem Hügel 6, 53111 Bonn, Germany, e-mail: whitney.cory@gmail.com

Can Subsistence Farming Help to Achieve Household Food Security? Evidence from Gurue, Central Mozambique

Custodio Matavel, Harry Hoffmann, Constance Rybak, Stefan Sieber

Leibniz Centre for Agric. Landscape Res. (ZALF), Sustainable Land Use in Developing Countries (SusLAND), Germany

In Mozambique, about 70 percent of population live in rural areas and depend directly on ecosystem services for their livelihoods. Subsistence agriculture has been given particular importance by the government of Mozambique as a strategy to fight food insecurity and poverty. Nonetheless, low soil quality, poor and irregular rainfall, low level of synthetic inputs, food losses, inadequate infrastructure and support services are the principal barriers to agricultural productivity. Therefore, the question is can subsistence farmers achieve some level of food security? To answer this question, we collected data from 300 households in Gurue district, central Mozambique, using a semi-structured questionnaire. The results revealed very limited income earning opportunities outside agriculture, therefore, undermining households' ability to buy food. Only 16 % of households had a non-farm income. The number of households with three or more meals a day grows during the harvesting period (March and July), reaching over 60 percent of the households, but this number decreases to 5 percent in January and February when the food reserves are already scarce. Two strategies stand out in alleviating food insecurity, particularly crop diversity and the use of processing and preservation techniques for agricultural products. The more crops a farmer produces, the more likely he/she is to harvest for a longer period throughout the year. The preservation of agricultural products helps to reduce post-harvest losses, while allowing food to be available for a longer period. However, farmers in the study area use traditional preservation techniques, mainly open sun drying, which may have negative impact on food quality, therefore creating a need for the implementation of innovative drying techniques that preserve the quality of food and use sustainable sources of energy.

Keywords: Crop diversity, food processing, food security, rural Mozambique

Contact Address: Custodio Matavel, Leibniz Centre for Agric. Landscape Res. (ZALF), Sustainable Land Use in Developing Countries (SusLAND), Eberswalder Str. 84, 15374 Müncheberg, Germany, e-mail: custodiomatavel@unilurio.ac.mz

Genetic resources and crop improvement

Pheno-Genetics of Local Apple Varieties in Northern Region of Pakistan: A Hidden Pool of Apple Diversity?

Martin Wiehle[1], Muhammad Arslan Nawaz[2], Richard Dahlem[3], Iftikhar Alam[4], Asif Ali Khan[5], Oliver Gailing[6], Markus Müller[6], Andreas Buerkert[2]

[1]*University of Kassel, Tropenzentrum / ICDD / Organic Plant Production and Agroecosystems Research in the Tropics and Subtropics (OPATS), Germany*

[2]*University of Kassel, Organic Plant Production and Agroecosystems Research in the Tropics and Subtropics, Germany*

[3]*Pomologenverein, Landesgruppe Rheinland-Pfalz/Saarland/Luxemburg, Germany*

[4]*Karakoram International University, Pakistan*

[5]*Muhammad Nawaz Shareef University of Agriculture, Plant Breeding and Genetics, Pakistan*

[6]*University of Goettingen, Inst. of Forest Genetics and Forest Tree Breeding, Germany*

As a main fruit crop of cold temperate regions, apple (*Malus × domestica*) is traded globally. As a well documented plant cultigen, traditional germplasm in remote regions is still unexplored. In Gilgit Baltistan, Pakistan's increasing market demands for modern varieties and farmers' preference for high yields increasingly lead to the loss of local/indigenous varieties. We therefore studied local apple diversity by assessing varietal richness and diversity, arboricultural activities as well as phenotypic and genotypic characters of apple germplasm. In total, 106 individual apple trees were sampled, and 35 tree owners were interviewed from seven villages of the valleys Ishkoman and Baltistan. Alpha- and beta-diversity, dendrometric parameters, and fruit traits (qualitative and quantitative) in each village were measured. In *Malus*, such alpha- and beta-diversity measurements were conducted for the first time. Alpha diversity was higher in Baltistan, where even a wild apple relative (*Malus baccata* (L.) Borkh.) was discovered. Beta-diversity revealed three compositionally distinct clusters among villages. Calyx depth (cm), fruit weight (g), and axis pit width (cm) showed highest variability among the assessed characters, highlighting those as distinguishable fruit characters for the accessions observed. Arboricultural measures were found to be comparatively low and only a few local varieties were marketed, while most varieties were used for home consumption only. Genetic data revealed a moderate to high diversity. Considering the high varietal, phenotypic, and genetic variability, Gilgit-Baltistan appears a promising source of apple germplasm for future breeding programs. Improved management and product diversification could open up new revenue sources for farmers.

Keywords: Alpha-diversity, beta-diversity, Gilgit-Baltistan, varietal diversity

Contact Address: Martin Wiehle, University of Kassel, Tropenzentrum / ICDD / Organic Plant Production and Agroecosystems Research in the Tropics and Subtropics (OPATS), Steinstraße 19, 37213 Witzenhausen, Germany, e-mail: wiehle@uni-kassel.de

Increasing Chickpea (*Cicer arietinum* L.) Tolerance to Soil Acidity: Screening Landraces for Tolerance to Aluminium and Manganese Toxicity in Solution Culture

Richard Onwonga[1], Karthika Pradeep[2], Wendy Vance[2], Richard Bell[2]

[1]*University of Nairobi, Dept. of Land Resource Management and Agricultural Technology, Kenya*

[2]*Murdoch University, College of Science, Health Engineering and Education, Australia*

Chickpea (*Cicer arietinum* L.) is a nutritious food security crop but its production in many parts of the world is constrained by soil acidity (Aluminium and Manganese toxicity). Selected chickpea landraces (90) from Ethiopia, Bangladesh, Nepal, India, Australia (Striker, Amber) and wild cicer (6) from Turkey were screened for tolerance to Al or Mn toxicity, in a plant growth room, using a hydroponic screening method. In each experiment, replicated thrice, main plots were 0, 15, and 60 μM Al or 2 and 150 μM Mn treatments, subplots were the chickpea accessions. The chickpea accessions were harvested at 10 (Al) and 26 (Mn) days after sowing (DAS). The longest root length (LRL; Al), shoot and root dry weight (Al, Mn) and visual (toxicity) symptom score (Mn), and shoot length (Mn) were measured. Relative Root Elongation index (EI) was derived from the LRL data. Chickpea landraces were classified as Tolerant (E1>70), Moderate (E1<=70, E1>50) or Sensitive (E1<=50). In 15µM Al treatment, wild cicer (wC) (499398 and 50011) were tolerant while landraces (41046, and 42395) were tolerant in 60µM Al. Most landraces and wC, in addition to Amber and Striker, were tolerant to Mn (150 μM) toxicity. Chickpea landrace (41046) was tolerant to both Al (15 and 60µM) and Mn toxicity while landrace (42400) was tolerant to Al (60µM) and Mn toxicity. Striker and wC 49938 were tolerant to Al (15µM Al) and Mn toxicity. Landraces; 41046 and 42400, and wC 49938 require further investigations in acid soils as they show promising tolerance to both Al and Mn toxicity. Further collection of chickpea landraces and wC grown on acid soils such as eastern Kenya and Ethiopia is recommended to extend the range of acid tolerant germplasm for screening and, to identify candidates for use in breeding programmes to produce acid tolerant chickpea.

Keywords: Acid tolerance, chickpea landraces, hydroponic screening, wild cicer

Contact Address: Richard Onwonga, University of Nairobi, Dept. of Land Resource Management and Agricultural Technology, Waiyaki Way, Nairobi, Kenya, e-mail: onwongarichard@yahoo.com

Delivery of Clean Planting Material of Orange-Fleshed Sweetpotato to Smallholder Farmers through Decentralised Vine Multipliers in Uganda

Joshua Okonya[1], Norman Kwikiriza[1], Frederick Grant[1], Joyce Maru[2], Simon Heck[2]

[1]*International Potato Center (CIP), Uganda*

[2]*International Potato Center (CIP), Kenya*

Crop biofortification is one of the most promising approaches being used in recent years to fight hidden hunger among resource-poor communities or persons in fragile environments. Vitamin A rich orange-fleshed sweetpotato (OFSP) has proven to be effective in improving the health of children below five years and pregnant women in many sub-Saharan countries. However, access to improved and clean OFSP varieties has limited realisation of the full benefits of this crop. To increase the availability of disease-free and high yielding planting material of OFSP, CIP together with partners have identified, recruited and trained more farmers in 12 districts of Uganda to start the production of pest and disease-free planting material of selected OFSP to increase its consumption within households in areas with the highest Vitamin A deficiency (VAD) in the country. Its believed that making the planting material of this nutrient-rich crop readily accessible to local communities will go a long way in increasing consumption of OFSP roots hence a reduction in VAD. The training was conducted in December 2019 and January 2020 and involved a total of 81 stakeholders and only 48 farmers were selected to become vine multipliers. During the training, perceptions of the stakeholders on effective ways of increasing OFSP consumption in their communities were also captured. Challenges, opportunities, and experiences of producing, marketing, and consumption of OFSP were also discussed. Community sensitisation, nutritional education, timely availability of improved OFSP varieties, access to irrigation, and market availability were reported to be among the key factors that influence the consumption of OFSP.

Keywords: Biofortified crops, scaling-up, stakeholder perceptions, sweet-potato, vine multiplication

Contact Address: Joshua Okonya, International Potato Center (CIP), Plot 47 Ntinda II Road, Naguru, Kampala, Uganda, e-mail: j.okonya@cgiar.org

In vitro Induced Polyploidy in *Celosia argentea* var. *plumosa*

Tomáš Thanh Nguyen Cong[1], Yamen Homaidan Shmeit[1], Jana Šedivá[2], Ingrid Melnikovová[1], Cusimamani Fernández Eloy[1]

[1]*Czech University of Life Sciences Prague, Fac. of Tropical AgriSciences, Czech Republic*

[2]*Silva Tarouca Research Institute for Landscape and Ornamental Gardening, Dept. of Plant Biotechnology, Czech Republic*

Celosia argentea L., Amaranthaceae, is an annual herb cultivated for its medicinal, nutritional and horticultural values. It is rich with minerals, proteins, vitamins and several other compounds such as saponins, alkaloids, peptides, glycosides, flavonoids, amino acids and fatty acids. Medicinal properties of this plant are applied in traditional medicine to treat tumors, jaundice, fever, diarrhea, inflammation and various other diseases. As an ornamental plant it is particularly grown for its flowers. Somatic induced polyploidisation *in vitro* is one of the methods used in breeding programs of ornamental plants. Chromosome doubling usually increases plant vigour and enhances morphological traits like flower shape and size.

The aim of this study was to obtain octoploid plants (2n = 72) from tetraploid *Celosia argentea* var. *plumosa* (2n = 36) by *in vitro* induced polyploidy. In total 320 nodal segments were treated with oryzalin at concentrations of 20, 40, 60 and 80 μM along two time intervals of 24 and 48 hours. The level of ploidy of the affected plants was determined using flow cytometry.

Four octoploid plants were obtained, of which three plants at the concentration of 40 μM for 24 hours and one plant at the concentration of 60 μM for 24 hours. *In vitro* morphological changes such as thicker stems, more compact growth and larger leaf area size were observed in newly acquired genotypes compared to control plants. In the second phase of the research, newly acquired genotypes will be evaluated in in vivo conditions for better analysis of morphological and biochemical properties.

Keywords: Amaranthaceae, *Celosia argentea* var. *plumosa*, flowcytometry, oryzalin, polyploidy

Contact Address: Yamen Homaidan Shmeit, Czech University of Life Sciences Prague, Fac. of Tropical AgriSciences, Dept. of Crop Sciences and Agroforestry, Kamycka 933, Prague, Czech Republic, e-mail: yamen.shmait@gmail.com

The Origin of Date Palm (*Phoenix dactylifera* L.) – State of the Art and Methods to Distinguish between Putatively Wild and Domesticated Populations

Martin Wiehle[1], Thomas Astor[2], Hannes Kahl[3], Valentin L. F. Wolf[4], Konstantin Krutovsky[5], Andreas Buerkert[6]

[1]*University of Kassel, Tropenzentrum / ICDD / Organic Plant Production and Agroecosystems Research in the Tropics and Subtropics (OPATS), Germany*

[2]*University of Kassel, Grassland Science and Renewable Plant Resources, Germany*

[3]*University of Leipzig, Ancient History, Germany*

[4]*University of Göttingen, Tropical Plant Production and Agricultural Systems Modelling , Germany*

[5]*University of Göttingen, Dept. of Forest Genetics and Forest Tree Breeding, Germany*

[6]*University of Kassel, Organic Plant Production and Agroecosystems Research in the Tropics and Subtropics, Germany*

Date palm (*Phoenix dactylifera* L., Arecaceae) is one of the few extratropical palms and a widely recognised keystone species of arid and semi-arid ecosystems across North Africa, Cape Horn, the Middle East, and the Indus Valley. According to FAO, date palm accounts for roughly 1 % of the global fruit production and is therefore an important agricultural commodity. It belongs to the few perennial and intensively domesticated species, whose wild relative(s) and putative centre(s) of domestication are not yet recognised. Hence, date palm's domestication process is considered one of the most difficult issues in agro-biodiversity contexts, but the search for its wild relative(s) has gained momentum in recent years. It is driven by (i) the need to increase the genetic pool needed for selection and adaptive management to mitigate climate change and (ii) scientific interest in the likely complex evolution of this species. In this study we review current trends in date palm research, consider climatic conditions between 130,000 BCE and today, and use archaeo-botanical (such as pyhtoliths and pollen) and historic (such as cave paintings and reliefs) records to unravel date palm's evolutionary process. We integrate data obtained from studies of other domesticated plants and propose approaches that allow to better predict putative regions of wild date palm. The data provide evidence for a three domestication centre theory (Persian Gulf as well as East Mediterranean and north-west African regions). The present study underlines the need for a more structured and collaborative research agenda with potential applications for other species of interest.

Keywords: Archaeology, cave painting, domestication, genetic diversity, palynology, phytoliths, private alleles

Contact Address: Martin Wiehle, University of Kassel, Tropenzentrum / ICDD / Organic Plant Production and Agroecosystems Research in the Tropics and Subtropics (OPATS), Steinstraße 19, 37213 Witzenhausen, Germany, e-mail: wiehle@uni-kassel.de

Phenotypic Plasticity of Fruits of *Acrocomia aculeata* in Western Minas Gerais, Brazil

Claudio E.M. Campos[1], Catherine Meyer[2], Thomas Hilger[2], Georg Cadisch[2], Sérgio Motoike[3]

[1]*Federal University of Viçosa, Department of Forestry, Brazil*
[2]*University of Hohenheim, Inst. of Agric. Sci. in the Tropics (Hans-Ruthenberg-Institute), Germany*
[3]*Federal Univerity of Viçosa, Department of Plant Science, Brazil*

Vegetable oils are an important international commodity in today's food and non-food industry. Nowadays, the oil palm (*Elaeis guineensis*) has the biggest share of this commodity. By the next few years, the demand of the cost-competitive palm oil is expected to rise up. An alternative to supply this demand is *Acrocomia aculeata*, an oilseed palm endemic to the Americas which has a productivity and oil composition comparable to *E. guineensis*. Acrocomia also shows a variety of agricultural advantages: being fire and drought tolerant, its possible cultivation on degraded soils and its environmental plasticity. Mature fruits of Acrocomia have a smooth hard exocarp, a fleshy mesocarp and a very hard, thick endocarp encompassing the endosperm. The mesocarp as the endosperm can be used for oil extraction whereas the endocarp and exocarp are promising for the production of bio-charcoal. This study aims to elaborate the variability in fruit phenotypic characteristics of ecotypes from different regions of Minas Gerais, Brazil, and to assess the difference between plants, originating from dry and humid regions. Mature fruits from different ecotypes were collected at the Macaúba Active Germplasm Bank, of the Universidade Federal de Viçosa in Araponga, MG, Brazil, in February 2020. The layers were separated. Their thickness, proportions, fresh and dry weight were determined. The mesocarp was dried for 48 hours so oil extraction could be done. The fruits are showing natural ecotypic differences. Fruit weight ranged from 20.4 to 41.3 g per fruit, where the predominant fruit mass was situated between 27.4–32.9 g and 32.2–36.5 g for the ecotypes from the dry northern and humid midwestern Minas Gerais, respectively. Fruits from dry regions tend to be smaller than fruits from humid regions. Exocarp thickness is between 0.5–1.5 mm independent of the ecotype. However, a difference between ecotypes can be found in the proportion of the exocarp to the total fruit. The majority of fruits show a proportion from 23.7–25.6 % and 18.5–23.3 % for the northern and midwestern ecotypes, respectively. Acrocomia ecotypes show a high fruit phenotypic plasticity, important for further development of the species as a crop.

Keywords: *Acrocomia aculeata*, Brazil, fruit characteristics, oilseed palm, phenotypic plasticity

Contact Address: Catherine Meyer, University of Hohenheim, Inst. of Agric. Sci. in the Tropics (Hans-Ruthenberg-Institute), Garbenstrasse 13, 70593 Stuttgart, Germany, e-mail: catherine.meyer@uni-hohenheim.de

Genomic Studies of Myanmar Rice (*Oryza sativa* L.) Varieties Using DArT and SNP Markers

Aye Aye Thant[1], Hein Zaw[2], Marie Kalousová[1], Rakesh Kumar Singh[3], Bohdan Lojka[1]

[1]*Czech University of Life Sciences Prague, Fac. of Tropical AgriSciences, Dept. of Crop Sciences and Agroforestry, Czech Republic*
[2]*Plant Biotechnology Center, Czech Republic*
[3]*The International Centre for BioSaline Agriculture, Crop Diversity and Genetics Section, United Arab Emirates*

Two ultra-high-throughput diversity array technology (DArT) markers (silicoDArT and SNP) were employed to investigate the genetic diversity and population structure of rice (*Oryza sativa* L.) varieties of Myanmar. The study was performed using 117 rice genotypes comprising 112 landraces and 5 improved (control) varieties with 4,064 silicoDArT and 7,643 SNPs derived from DArT platform. Quality control parameters included > 95 % call rate, > 95 % reproducibility, and minor allele frequency (MAF) >0.1 for screening. Polymorphic information content (PIC) values for silicoDArT ranged from 0.02 to 0.5 with an average of 0.37. In the case of SNP markers, PIC values ranged from 0 to 0.5 with an average of 0.41. Genetic variance among the genotypes ranged from 0.001 to 0.954 in silicoDArT and 0 to 0.753 in SNP markers. Genetic relationships among the genotypes were identified utilizing weighted neighbor-joining dendrograms. All the genotypes were grouped into two major clusters with both silicoDArT and SNP markers. Population structure were tested using K values from 1 to 5, maximum Δk was found at $K = 2$, confirming that the structure analysis also revealed two distinct genetic clusters. Analysis of Molecular Variance (AMOVA) with SNP markers showed that within individuals only 4% diversity existed whereas among individuals it was 22 %. Maximum diversity has been observed at population level (74 %). This study demonstrated that DArT markers are a useful tool for the genomic studies with regard to rice. It will support researchers to identify useful DNA polymorphisms in genes and germplasm of interest and apply that information for rice varietal development and release.

Keywords: DArT markers, genetic diversity, Myanmar, rice landraces, Silico-DArT, SNP

Contact Address: Aye Aye Thant, Czech University of Life Sciences Prague, Fac. of Tropical AgriSciences, Dept. of Crop Sciences and Agroforestry, Prague, Czech Republic, e-mail: aye18988@gmail.com

Characteristics of PBA Profiling Markers in the Analysis of *Arachis hypogaea* L. Genome

Julio Montero-Torres[1], Michal Droppa[2], Tania Pozzo[3], Roberto Acebey-Aldunate[1], Eloy Fernández Cusimamani[4], Jana Žiarovská[2]

[1]*Universidad Mayor, Real y Pontificia de San Francisco Xavier de Chuquisaca, Fac. of Agricultural Sciences, Bolivia*

[2]*Slovak University of Agriculture, Dept. of Genetics and Plant Breeding, Slovakia*

[3]*University of California, Dept. of Plant Sciences, United States*

[4]*Czech University of Life Sciences Prague, Fac. Trop. AgriSci., Dept. of Crop Sci. and Agroforestry, Czech Republic*

In this study, the PBA technique was utilised to characterise its effectivity to be used for the possible analysis of *Arachis hypogaea* L. genom polymorphism. Three different peanut genotypes were chosen that were characterised previously to have different profiles in iPBS fingerprints. All the genotypes were collected from Chuquisaca Department, Bolivia as original plant sources. The seeds were transferred to the Faculty of Tropical Agrisciences, CULS in Prague, Czech Republic and planted in pots. Young plants were transferred to the AgroBioTech Research Centre; SUA in Nitra, Slovak Republic where biological material was analysed. Three different PBA primer combinations were used in PCRs with the result of generating different PBA fingerprint profiles – CYP1A, CYP2B, and CYP2C. A total of 83 amplicons were generated for the analysed peanut accessions with the highest number of 33 amplicons for marker CYP2C, but for this marker, the lowest percentage of polymorphism was obtained on the level of 60 %. CYP1A marker achieved the polymorphism on the level of 63 % and CYP2B marker on the level of 79 %. CYP1A marker achieved the value of effective number of alleles 1.7634 and the Shannon's Information index 0.6245. CYP2B marker achieved the value of effective number of alleles 1.6500 and the Shannon's Information index 0.5830. CYP2C marker achieved the value of effective number of alleles 1.9780 and the Shannon's Information index 0.6876. None of the markers used in this study has generated the same profile for any of the analysed peanut accessions, that is why all of them should be useful for DNA based profiling of *Arachis hypogaea* L. germplasm, but CYP2B should be used preferably.

Keywords: *Arachis hypogaea*, genome, germplasm, PBA markers, polymorphism

---

Contact Address: Jana Žiarovská, University of Agriculture in Nitra, Dept. of Genetics and Plant Breeding, Nitra, Slovakia, e-mail: jana.ziarovska@uniag.sk

Zambian Neglected Species: Oils and Cakes Composition of Traditional Oil-Bearing Trees

Anna Manourova[1], Jan Tauchen[2], Adela Frankova[2], Ondrej Drabek[2], Vladimir Verner[1], Mukelabai Ndiyoi[3]

[1]*Czech University of Life Sciences Prague, Fac. of Tropical AgriSciences, Dept. of Crop Sciences and Agroforestry, Czech Republic*

[2]*Czech University of Life Sciences Prague, Fac. of Agrobiology, Food and Natural Resources, Dept. of Food Science, Czech Republic*

[3]*University of Barotseland, School of Natural Resources, Zambia*

Due to the increased contact between disparate human populations and the development of a global trading system, the number of species upon which global food security and agricultural incomes depend narrowed drastically. More than half of today's needs for energy are met by maize, wheat and rice. The impact of the species base narrowing is mostly felt most by the rural communities in marginal areas, where the set of livelihood options is limited. Many neglected/underutilised species occupy important niches, perfectly adapted to the local fragile conditions, contributing to sustainable production with minimal inputs as well as to the diversity and the stability of agro-ecosystems. The scientific community, as well as the food industry, know the only a minimal share of Zambian food plants. Therefore, our study focused on the chemical composition of oils and cakes of three neglected but traditionally important oil-bearing plants, namely: *Parinari curatellifolia*, *Schinziophyton rautanenii* and *Ochna serrulata*, in surroundings of Mongu, Western Province, Zambia. *P. curatellifolia* and *S. rautanenii* oils were chiefly composed of α-eleostearic acid (28.58–55.96 %), linoleic acid (9.78–40.18 %), and oleic acid (15.26–24.07 %), whereas *O. serrulata* contained mainly palmitic (35.62–37.31 %), oleic (37.31–46.80 %), and linoleic acid (10.61–18.66 %). Vitamin E of *S. rautanenii* oil was mainly composed of γ-tocopherol (3236.18 µg/g); *O. serrulata* contained similar proportions of α- (287.37 µg g^{-1}) and γ-tocopherol (361.11 µg g^{-1}), whereas *P. curatellifolia* had negligible levels of vitamin E. All three species can be considered as a good source of essential minerals. The results suggest great potential of the traditional Zambian species to be introduced into food, technical and/or pharmaceutical industry. Especially *O. serrulata* deserves deeper research attention, because of the considerable quantities of α-tocopherol in its oil which exhibits non-rancid properties. Due to the nutrient-rich cakes, the tested species might be also promising for animal fodder fortification.

Keywords: Cooking oils, neglected crops, sustainable diet, underutilised species, Zambia, α-eleostearic acid

---

Contact Address: Anna Manourova, Czech University of Life Sciences Prague, Fac. of Tropical AgriSciences, Dept. of Crop Sciences and Agroforestry, Prague, Czech Republic, e-mail: manourova@ftz.czu.cz

Biotic stresses, crop protection and multifunctional organisms

Effect of Biopesticides Neem Extract (*Azadirachta indica*) treatments on Soil Biochemical Properties and Plant Growth Promoting Rhizobacteria Viabilities

Sarjiya Antonius[1], Lita Nurliana Ulfa[2], Tirta Kumala Dewi[1]

[1]*Research Center for Biology, Indonesian Institute of Sciences, Indonesia*
[2]*Brawijaya University, Indonesia*

Recently the use of biopesticide is becoming popular. The neem seed contains Azadirachtin which is the most important component as a biopesticide. However, it is not known how the effect of this compound on PGPR and soil biochemical properties. This study aims to determine the impact of neem extract biopesticides (*Azadirachta indica*) on the viability and activity of plant growth-promoting rhizobacteria (PGPR). The method used is *in vitro* assay, by implanting disc blank which has been treated with the variation of biopesticide concentration of neem extract by 0 % (control), 2.5 %, 5 % and 10 % on 11 isolates of PGPR ie 4a, 4e, III b, PIKO, PP2, AD71, *Pseudomonas*, L7TO3, 140B, AA2, and AA1. In Vivo assay was conducted by applying biopesticide neem extract directly into soil with different concentrations. Soil biochemical properties monitored include soil respiration, PME-ase, and urease in soil samples that have been treated with biopesticides. The results showed that the neem extract biopesticide (*Azadirachta indica*) was able to inhibit eight of 11 PGPR isolates. The eight isolates were 4a, 4e, III b, PIKO, L7TO3, 140B, AA2, and AA1. Meanwhile, PP2, AD71, and Pseudomonas isolates were not inhibited. The value of the soil respiration test was proportional to the PGPR population number. In the PME-ase activity, measurement decreased as biopesticide extract treatments at higher concentrations. Similar results were observed in the Urease activity. From this work, it must be considered on the application of biopesticide neem extract in the plantations, because it will produce a negative effect on PGPR and soil biochemical properties.

Keywords: *Azadirachta indica*, PGPR, PME-ase, soil respiration, urease

Contact Address: Sarjiya Antonius, Research Center for Biology, Indonesian Institute of Sciences, Cibinong, Indonesia, e-mail: sarj.antonius@gmail.com

Impact and the Control of Root-Knot Nematodes in Tomato Production in Nepal

Jenish Nakarmi[1], Rabin Rijal[2], Shashank Kafle[2], Gopal Bahadur K.C.[2], Florian M.W. Grundler[1]

[1] *University of Bonn, Inst. of Crop Science and Resource Conservation; Molecular Phytomedicine, Germany*

[2] *Tribhuwan University, Institute of Agriculture and Animal Science (IAAS), Nepal*

Root-knot nematodes (RKN), one of the economically most important plant parasitic nematodes, cause severe losses all over the world. In Nepal, devastation caused by RKN is increasing rapidly especially in tomato production. Efficient management strategies against RKN are very limited. Therefore we studied the efficacy of different biological, botanical and synthetic agents in field trials performed in polyethylene houses at Bhaktapur and Kavre districts of Nepal simultaneously. Using randomised complete block design (RCBD), experiments were carried with six different treatments including untreated control with four replications during March to November 2019. Hybrid tomato (cv. Srijana), one of the most popular tomato varieties among the farmers in Nepal was used as model plant. The treatments included BioAct Prime (*Purpureocillium lilacinum*), Serenade ASO (*Bacillus subtilis* strain QST 713), and Velum Prime (Fluopyram), and Neem extract (Azadirachtin), combination of the BioAct Prime and Serenade ASO along with control (without any treatment). Observations were obtained on the plant growth, total yield and the disease parameters from six randomly selected plants by systematic sampling at certain intervals. The treatments varied significantly in the total yield, single fruit weight, egg population density in the root per system (RPS) of tomato plants, root length and fresh plant height taken during final harvest. Total yield was highest in Serenade ASO (112.23 kg plot^{-1}), followed by Velum Prime (105.27 kg plot^{-1}) with significant difference between two treatments, while average single fruit weight was maximum in Velum Prime (47.60 g) and Serenade ASO (47.58 g). Galling Index (GI) in RPS was observed lowest in Serenade ASO (3.083), accompanied by Velum Prime (3.042). Similarly, the lowest no. of eggs RPS was also observed in Serenade ASO (69687.5), followed by Velum (78250). Serenade ASO, a bio-control agent, appeared to be a good substitute of Velum Prime, a chemical nematicide, in suppression of RKN in the tomato poly house and high yield of tomato with the safe environment.

Keywords: Bio-control agents, Nepal, root-knot nematodes, *Solanum lycopersicum*, Velum Prime

Contact Address: Jenish Nakarmi, University of Bonn, Inst. of Crop Science and Resource Conservation; Molecular Phytomedicine, Jagdweg 1a, 53115 Bonn, Germany, e-mail: jnak@uni-bonn.de

Evaluation of Selected Common Bean Genotypes for their Reaction to *Xanthomonas campestris* pv. *phaseoli* in Kakamega County, Kenya

ANGELINE CHEPKEMBOI[1], JOHN MAINGI[2], SHEM NCHORE[3]

[1]*Kenyatta University, Department of Plant Sciences, Kenya*
[2]*Kenyatta University, Department of Biochemistry, Microbiology and Biotechnology, Kenya*
[3]*Kenyatta University, School of Agriculture and Enterprise Development, Kenya*

Common bean (*Phaseolus vulgaris* L.), plays a significant role in food security owing to its nutritional value and generation of income. However, output of dry beans in Western Kenya is constrained with a myriad of biotic and abiotic challenges including; diseases, pests, soil infertility and unfavourable weather conditions resulting to low productivity. Of the many diseases of beans, common bacterial blight (CBB) caused by *Xanthomonas campestris* pv. *phaseoli* (Xap) is a major biotic constraint to bean production causing up to 80 % yield losses on farmers fields when severe. Due to the fact that most agro-chemicals have not been effective against CBB and limited information on pathogen distribution in Kakamega County, the use of resistant genotypes is therefore a central management strategy. The aim of this study was to screen the available nine bean genotypes for resistance to CBB disease in both greenhouse and field conditions. Experiments were conducted in randomised complete block design with three replications in a factorial factor of 9×2×2 during the greenhouse and 9×2 field screening. During growth, data on plant height number of pods/plant, length of pods and size and number of CBB spots was taken. Yield parameters were also assessed. The findings from the experiment revealed a significant variation ($p < 0.05$) on the entire traits studied among the nine bean genotypes. Data from the field and greenhouse experiments were in conformity. None of the evaluated genotype was immune to CBB. In the green house, it was observed that disease symptoms were significantly severe ($p < 0.05$) in beans planted in non-sterile soil and inoculated with Xap compared to those planted in sterile soil and non-inoculated respectively. Cal77 and Cal156A genotypes exhibited high level of resistance to CBB while seven genotypes namely Cal285, Cal256, Cal271A, Cal274, KK8 and Cal87 showed moderate resistance. Further evaluation and screening needs to be done and the susceptible genotypes be tried in other locations.

Keywords: Common bean, immune, response, *Xanthomonas*

Contact Address: Shem Nchore, Kenyatta University, School of Agriculture and Enterprise Development, P.O.Box 43844 , 00100 Nairobi, Kenya, e-mail: nchoree@gmail.com

Adoption and Dis-Adoption of Sustainable Agriculture: A Case of Farmers' innovation and IPM Technologies for Suppressing Fruit Flies in the Kenyan Mango Farming Systems

Charity Wangithi[1], Beatrice Muriithi[1], Raphael Belmin[2]

[1]*International Centre for Insect Physiology and Ecology, Social Science and Impact Assessment Unit, Kenya*

[2]*Agricultural Research for Development (CIRAD), Senegal*

The invasive fruit fly *Bactrocera dorsalis* poses a major threat to the production and trade of mango in sub-Saharan Africa. The loss of market opportunities due to stringent phytosanitary measures makes the livelihoods of many small scale farmers who dominate the mango sub-sector vulnerable. In attempt to minimise the yield losses and production costs while maximising revenues, farmers devise different innovations to manage the pest challenges at the farm level. Most of these innovations guide research and development of modern bio-technologies. Using multi stage sampling technique,two sub counties in Embu county, Kenya were selected where mango is a predominant economically important fruit crop. Household survey data was collected after 2019 mango season. Our study tracked farmers' innovations in the management of the invasive fruit fly, analysed farmers' knowledge, perception and practices on the management of Bactrocera dorsalis and determinants of adoption and dis-adoption decisions on sustainable pest management strategies. Results reveal that farmers consider fruit flies as the major threat to the productivity of the mango sub-sector (98.8 %) and hence depend heavily on pesticides (89.7 %). Some farmers (34.6 %) use indigenous methods to manage the pest. Though farmers possess good knowledge of different non-pesticides strategies, uptake is relatively low. Results from the multinomial Logistic Regression model reveal that experience in mango farming, the number of mature mango trees owned, land allocated to mango production, access, and availability of the market determines the likelihood of farmers to adopt, dis-adopt or not adopt a non-pesticide fruit fly management strategy. We recommend strengthening information exchange networks to farmers for sustainable adoption of biocontrol technologies in the management of *B. dorsalis*.

Keywords: *Bactrocera dorsalis*, farmers' innovations, Kenya, knowledge, mango

---

Contact Address: Charity Wangithi, International Centre for Insect Physiology and Ecology, Social Science and Impact Assessment Unit, 30772, 00100 Nairobi, Kenya, e-mail: wangithimuthoni@gmail.com

Eiphosoma laphygmae, a Classical Solution for the Biocontrol of the Fall Armyworm, *Spodoptera frugiperda*?

Tabea Allen[1], Marc Kenis[2], Lindsey Norgrove[1]

[1]*Bern University of Applied Sciences, School of Agricultural, Forest and Food Sciences, Switzerland*

[2]*CABI, Switzerland*

The fall armyworm. *Spodoptera frugiperda* (J E Smith, 1757) is an invasive Lepidoptera and one of the most damaging cereal pests in the tropics, having arrived in Africa in 2016, spreading through the continent, then on to Asia. Current control methods rely on insecticides whereas biological control might offer a more sustainable solution. The parasitoids, *Eiphosoma laphygmae* and *E. vitticole* (Hymenoptera: Ichneumonidae), previously considered as synonyms, are potential classical biological control agents, yet knowledge on their biology needs to be collated and specificity assessments conducted. We aimed to assess existing knowledge on biology, identify their natural distribution, collate reported parasitism rates from field studies and determine which other parasitoids co-occur with them. We conducted a systematic literature review using the keyword "*Eiphosoma*" on 11.11.2019 in Web of Science, Agricola, CAB-Abstracts, and Food Science and Technology Abstracts. On 12.11.2019, we searched using the search string ("*Eiphosoma vitticole*" OR "*Eiphosoma laphygmae*") in full text in googlescholar. We had 121 initial hits. We then excluded papers from outside the topic areas and three for which we had no access, retaining 44 papers in English, Portuguese and Spanish.

Reports on the natural distribution of *E. laphygmae* were restricted to the American tropics with the most northerly record from Northern Mexico and the most southerly from the state of Sao Paulo, Brazil. In fields where E. laphygmae naturally occurred, it was the second most important contributor to the mortality of the fall armyworm, after *Chelonus insularis*, another hymenopteran parasitoid. On average, *E. laphygmae* parasitized 4.3 % of fall armyworm in field studies. The highest parasitism rates were observed in Costa Rica (13 %) and Minas Gerais Brazil (14.5 %). *E. laphygmae* appeared to establish better in more diverse systems with weeds. Given that it is assumed to be synovigenic, it is dependent on protein and nectar from wild flowers for egg production. As African farming systems often have high diversity, this may favour the establishment of *E. laphygmae* if eventually introduced as a classical biological control agent.

Keywords: Biological control, *Eiphosoma laphygmae*, *Eiphosoma vitticole*, fall armyworm, larval parasitism, maize, *Spodoptera frugiperda*

Contact Address: Lindsey Norgrove, Bern University of Applied Sciences, School of Agricultural, Forest and Food Sciences, Laenggasse 85, 3052 Zollikofen, Switzerland, e-mail: lindsey.norgrove@bfh.ch

Response of *Solanum lycopersicum* L. (Tomato) to *Tuta absoluta* and *Glomus clarum* Using SSR Marker

Odunayo Olawuyi, Itorobong Nsi

University of Ibadan, Dept. of Botany, Nigeria

Tomato (*Solanum lycopersicum* L.) is an annual crop and tropical nutritious vegetable that is sensitive to both abiotic and biotic stresses. *Tuta absoluta* is one of the most devastating pests resulting to economic loss in tomato globally. The use of chemicals and other biological methods have been employed in the management of *T. absoluta*, but there are limited information on the use of arbuscular mycorrhizal fungus (*Glomus clarum*) which is environmentally friendly, and the molecular primers associated with resistance of tomato to *T. absoluta*. Screen house experiment was conducted at the nursery farm of the Department of Botany, University of Ibadan to investigate the response of tomato to *T. absoluta* and *G. clarum* using morphological and molecular techniques. A total of eight tomato varieties which consisted of four varieties from TechnoServe Nigeria Limited, Katsina State, three from National Centre for Genetic Resources and Biotechnology (NACGRAB), Ibadan and one from Ojoo market, Ibadan in Nigeria were evaluated using Complete randomised design with three replicates.The three treatments comprised of *G. clarum* + *T. absoluta*, *T. absoluta* alone and Control (uninoculated). Growth, agronomic, yield and infestation parameters were evaluated,while DNA extraction and amplification of young apical leaves from treated and untreated (control) tomato varieties were carried out using four Simple Sequence Repeat (SSR) primers. Results showed that there was positive interaction between *G. clarum* and *T. absoluta* on growth, agronomic, yield and infestation characters of tomato. Plants inoculated with *G. clarum* had reduced percentage incidence (13 %) and severity of infestation (27 %) compared to control (23 % and 40 %) respectively. This could have been due to *G. clarum* which induced the activation of tomato defence mechanism and optimum nutrient uptake. There are variations in genetic resistance and susceptibility among the tomato varieties in treatment combinations of *T. absoluta* and *G. clarum*. Primers; LEAAT005 (0.67) and LECAA001 (0.57) detected the highest genetic polymorphism and could be used in marker assisted breeding of other *Solanum* spp. The tomato varieties; NGB00717 (NACGRAB), NGB00725 (NACGRAB) and Dan Eka Jibia (Technoserve) were highly tolerant to *T. absoluta* and could be recommended for breeding of tomato against *T. absoluta*.

Keywords: Genetic resistance, *Glomus clarum*, SSR marker, tomato, *Tuta absoluta*

Contact Address: Odunayo Olawuyi, University of Ibadan, Dept. of Botany, Ibadan, Nigeria, e-mail: olawuyiodunayo@yahoo.com

The Use of Agrochemicals and Mortality by Stomach Cancer in Brazil Between 1979 and 2015

Thiago Henrique Costa Silva[1], Nara Rabia Rodrigues do Nascimento-Silva[2], Luciana Ramos Jordão[1], Dinalva Donizete Ribeiro[3]

[1]*Federal University of Goiás (UFG), Law College, Brazil*
[2]*Federal University of Goiás (UFG), Nutrition, Brazil*
[3]*Federal University of Goiás (UFG), Geography, Brazil*

The intensive use of agrochemicals in Brazilian agriculture, typical of productivist production model and based on commodities for export, has been correlated to several environmental problems and to human health, among them stomach cancer. This study seeks to analyse the correlation between the use of pesticides per area harvested from the three main crops of the country (soybean, corn and sugarcane), which together correspond to more than 70 % of the planted area of the Brazil, and the annual mortality of stomach cancer. Through statistical analyses in an ecological study, with a qualitative and quantitative approach, from the data from 1979 to 2015, the correlation of the variables is inferred and the need for public policies that allow to make food sovereignty feasible, in order to allow the production of food that takes into account the respect for nature and for human life. The Brazilian dependence on pesticide use itself as an obstacle to the exercise of food security and sovereignty, curbing industrialisation, and imposing the subordination of the State to the power of multinationals. The use of pesticides in sugarcane, soybeans, and corn increased from 100,476,688 liters per harvested area in 1979 to 535,177,688 in 2015, while mortality from stomach cancer in Brazil increased from 8,602 cases to 14,265 in the same period. By correlating these data, this study concluded that the variables have a strong positive correlation. Consequently, the use of pesticides can be considered a risk factor for stomach cancer, setting up further evidence that the super dependent chemical production model needs to be rethought. For this reason, the pesticide needs to stop being just part of an (Agri)business to be analysed under the focus that led to its use: healthily maintaining life. Therefore, the government needs to develop public policies that guarantee constitutional rights to an ecologically balanced environment and the health and life of society. However, what has been verified in recent years, mainly in the Jair Bolsonaro government, is an increase in the use of pesticides in the country and in the quantity of authorised substances, totaling 475 in 2019 and 150 until May 2020.

Keywords: Agrochemicals, commoditisation, food sovereignty, stomach neoplasms

Contact Address: Luciana Ramos Jordão, State University of Goias, Law College, Av. Perimetral Norte N. 4356 C. 88b Cond. Alto da Boa Vista, 74445500 Goiais, Brazil, e-mail: lr.jordao@me.com

Pesticides Residue on Brazilian Tomatoes, Food Safety and Regulatory Framework

Rebeca Barbosa Moura, Luciana Ramos Jordão, Barbara Luiza Rodrigues, Niury Ohan Pereira Magno, Igor Gomes de Araújo, Letícia Versiane Arantes Dantas, Thays Dias Silva, Victória Cardoso Carrijo

State University of Goias, Law College (Southeast Campus: Morrinhos), Brazil

Brazilian economy is especially focused on agriculture. Even though the Constitution stablishes individual rights and human dignity as a priority, economic forces generally surpasses individual wellness on day to day issues. The current development on policies concerning pesticides exemplify this. In order to analyse changes on allowed pesticide residues and food regulations, this research evaluates one of the most consumed vegetable in Brazil, the tomatoes. Brazilians eat 21 tomatoes per capita early. The item is consumed by 205 million people, since tomatoes are one of the most important vegetable on everyday meals. However, it also has one of the highest scores on pesticide residue levels according to ANVISA, the Brazilian health surveillance agency. From 2013 to 2015, ANVISA analysed 730 tomato samples. Unauthorised pesticides were identified in 200 of them. Even knowing each Brazilian consumes 7 to 9 liters of pesticides, the government discusses new regulations to make new products approval faster by lowering criteria and creating a committee linked to Agriculture Ministry which would have no direct interference of environmental public agencies. During the first year and half of President Bolsonaro, more than 600 pesticides were authorised, 150 of them up to May 2020 and 474 in 2019. There is a clear incentive on pesticide usage whereas residue control is put aside just as discussions about healthier nutrition, working conditions and environmental impacts. By analysing one of the most important vegetables in everyday consumption in Brazil, especially the regulations on pesticide residue and statistic data on its impacts on health, nutrition and environmental preservation, one is led to the conclusion that residue control can cause not even health damage to people, but will also reduce soil quality and bring negative economic consequences in the long run. Thus, this research, based on deductive method, follows the premise in which expanding pesticides usage needs stronger regulations on residue control as it is determined by the Constitution.

Keywords: Food security, human rights, rural development, sustainable development

Contact Address: Luciana Ramos Jordão, State University of Goiás, Law College (Southeast Campus: Morrinhos), Av. Perimetral Norte, n. 4356, c. 88b, Alto da Boa Vista, 74445500 Goiânia, Brazil, e-mail: luciana.jordao@ueg.br

Bacillus-mediated Changes in Iron Partitioning and Sequestration in Lowland Rice under Iron Toxic Conditions

Tanja Weinand, Islam Gul Zeeshan, Julia Hartmann, Folkard Asch
University of Hohenheim, Inst. of Agric. Sci. in the Tropics (Hans-Ruthenberg-Institute), Germany

Iron toxicity is the result of large concentrations of reduced iron (Fe II) in the soil solution, which occurs in flooded (arable) soils. It constitutes a severe stress in lowland rice cultivation and can lead to significant economic losses. Despite the widespread occurrence of iron toxicity in many rice producing areas, little is known about the impact of rhizobacteria on rice genotypic responses.

Rice cultivars tolerant to iron toxicity have been shown to employ different tolerance strategies and we hypothesise that the extent of iron toxicity symptom mitigation or aggravation upon *Bacillus* spp. inoculation is partly dependent on the genotypic adaptation mechanism. This study investigates *Bacillus*-mediated changes in iron sequestration in lowland rice and their impact on the expression of iron toxicity symptoms.

Six rice (*Oryza sativa*) cultivars (IR75866–17-B-12-WAB1, Suakoko 8, Nipponbare, IR31785–58-1–2-3–3, TOX 4004–8-1–2-3, Sahel 108), differing in their tolerance to iron toxicity were subjected to inoculation with three different *Bacillus* isolates (*B. pumilus* and *B. megaterium*) and subsequently exposed to 1000 ppm Fe (II) over the course of one week. The effects of bacteria treatment were evaluated by symptom scoring, dry weight determination and measurement of plant iron content.

Screening of leaf scores showed that the impact of *Bacillus* isolates on iron toxicity symptoms significantly differs between rice cultivars. Further, iron uptake and shoot iron concentration do not always correlate with severity of leaf symptoms.

Iron partitioning in rice tissues will be further analysed. Varietal differences in symptom expression changes following bacteria treatment will be related to altered iron concentrations in leaf blades, leaf sheaths and roots. Further, a possible impact of *Bacillus* inoculation on iron sequestration within the plant cell will be investigated.

Keywords: *Bacillus* spp, iron toxicity, *Oryza sativa*

Contact Address: Tanja Weinand, Institute of Agricultural Science in the Tropics (Hans-Ruthenberg-Institute), Management of Crop Water Stress in the Tropics and Subtropics, Garbenstr. 13, 70599 Stuttgart, Germany, e-mail: tanja.weinand@uni-hohenheim.de

Abiotic stresses in crops

Oral Presentations

Posters

Reserve Mobilisation, Dry Matter Partitioning, and K/Na Ratio of Rice (*Oryza sativa*) Seedlings in Response to Varying Levels of Salt Stress

Kristian Johnson, Folkard Asch

University of Hohenheim, Inst. of Agric. Sci. in the Tropics (Hans-Ruthenberg-Institute), Germany

Across South and Southeast Asia millions of hectares of arable land are left uncultivated or only support low yields due to salinity. This will likely worsen through a combination of changes in land use and climate, such as urbanisation, intensification of agriculture, unpredictable wet seasons, irregular rainfall, and rising sea levels. Greater insight into the underlying physiological mechanisms of varietal tolerance in rice (*Oryza sativa*) to salt stress across development stages, in particular germination and early seedling stages, will improve management practices and varietal screening, ultimately leading to more targeted breeding.

This study investigates morphophysiological characteristics related to rice seedling vigour and salt stress tolerance, such as the kinetics of reserve mobilisation, dry matter partitioning among organs, the K/Na ratio, and early seedling growth in seedlings subjected to salt stress . Five rice cultivars ranging in their sensitivity to salt stress (I Kong Pao, IR4630–22-2–5-1–3, OM5451, IR31785–58-2–3-3, and IR64) were germinated in sand and subjected to three levels of salinity (0, 50, 100 mM) over the course of 17 days in a greenhouse. Dry matter fractions were measured from samples taken every other day. Sodium and potassium concentrations within the seedling biomass were determined on a whole seedling basis for each sampling period, except for the final sampling date, during which above and below-ground biomass were analysed separately. Varietal differences in early growth vigour across salt stress treatments will be discussed in relation to genotypic partitioning strategies among organs, growth respiration, transition to autotrophy, and the sodium and potassium concentrations within the seedling biomass.

Keywords: K/Na ratio, rice (*Oryza sativa*), salinity, salt stress, seedling vigour

Contact Address: Kristian Johnson, University of Hohenheim, Inst. of Agric. Sci. in the Tropics (Hans-Ruthenberg-Institute), Brühlstrasse 9, 70563 Stuttgart, Germany, e-mail: k.johnson@uni-hohenheim.de

Root Architecture of Rice as Affected by Phosphorus Starvation and Salt Stress

Alenna Vazquez-Glaría[1], Mareike Kavka[2], Loiret Fernández[1], Eduardo Ortega[1], Bettina Eichler-Löbermann[2]

[1]*University of Havana, Plant Physiology, Cuba*

[2]*Rostock University, Agricultural and Environmental Faculty, Germany*

Salt stress is one of the most harmful environmental factors affecting production of crops such as rice (*Oryza sativa*). In Cuba about 15 % of the agricultural land is affected. Plants exposed to high salinity often also suffer from insufficient nutrient supply. It is known that salinity and phosphorus (P) availability in soil can modify the root morphology, but only little information is available regarding their combined effects on root-system architecture (RSA). In our study we quantified the variation on RSA of two rice genotypes (INCA LP-5 and Perla de Cuba) in dependence of salt level (0 and 50 mM) and P availability (1 and 10 ppm) in mini-rhizotrons under controlled conditions. After 21 days of growth the root systems were carefully placed on a scanner and analysed with the free software GiA Roots. The results showed significant differences of the root system between both rice cultivars mainly regarding the total length, network surface area, network volume and root system extension, which all were higher in INCA LP-5 than Perla de Cuba. Phosphorus starvation and salt stress resulted in a decrease of these characteristics. A combined stress of P shortage and salinity, however, did not further alter the traits except the total root size, which was more decreased when both stress factors occurred together. Both, P starvation and salt stress also reduced considerably the shoot biomass by more than 50 %. The P concentration was rarely affected. The root architecture at early growth stage could possibly affect the adaptation to salt stress and P deficiency at later growth periods, and might be used as an indicator to select adapted genotypes.

Keywords: Nutrients, rhizotron, *Oryza sativa*, root size, salt stress

---

Contact Address: Bettina Eichler-Löbermann, Rostock University, Agricultural and Environmental Faculty, Justus-von-Liebig-Weg 6, 18059 Rostock, Germany, e-mail: bettina.eichler@uni-rostock.de

Phosphorus Acquisition Efficiency in Rice: Does Carbon Supply Limit Root Growth under Extreme P Deficiency?

Alejandro Pieters, Sabine Stürz, Folkard Asch

University of Hohenheim, Inst. of Agric. Sci. in the Tropics (Hans-Ruthenberg-Institute), Germany

Phosphorus (P) acquisition efficiency (PAE) depends largely on a large root system. For high PEA varieties, enlarging the root system under P deficiency requires increased carbon partitioning towards the roots under conditions that also limit photosynthesis. Thus, root growth under P deficiency could be carbon limited. To test this hypothesis, we grew four rice genotypes (DJ123, Kasalath, Santhi Sufaid, high PAE, and IR64 a low PAE check), in hydroponics under glasshouse conditions and provided them with either high (0.32 mM) or low (0.0032 mM) P in the nutrient solution for 50 days. Photosynthetic rate was measured, plant dry weight and carbohydrate concentrations in leaves and roots were determined. Under high P supply shoot and root dry weight was different amongst varieties ($p < 0.001$). P deficiency decreased total and organ biomass but the impact was genotype specific as depicted by a significant PxG interaction ($p < 0.001$). P deficiency increased the root to shoot ratio and high PAE genotypes showed the largest increases compared to IR64. At high P supply photosynthesis rate differed amongst varieties and decreased under P deficiency ($p < 0.001$), the most affected variety was Kasalath and the least affected was Santhi Sufaid, both high PAE varieties. Sucrose concentrations in leaf blades and roots were lower in P starved plants than in high P plants. Starch concentrations in leaves increased under P deficiency but decreased in roots. Root dry weight was hyperbolically related to photosynthesis and linearly related to starch concentration in roots. Our results suggest that root growth under P deficiency could be limited by carbon supply, as leaf sucrose concentrations decreased importantly under P deficiency. However, whether the increase of leaf starch under P deficiency represents unused carbon due to restricted growth or contributes to maintain P homeostasis within the chloroplast allowing photosynthesis to continue, remains to be elucidated. Under these circumstances, selecting for high photosynthetic rates under P deficiency, coupled to an efficient translocation of assimilates to roots could contribute to breeding for high PAE in rice.

Keywords: Carbohydrates, carbon partitioning, IR64, *Oryza sativa*, photosynthesis

Contact Address: Alejandro Pieters, University of Hohenheim, Inst. of Agric. Sci. in the Tropics (Hans-Ruthenberg-Institute), Garbenstr. 13, 70599 Stuttgart, Germany, e-mail: alejandro.pieters@uni-hohenheim.de

How Drought Affects Water Use Efficiency and Photosynthesis in the Neotropical Oilseed Palm *Acrocomia aculeata*

Catherine Meyer[1], Thomas Hilger[1], Sérgio Motoike[2], Georg Cadisch[1]

[1]*University of Hohenheim, Inst. of Agric. Sci. in the Tropics (Hans-Ruthenberg-Institute), Germany*

[2]*Federal Univerity of Viçosa, Department of Plant Science, Brazil*

Drought affects a wide range of morphoanatomical, physiological and biochemical processes in plants and has a major impact on their growth and development. Hence, it is a substantial limiting factor for crop expansion, further being intensified in the future by global climate change. An immediate plant response to water deficit is the closure of stomata, so decreasing the leaf gas exchange and subsequently the photosynthetic rate.

This can lead in the long run to a decrease in biomass accumulation and yield. Water use efficiency (WUE) is the relationship between accumulated biomass or assimilated carbon and the amount of transpired water. It is so an important ecophysiological indicator to drought stress adaptation of plants, as plants with a high WUE are more resilient to water deficits. *Acrocomia aculeata*, endemic to semi-arid and arid regions of Central and South America, is an agricultural and industrial promising oilseed palm with a tolerance to prolonged drought up to six months. This study aimed to assess water use efficiency, stomatal conductance and photosynthetic rate of Acrocomia ecotypes originating from different climatic regions of Brazil in the dry and rainy season. The study was conducted at the Macaúba Active Germplasm Bank of the Universidade Federal de Viçosa in Araponga, MG, Brazil. Five ecotypes originating from Cerrado, Mata Atlântica and Caatinga regions were selected. Leaf gas exchange and relative leaf water status (RWC) were measured in September 2019 (dry season) and in January 2020 (rainy season). Additionally, climate data and soil moisture content were recorded. Photosynthetic rate and stomatal conductance were reduced in the dry season where humid adapted ecotypes were more affected than arid adapted ecotypes. WUE was 1.5–2 times higher under drought conditions, suggesting an efficient stomatal control of transpiration by *A. aculeata*. This is also reflected in a high RWC during the dry season, being above 85 % in both seasons, considering that soil moisture content was 0.156 m^3 m^{-3} and 0.261 m^3 m^{-3} in September and January, respectively. This suggests that *A. aculeata* is able to acclimate to drought events, however, ecotype differences need to be taken into account.

Keywords: *Acrocomia aculeata*, Brazil, drought, photosynthesis, water use efficiency

---

Contact Address: Catherine Meyer, University of Hohenheim, Inst. of Agric. Sci. in the Tropics (Hans-Ruthenberg-Institute), Garbenstrasse 13, 70593 Stuttgart, Germany, e-mail: catherine.meyer@uni-hohenheim.de

Photosynthesis, Dry Matter Production and Yield Performance of Lentil Varieties under High Temperature

Md. Tariqul Islam

Bangladesh Institute of Nuclear Agriculture, Crop Physiology Division, Bangladesh

Climate is changing and air temperature is raising due to increasing concentration of CO_2 and other atmospheric greenhouse gases. The rise in atmospheric temperature causes detrimental effects on growth, yield, and quality of the crop varieties by affecting their phenology, physiology, and yield components. Lentil (*Lens esculenta* Medik.) is an important pulse crop with high protein content, has the potential capacity to combat nutritional deficiencies in developing regions and countries. It occasionally faces high temperature at its reproductive stage in February-March due to late sowing after Transplanting Aman rice harvest. Efforts can be made to increase area as well as yield of lentil crops by the use of temperature stress tolerant varieties. A pot experiment was carried out with ten high yielding lentil varieties cultivating by the farmers to assess the effects of high temperature (34°C) on photosynthesis, chlorophyll content (SPAD reading) in leaves, dry mass production, yield attributes and yield and to find out temperature stress tolerant varieties. The experiment was conducted from 21 November 2017 to 15 March 2018 at BINA, Mymensingh, Bangladesh. High temperature (34°C) at two growth stages like flower initiation stage and pod growth stage of the lentil varieties *viz.* B-2, B-3, B-4, B-5, B-6, B-7, B-8, B-9, B-10 and BARI masur-5 was imposed separately for 7 days in plant growth chamber. The experiment was laid out in a completely randomized design with three replications. The nitrate reductase activity, chlorophyll content and photosynthetic rate of leaves were determined during temperature imposed. At harvest, total dry matter, seed yield and yield related parameters were recorded. Photosynthesis, chlorophyll content in leaves, total dry matter and yield attributes were highest in control plants and decreased under high temperature. High temperature imposed either at flower initiation stage or pod growth stage had significantly negative influence on plant parameters but temperature imposed at pod growth stage had greater negative effect. Seed yield drastically reduced under high temperature in all varieties at any growth stage compared to control.. The higher yield reduction was recorded in Binamasur-2 and Binamasur-3. But the yield loss under high temperature was less in two varieties *viz.* Binamasur-6 and Binamasur-8 and showed tolerance to high temperature.

Keywords: Dry matter yield, high temperature, lentil, photosynthesis

Contact Address: Md. Tariqul Islam, Bangladesh Institute of Nuclear Agriculture, Crop Physiology Division, Bangladesh Agricultural University Campus, 2202 Mymensingh, Bangladesh, e-mail: islamtariqul05@yahoo.com

Effects of Soils and Droughts on Yield, Water Use Efficiency and Thiamine of Leafy Vegetables

TAEWAN PARK[1], SAHRAH FISCHER[1], CHRISTINE LAMBERT[2], THOMAS HILGER[1], GEORG CADISCH[1]

[1]*University of Hohenheim, Inst. of Agric. Sci. in the Tropics (Hans-Ruthenberg-Institute), Germany*

[2]*University of Hohenheim, Inst. of Biological Chemistry and Nutrition, Germany*

Most research on crops, and interactions with soil fertility and water availability, has focussed on staple and cash crops. Green leafy vegetables, vital for food and nutrition security and human health, have been neglected. In a time when intensive agriculture is degrading soils, and climate change is causing increasing cases of erratic weather, this lack of knowledge can be harmful. The main aim of this research was to evaluate the effect of (i) soil fertility and (ii) drought on yield, water use and thiamine (Vitamin B1) content of green leafy vegetables. For this purpose, the green leafy vegetables cowpea (*Vigna unguiculata*), black nightshade (*Solanum nigrum*), and collard greens (*Brassica oleracea* var. *viridis*) were cultivated in a greenhouse trial. The vegetables were subjected to three watering regimes, i.e. 25 % (severe drought), 50 % (mild drought) and 75 % pot capacity (control), and cultivated on two soils (low vs. high fertility). The vegetables were evaluated on above- and belowground biomass, yield, nodulation in cowpea, total water use, water use efficiency of yield (WUEY) and thiamine. The yield of cowpea and nightshade was higher in the fertile soil under all watering regimes than in the infertile soil. Fertile soil had a higher WUEY than infertile soil in all three vegetables. Severe drought resulted in the highest WUEY of all vegetables in fertile soil (cowpea: 15.0 g L^{-1}, nightshade: 16.7 g L^{-1} and collards: 37.3 g L^{-1}). Collards had the highest WUEY in all treatments. Thiamine of collards and cowpea was significantly increased in infertile soil under mild and severe drought (collards: 141 – 305 % and cowpea: 133 – 185 % to fertile soil), however the absolute thiamine amount was lower due to the decreased yield. In conclusion, cowpea and nightshade were more dependent on soil fertility than collards in terms of yield, and collards was the most productive vegetable under drought conditions. In rural areas, green leafy vegetables often represent the main source of nutrients in the diet. The results can be used to suggest better-quality and -quantity diets in rural areas and understand the effects of drought and soil fertility on food and nutrition security.

Keywords: Drought, food quality, green leafy vegetables, nutrition, soil fertility, vitamin

Contact Address: Taewan Park, University of Hohenheim, Inst. of Agric. Sci. in the Tropics (Hans-Ruthenberg-Institute), Garbenstr. 13, 70599 Stuttgart, Germany, e-mail: taewan.park@uni-hohenheim.de

Short-Term Memory Induction, a Method for Improving Drought Stress Tolerance in Sweetpotato Crop Wild Relatives

Fernando Guerrero-Zurita[1], David A. Ramírez[1], Javier Rinza[2], Johan Ninanya[2], Bettina Heider[1]

[1]*International Potato Centre (CIP), Genetics, Genomics and Crop Improvement, Peru*
[2]*International Potato Centre (CIP), Crop and Systems Science Division, Peru*

Sweetpotato crop wild relatives (*Ipomoea* series Batatas (Choisy) D. F. Austin) are a group of species whose physiological potential for drought stress tolerance is very high but poorly studied. In the present study, short-term memory induction was tested in 59 sweetpotato crop wild relatives (SP-CWR) accessions with the aim of improving drought stress tolerance. For this purpose, accessions were subjected to two treatments, i) non-priming: full irrigation up to field capacity, and ii) priming: three drought stress periods with no irrigation of increasing length. The priming process started after flowering onset with drought stress periods of 8, 11, and 14 days followed each by 14 days of recovery with full irrigation. Senescence delay (S), foliar area (FA), leaf-minus-air temperature, and leaf carbon isotopic discrimination were the eco-physiological indicators measured to determine short-term memory induction in all accessions. Drought stress tolerance was evaluated in terms of resilience capacity index, and production capacity index; both were calculated per accession based on yield performance. An increase in S, improved leaf photosynthetic performance, efficient leaf thermoregulation and increased FA were the memory mechanisms identified in 81.4, 50.8, 28.8 and 23.7 % of the total number of accessions, respectively. Under a severe drought stress scenario, a resilient response included more long-lived green leaf area with a concomitant higher aerial biomass development while a productive response was related to optimized leaf thermoregulation and gas exchange. The results of this study highlight the potential to improve sweetpotato resilience in dry environments in introgression breeding programs that include *I. triloba* and *I. trifida*. Moreover, since *I. splendor-sylvae, I. ramosissima, I. tiliacea*, and wild *I. batatas* were the most productive species in this study, further genetic and metabolic studies should be conducted in this species looking forward to increasing sweetpotato drought stress tolerance. This study proposes a new model of drought stress tolerance improvement based on short-term memory induction. It highlights the importance of including crop wild relatives due to their outstanding physiological response (higher than sweetpotato cultivars) in limited water conditions.

Keywords: Carbon isotopic discrimination, foliar area, leaf temperature, productivity, resilience, senescence delay

Contact Address: Fernando Guerrero-Zurita, International Potato Centre (CIP), Genetics, Genomics and Crop Improvement, Av. La Molina 1895, Lima 51 La Molina, Peru, e-mail: f.guerreroz13@gmail.com

The Effect of Drought on Leaf Emergence and Development in the Oilseed Palm *Acrocomia aculeata*

Catherine Meyer[1], Thomas Hilger[1], Sérgio Motoike[2], Georg Cadisch[1]

[1]*University of Hohenheim, Inst. of Agric. Sci. in the Tropics (Hans-Ruthenberg-Institute), Germany*

[2]*Federal Univerity of Viçosa, Department of Plant Science, Brazil*

Acrocomia aculeata is an oilseed palm native to semi-arid and arid regions of Central and South America. They are solitary palms with pinnate leaf blades of around two to three metres in length. Young leaves are emerging tightly furled, reminding of a spear, thus named spear leaves. The leaves unfurl and expand when completely emerged. *Acrocomia* shows a resilience to drought, however knowledge is lacking on how water deficit affects the leaf formation and development. This study aimed to determine the monthly leaf emergence and unfurling in different *Acrocomia* ecotypes and to evaluate the effect of water deficit on leaf development using specific leaf area (SLA), leaf dry matter (LDM) and leaf thickness (LT) as indicators. It was hypothesised that leaf emergence, SLA and LDM decrease during the dry season whereas LT increases. Data collection was done in the Macaúba Germplasm Bank of the Universidade Federal de Viçosa in Araponga, MG, Brazil. Selection of the Acrocomia ecotypes was based on their climatic origin. From April 2019 to March 2020, the number of newly formed leaves was counted monthly. Leaflets from the petiole middle of one-month old leaves were sampled. The fresh and dry weight, and the leaf length and width were determined. Additionally, data on climate and soil moisture content were collected. In the dry season, the leaf unfurling was slowed down, leading to an accumulation of spear leaves. Furthermore, only zero to one leaf emerged versus two to three leaves in the rainy season. Consequently, water deficit caused a decelerated vegetative growth. A difference between the ecotypes could be observed, where ecotypes coming from drier regions used the strategy of reduced leaf unfurling and ecotypes from humid regions showed a decreased leaf emergence. Contrary to expectations, the SLA, LT and LDM showed no clear response to water availability. SLA ranged between 4.2–7.9 $m^2\ kg^{-1}$ and 4.0–6.7 $m^2\ kg^{-1}$, LDM between 26.2–45.8 % and 34.5–45.8 %, and LT between 380–596 μm and 379–653 μm for the dry and rainy seasons, respectively. This could be explained by the faster leaf development in the rainy season, which has a decreasing impact on SLA and LDM.

Keywords: Brazil, leaf development, leaf emergence, specific leaf area, water availability

Contact Address: Catherine Meyer, University of Hohenheim, Inst. of Agric. Sci. in the Tropics (Hans-Ruthenberg-Institute), Garbenstrasse 13, 70593 Stuttgart, Germany, e-mail: catherine.meyer@uni-hohenheim.de

Salinity Effects on the Activities of ROS Scavenging Enzymes in Leaves of two Sweet Potato Clones

Shimul Mondal[1], Susanne Burgert[1], Julia Hartmann[1], Ebna Habib Md Sofiur Rahaman[2], Folkard Asch[1]

[1]*University of Hohenheim, Inst. of Agric. Sci. in the Tropics (Hans-Ruthenberg-Institute), Germany*

[2]*International Potato Center (CIP), Sweet potato breeding in the tropics, Bangladesh*

In the Ganges-Brahmaputra delta of Bangladesh sweet potato significantly contributes to food security, but seasonal differences in river water levels combined with tidal effects, render the sweet potato production prone to salinity. Therefore, salt tolerant clones would be needed to maintain production, but to date little is known about salt tolerance traits in sweet potato. Salt (NaCl) stress in sweet potato may result in excessive uptake of sodium or chlorine into plant tissue leading to the formation of reactive oxygen species (ROS), which in turn may destroy chloroplasts' thylakoid membranes and reduce photosynthesis. Plant defensive mechanism include antioxidant enzymes such as glutathione reductase (GR), catalase (CAT), super oxide dismutase (SOD), ascorbate peroxidase (APX) and peroxidase (POX) whose levels may be increased in clones possessing a high tissue tolerance to NaCl. Cuttings of two contrasting cultivars of sweet potato, BARI SP 8 (tolerant) and BARI SP 4 (sensitive), were greenhouse-cultivated in nutrient solution for 21 days and then exposed to 100 mmol NaCl for 7 days. 3, 5, and 7 days after salt application the youngest leaves were sampled individually and enzyme activities, K and Na concentrations, chlorophyll content (SPAD), and dry mater (DM) determined. Leaf DM and leaf K/Na ratio in general, and GR activities in BARI SP 4 decreased for salt stressed plants whereas SPAD and CAT activities in general, and GR activities in BARI SP 8 increased. Varieties differed strongly in their responses with BARI SP 8 always showing more severe effects. Activity levels of SOD, APX and POX were not affected by salinity in neither variety. We conclude that salinity does not lead to increased levels of ROS in sweet potato under salt stress. Further studies are needed on the regulation of GR, sodium, chlorine, and potassium uptake as salt tolerance traits uptake in sweet potato.

Keywords: Antioxidant enzyme, dry matter, K/ Na ratio, salt tolerant

Contact Address: Shimul Mondal, University of Hohenheim, Inst. of Agric. Sci. in the Tropics (Hans-Ruthenberg-Institute), Management of Crop Water Stress in the Tropics and Subtropics, Garbenstr. 13, 70599 Stuttgart, Germany, e-mail: shimul.mondal@uni-hohenheim.de

Comparative Study on Water Use Efficiency of Cotton in Sole and Relay Intercropping Wheat-Cotton Production Systems

Uğur Çakaloğullari, Özgür Tatar
Ege University, Fac. of Agriculture, Dept. of Field Crops, Turkey

Climate change induced water scarcity is one of the most important challenges for agricultural production. Therefore, agricultural lands are expected to be limited especially in arid and semi-arid areas due to restricted water resources. Furthermore, land-use competition between staple crops such as wheat and cash crops such as cotton will inevitably rise. In the present study, we aimed to evaluate water use efficiency (WUE) in relay-intercropping system which increases land-use efficiency and allows to cultivate wheat and cotton in a sam. land and growing season. A field experiment during 2017/18 was carried out with two different water regimes (deficit-watered and well-watered) and two different cropping systems (sole and relay strip intercropping of wheat-cotton). Irrigation amount in well-watered treatment was adjusted according to local farmer practices in the experimental site (480 mm) while 13 % reduced water applied in deficit watered treatment to simulate moderate water limited conditions. Our results indicated that WUE of cotton and wheat were affected significantly by cropping systems ($p < 0.05$). Sole cropping of cotton had higher WUE (0.209 kg ton^{-1}) than relay strip intercropping with wheat (0.142 kg ton^{-1}). Similarly, sole cropping of wheat had higher WUE (0.554 kg ton^{-1}) than relay strip intercropping with cotton (0.388). These results were found by evaluated of each individual crops separately in cropping systems. However, it is possible to revealing synergistic effect of relay strip intercropping by calculating monetary water use efficiency (WUEM). According to monetary comparison based on cropping systems, sole cotton (0.170 ton^{-1}) and relay strip intercrop (0.165 ton^{-1}) had almost same WUEM and also higher than WUEM of sole wheat (0.146 ton^{-1}). Consequently, while well watering conditions lead to increase (17 %) in WUEM of sole cotton, the increase was more pronounced (45 %) in relay strip intercropping syste*m*. According to our first findings, wheat-cotton production by relay strip intercropping system could be suggested as alternative system to sole cotton in terms of water use efficiency.

Keywords: Cotton, relay strip intercropping, water scarcity, water use efficiency, wheat

Contact Address: Özgür Tatar, Ege University, Fac. of Agriculture, Dept. of Field Crops, İzmir, Turkey, e-mail: tatar.ozgur@gmail.com

Effect of Awn on Water Use Efficiency of Wheat

Recep Karakoç*, Özgür Tatar

Ege University, Fac. of Agriculture, Dept. of Field Crops, Turkey

Increasing frequency of drought conditions comprising grain filling period of wheat in Mediterranean region more often limits grain yield. Although several phenotypic traits to increase grain yield during terminal drought conditions have been reported, effects of awn, the thread-like extension of lemma, on dry matter assimilation during grain filling stage under limited water conditions has been still not clear. In the present study, awns of bread wheat (*Triticum aestivum* L.) and durum wheat (*Triticum durum* L) were removed at post-anthesis stage to assess their effects on water use efficiency and dry matter assimilation under rainfed and irrigated conditions. Grain yield was higher in awned than awnless bread wheat in both rainfed (16 %) and irrigated (9 %) conditions. However, there were no significant changes in grain yield of awned and awnless durum wheat under rainfed conditions while awns had positive effects on grain filling under irrigated conditions (17 %). Awned bread wheat had higher water use efficiency (WUE) in both conditions (19 % and 12 %) while WUE of awned drum wheat was only higher under irrigated conditions (21 %). Our findings suggested that xeromorphic structure of awns could be an advantageous for grain yield under water deficit conditions. Similar responses of awned and awnless durum wheat under rainfed conditions were perceived as the different adaptation abilities of drum wheat to limited water conditions.

Keywords: Awn, *triticum aestivum*, *Triticum durum*, water use efficiency, wheat

Contact Address: Özgür Tatar, Ege University, Fac. of Agriculture, Dept. of Field Crops, İzmir, Turkey, e-mail: tatar.ozgur@gmail.com

Soils and fertilisers

Soils

Oral Presentations

Posters

The Effects of Climate on Decomposition of Cattle, Sheep and Goat Manure in Kenyan Tropical Pastures

Yuhao Zhu[1], Lutz Merbold[2], Sonja Leitner[2], Klaus Butterbach-Bahl[1]
[1]*Karlsruhe Institute of Technology (KIT), Inst. of Meteorology and Climate Research, Atmospheric Environmental Research, Germany*
[2]*International Livestock Research Institute (ILRI), Mazingira Centre, Kenya*

Decomposition of manure deposited onto pasture from grazing animals represents an important process for carbon (C) and nitrogen (N) cycles in grassland systems. However, most decomposition studies focus on plant litter while studies investigating manure decomposition are scarce; especially in sub-Saharan Africa (SSA). In this study, we measured decomposition of three types of animal manure (cattle, sheep, goat) over > 1 year using litter bags at four climatically different sites across Kenya. Manure dry matter, total C, total N and ammonium concentrations decreased exponentially, with the most rapid decrease occurring during the first few weeks following application, followed by slower changes during the following 2–3 months. Rates of N mineralisation were lower than those for C mineralisation, resulting in decreasing C/N ratios over time, indicating N retention and immobilisation by the decomposing manure. Generally, cattle manure decomposed faster than sheep or goat manure despite having a higher initial C/N ratio and lower N concentration, with decomposition rates for dry matter ranging from 0.200 to 0.989 k $year^{-1}$. Overall, we found that >30 % of cattle manure mass and >50 % of sheep and goat manure mass remained after 378 days. Cellulose decomposed first, while lignin concentrations increased among all manure types and at all sites. We found that total manure decomposition rates were positively correlated with cumulative precipitation and aridity index, but negatively correlated with mean temperature. Our results show much slower decomposition rates of manures in semi-arid tropical environments of East Africa as compared to the few previous studies in temperate climates.

Keywords: Cellulose, climatic conditions, lignin, litter bags, manure, mineralisation, sub-Saharan Africa (SSA)

Contact Address: Yuhao Zhu, Karlsruhe Institute of Technology (KIT), Inst. of Meteorology and Climate Research, Atmospheric Environmental Research, Kreuzeckbahnstr. 19, 82467 Garmisch-partenkirchen, Germany, e-mail: yuhao.zhu@kit.edu

Patterns and Drivers of Soil Carbon Dioxide Emissions from Well-Watered and Drought-Stressed Silage Maize Production

Khatab Abdalla[1], Mutez Ahmed[2], Johanna Pausch[1]
[1]*University of Bayreuth, Agroecology, Germany*
[2]*University of Bayreuth, Soil Physics, Germany*

The projected global warming risks due to high emissions of greenhouse gases, mainly from anthropogenic activities, increases the need for an agricultural practice with high carbon sink capacity and water use efficiency without compromising on food security. On one hand, it's well accepted that soil temperature and moisture content directly affect microbial activity, whereas on the other hand, drought stress was recently postulated to increase root exudates, which in turn will accelerate soil organic matter mineralisation. Thus, the objective of this study was to investigate the interplay between soil moisture (well-watered and drought stressed) and maize (*Zea mays* L.) root exudates on soil CO_2 efflux. The experiment consisted of three treatments, which are well-watered, drought stressed maize plus a control (without plants). Soil CO_2 efflux, soil temperature and moisture content were measured weekly during the growing season (April to September) and monthly in the fallow period. Under well-watered conditions, the annual average of CO_2 efflux was 0.12 g CO_2-C m^{-2} hr^{-1}, which was 24.5% and 20 % significantly higher than under drought stressed and the control, respectively. Moreover, well-watered treatment had significantly greater primed carbon than drought stressed maize. Soil temperature in deeper soil layers (25, 50 and 75 cm) correlated positively with the CO_2 efflux, while soil moisture correlated negatively at the 5 cm and 25 cm. In addition, above and below ground biomass correlated significantly to soil CO_2 efflux. Overall, these results suggested that the root exudates decreased under drought conditions, which decreasing soil respiration. Drought tolerance varieties could be an option to decrease soil respiration and maintain productivity.

Keywords: Carbon sequestration, ecosystem, microbial biomass, soil respiration

Contact Address: Khatab Abdalla, University of Bayreuth, Agroecology, Universitätsstraße 40, 95447 Bayreuth, Germany, e-mail: khatab.abdalla@uni-bayreuth.de

Physicochemical and Microbiological Properties of Amazonian Soils under Intensive Crop System

Roldán Torres-Gutiérrez[1], Carlos Alfredo Bravo Medina[2], Laura Esthela Cruz Samaniego[3], Karen Daniela Aguirre-Sánchez[3], Reinaldo Demesio Alemán Pérez[2], Bettina Eichler-Löbermann[4]

[1]*Regional Amazon University Ikiam, Live Science Faculty, Ecuador*

[2]*Amazon State University, Earth Science Faculty, Ecuador*

[3]*Amazon State University, Live Science Faculty, Ecuador*

[4]*Rostock University, Agricultural and Environmental Faculty, Germany*

Pitahaya (*Selenicereus megalanthus*) is among the promising crops that have gained an important space in international markets during the last years, becoming a source of incom. for farmers worldwide. However, the intensive production system of this crop is poorly studied until now. Having this background, our research had as objectives, to determine the influence of the intensive crop system of pitahaya on physicochemical and microbiological parameters, as well as the diversity of microorganisms of soils. Soil sampling was carried out in two farms, one with pitahaya and another with tea, as a representative non-intensive crop system. In all farms, two soil depths were assessed (0–10 cm and 10–30 cm). Composite soil samples were processed for chemical and physical properties analysis, as well as the texture of the soils. Bacteria and fungi communities were determined by colony forming units (CFU) and the Most Probable Number method. The growth kinetic was performed for each microbial group, counting the colonies at several times points. Subsequently, the morpho-cultural characterisation of the colonies was carried out to determine the diversity of both microbial groups using the Shannon and Simpson indices. The results for the chemical soil properties showed significant differences for % of total nitrogen (%NT) and organic matter in tea crop compared with pitahaya at 0–10 cm of depth. From 10–30 cm pitahaya crop increased the %NT. The physical properties and texture of the soils showed no significant differences between both crops. Regarding the depths, from 0–10 cm improved the aeration porosity for tea and hydraulic conductivity for pitahaya, while from 10–30 cm, the retention porosity increased significantly for tea. When comparing the microbial communities between both systems, a positive influence of tea was shown, having an increase of 67 % for bacteria and 52 % for fungi compared to pitahaya. However, diversity indices showed that although there was a high microbial diversity for both crop systems, there are no remarkable differences between them. These results open the gap to seek alternative systems for pitahaya cultivation and improve essential elements of agroecosystems, such as, organic matter and microbial communities.

Keywords: Amazon soils, bacteria, colonies forming units, diversity, edaphology, fungi

Contact Address: Roldán Torres-Gutiérrez, Regional Amazon University Ikiam, Live Science Faculty, Parroquia Muyuna Kilómetro 7 Vía a Alto Tena, 150150 Tena, Ecuador, e-mail: roldan.torres@ikiam.edu.ec

Factors that Influence Farmers' Perception of the Causes and Indicators of Land Degradation in Cameroon

Asheri Yayu Nathalie, Miroslava Bavorova, William Nkomoki

Czech University of Life Sciences Prague, Fac. of Tropical AgriSciences - Dept. of Economics and Development, Czech Republic

Land degradation is regarded as a major issue in developing countries such as Cameroon, yet there is still a paucity of information on the smallholder farmers' perception of land degradation. The objectives of this study were to examine farmers' perception of the causes and indicators of land degradation and to analyse factors that influence their perception of land degradation in Tiko municipality, Cameroon. The data were collected using focus group discussion and a semi-structured questionnaire involving 122 respondents. The data analysis was done using descriptive statistics and binary logistic regression. The results of this study showed that 89.3 % of the farmers were aware of the land degradation problems on their farms. A considerable number of the respondents identified water erosion (50.8 %) and low soil nutrients (44.3 %) as very serious causes of land degradation, while wind erosion was the least identified (33.6 %). In addition, most of the participants identified lower yields (72.2 %) and decreased vegetation (36.9 %) as the most important indicators of land degradation. The binary logistic regression model results indicated that socio-demographic characteristics (total household members, education), farm attributes (farm size), biophysical attributes (lower yields, soil texture, wind erosion), institutional characteristics (group membership), and information sources (radio) increased the likelihood of farmers perceiving their land as degraded. Furthermore, the model indicates that having more education, having a large farm, being a group member, an increase in household size, and the more severe the problem of wind erosion is, the higher probability of farmers' perceptions of their land as degraded.

Keywords: Cameroon, land degradation, perception, smallholder farmers

---

Contact Address: Asheri Yayu Nathalie, Czech University of Life Sciences Prague, Fac. of Tropical AgriSciences; Dept. of Economics and Development, Na hroudě 39, 100 00 Prague, Czech Republic, e-mail: asherinatalie67@gmail.com

Gravel Content of Soil Impact on Water Use, Growth and Yield of Pearl Millet

Sebastian Ayinbisa Yanore[1], Samuel Kwasi Godfried Adiku[1], Dilys Sefakor MacCarthy[2], Daniel Etsey Dodor[1], John Bright Amoah Nyasapoh[1]

[1]*University of Ghana, Dept. of Soil Science, Ghana*

[2]*University of Ghana, Soil and Irrigation Research Centre, Ghana*

The effect of gravel content (GC) of soil on water use efficiency (WUE), growth and yield of pearl millet (*Pennisetum glaucum*) was investigated under greenhouse conditions at the University of Ghana. Pearl millet was grown either in repacked soil columns with varying gravel contents (0, 10 %, 20 %, 30 %, 40 %, 50 % 60 % and 70 %), and in undisturbed monoliths which had 10 %, 34 %, 64 % and 72 % GC (w/w). The plants were irrigated following rainfall pattern in the Upper East Region of Ghana from where the soils were sampled and noted for pearl millet production. All the trials received fertiliser application of 100 kg N ha^{-1}, 20 kg P ha^{-1} and 20 kg K ha^{-1}. Data collected during the plant growth included runoff, drainage, plant evapotranspiration (ETa), and nutrient leaching. Plant data included plant height, number of leaves, chlorophyll, biomass, and grain yield at maturity. The WUE was determined as the ratio of grain yield to ETa. The results indicated that there was no significant difference between the ETa of the repacked soil columns and undisturbed monoliths. In soil columns with more than 40 % GC, appreciable (64 %) of irrigation was lost via runoff and drainage. On the contrary, soils with less than 40 % gravel lost on average 43 % of irrigation via runoff and drainage. Nutrient leaching, estimated as the Total Dissolved Salts in the drained water (TDS) showed that soils columns with more than 40 % GC lost at least 40 % of the applied nutrients and this was significantly ($p = 0.001$) higher than nutrient lost in columns with 0 % GC. Chlorophyll and leaf number were significantly higher ($p = 0.001$) when the GC was between 0 % and 40 %. Shoot, root, grain yield, and WUE were significantly higher ($p = 0.001$) at GC = 40 % compared to GC = 0 % and declined as the GC increased. As the GC increased (>50 %), soil moisture content decreased, and nutrient leaching increased. The study showed that increasing GC reduces soil and plant productivity, and such soils would require specialised amendments to offset the adverse impact of gravels on plant growth.

Keywords: Gravel content, nutrient leaching, plant growth, water use efficiency, yield

Contact Address: Sebastian Ayinbisa Yanore, University of Ghana, Dept. of Soil Science, Oppsite Akuafo Main Hall, GA-455-580 Accra, Ghana, e-mail: syanore_ayinbisa@st.ug.edu.gh

Applying Electrical and Electromagnetic Methods to Investigate Soil Salinity in Rice Production Systems in the Vietnam Mekong Delta

Van Hong Nguyen[1], Folkard Asch[1], Jörn Germer[1], Nha Duong Van[2]

[1]*University of Hohenheim, Inst. of Agric. Sci. in the Tropics (Hans-Ruthenberg-Institute), Germany*

[2]*Kien Giang University, Vietnam, Agriculture and Rural Development Faculty, Vietnam*

Rice is a major staple in Vietnam and more than half of its production is located in the Mekong River Delta (MRD). Due to sea level rise and seasonal fluctuations in river water levels, rice production in the MRD is threatened by salinity. Sea level rise may result in sea water intrusion into ground water tables, potentially allowing capillary rise of salt into the top soil. Low river water levels allow tidal intrusion of seawater into the irrigation canal system rendering the water unusable for irrigation. This study was conducted to investigate the suitability of two geo-electric methods to assess soil salinity in rice production systems. ARESII (electrical resistivity) and EM38 (electromagnetic conductivity) were employed at six benchmark sites of roughly 6 ha each in the Tra Vinh province of MRD to explore soil salinity as related to the cropping systems (2 or 3 rice crops per year) and distance to the sea defined via Northwest – Southeast and Southwest – Northeast transects along the peninsula. EM38, with a maximum penetration depth of 1 m, was used to characterise effects of irrigation management on top soil properties (EC in 0–1.5 m depth). ARESII, with a maximum penetration depth of 48 m was used to explore resistivity of sub-soil layers down to 40 m with different spatial resolutions (Ω in 1 m – 48 m depth). Top soil salinity was low ($< 4\ dSm^{-1}$), for all measured sites with EC reflecting water fluxes and irrigation management. Resistivity results show high values in the upper soil layers (> 50 Ωm), and low resistivity values (0–2 Ωm) vary from 7 m – 20 m depth at all sites. An increasing trend of deep soil resistivity is evident from the coast to inland and from south to north of the peninsula. Fields with three rice crops show a lower resistivity than those that are cropped twice. Although, the electrical resistivity demonstrated potential variations in depths and space of soil salinity among rice systems, geological information of measured areas is necessary for validating results. The suitability of the two methods to assess soil salinity in MRD is discussed.

Keywords: Conductivity, *Oryza sativa*, deep soil, resistivity, top soil

---

Contact Address: Van Hong Nguyen, University of Hohenheim, Inst. of Agric. Sci. in the Tropics (Hans-Ruthenberg-Institute), Garbenstr. 13, 70599 Stuttgart, Germany, e-mail: van.nguyen@uni-hohenheim.de

Soil Phosphorus Availability in Manganese-Rich Waste Amended Soils

Mohammed Abdalla Elsheikh[1,4], Khatab Abdalla[2,4], Louis Titshall[3], Pardon Muchaonyerwa[4]

[1]*University of Khartoum, Inst. of Desertification and Desert Cultivation Studies, Sudan*

[2]*University of Bayreuth, Agroecology, Germany*

[3]*South African Sugarcane Research Institute (SASRI), South Africa*

[4]*University of Kwazulu-Natal, School of Agricultural, Earth and Environmental Sciences, South Africa*

Mining, processing and smelting activities could pollute soil and groundwater resources with heavy metals, which could seriously affect the environment and ecosystems functioning. Land application of industrial and mining waste could increase supply and availability of essential nutrients for plant growth; however, it is negative consequences on plant growth, through toxicities of pollutant elements is well-known. In this regards, the interactions of the constituent elements in the soil with applied nutrients could reduce the effectiveness of fertilisers. Understanding the behaviour of phosphate in Manganese rich-wastes amended soils, will be a key for sustainable phosphorus management in the soils, e.g. optimisation of phosphorus fertilisers inputs for crop production on manganese waste treated soils. Therefore, the objective of this study was to determine the effects of phosphorus sources (i.e. phosphate rock and KH_2PO_4) and application rate on phosphorus and manganese relationship in a soil amended with manganese-rich waste. To achieve this purpose, an incubation experiment under room conditions was carried out. The changes in phosphorus, manganese, electrical conductivity and pH were determined at 0, 30 and 60 days of the incubation time. The results of this research showed that phosphorus concentrations were decreased in sandy soil for the highest application of manganese by 31.1, 40.2, and 41.9 % for 0, 30, and 60 days of incubation, respectively. While, the reduction in clay soil was 16.5, 27.0, and 42.0 % for 0, 30, and 60 days of the incubation period, respectively. Addition of phosphorus to rich- manganese amended soil works better under sandy soil, while clay soil has better performance concerning the reclamation of Manganese rich-waste. Manganese -rich waste reduced available soil phosphorus concentrations, meaning low efficiency of phosphorus fertilisation and less plant productivity. Further research is needed to mitigate manganese concentration in the soil, phytoremediation technique could be an option.

Keywords: Heavy metals, Manganese rich-wastes, soil pollution

Contact Address: Khatab Abdalla, University of Bayreuth, Agroecology, Universitätsstraße 40, 95447 Bayreuth, Germany, e-mail: khatab.abdalla@uni-bayreuth.de

Phosphate Solubilizing Activity of Native Guatemalan Isolates of *Pseudomonas fluorescens*

José Alejandro Ruiz-Chután[1], Julio Ernesto Berdúo-Sandoval[2], Aníbal Sacbajá[2], Bohdan Lojka[1], Marie Kalousová[1], Jana Žiarovská[3], Cusimamani Fernández Eloy[1], Amílcar Sánchez-Pérez[2]

[1]*Czech University of Life Sciences Prague, Fac. of Tropical AgriSciences, Dept. of Crop Sciences and Agroforestry, Guatemala*
[2]*Universidad de San Carlos de Guatemala, Facultad de Agronomía, Guatemala*
[3]*University of Agriculture in Nitra, Dept. of Genetics and Plant Breeding, Slovakia*

Phosphorus (P) is an essential element in agricultural production. However, due to its complex dynamics, only a small portion of the total present in the soil can be used by plants. This is because most are found in insoluble forms, especially in Andisol soils of volcanic origin. Phosphate solubilising microbes (PSMs) are an alternative to transform this element into soluble forms that can be used by plants; in addition to not generating environmental pollution and representing a low cost compared to the production of fertilisers. The main objective of this research was to identify and evaluate *in vitro*, the phosphate solubilising activity, and stability of native Guatemalan isolates of *Pseudomonas fluorescens* bacteria and its relation with its genetic diversity. The isolates were obtained from regions of Guatemala where Andisol soils represent a limitation to agricultural production due to the high P fixation phenomenon. A total of 35 *P. fluorescens* isolates were identified and confirmed through specific PCR. Subsequently, the genetic diversity of the isolates was analysed with the molecular marker AFLP. Phosphate solubilising activity and stability were evaluated by relating the solubilisation halo to the bacterial growth halo through the solubilisation index. For this, *in vitro* cultures of the isolates were performed in NBRIP medium taking tricalcium phosphate $Ca_3(PO_4)_2$ as the source of insoluble phosphorus. High genetic diversity was identified among the *P. fluorescens* isolates, as well as the isolates with the highest solubilisation index and stability showed greater genetic similarity. The Pf 30 isolate showed the highest solubilisation index, while the Pf 7 isolate was the most stable. Finally, due to the solubilisation potential of *P. fluorescens* isolates observed *in vitro*, future evaluation of the solubilising activity is suggested under field conditions with different soil types and in association with different crops. This in order to identify the isolates with the best solubilisation potential and thus offer a viable solution to agricultural producers affected by phosphorous fixation.

Keywords: Andisols, genetic diversity, phosphate solubilising activity, *Pseudomonas fluorescens*

Contact Address: José Alejandro Ruiz-Chután, Czech University of Life Sciences Prague, Fac. of Tropical AgriSciences, Dept. of Crop Sciences and Agroforestry, 15 Calle B 16-61 Zona 6, 01006 Guatemala, Guatemala, e-mail: josealejandro.ruiz@icloud.com

Possible Phosphate Solubilisation Mechanism and Growth Promotion of Wheat through *Bacillus megaterium* and *Bacillus subtilis*

Zafar Iqbal[1,2], Frank Rasche[2], Maqshoof Ahmad[1], Thomas Hilger[2]

[1]*The Islamia University of Bahawalpur, Department of Soil Science, Pakistan*
[2]*University of Hohenheim, Inst. of Agric. Sci. in the Tropics (Hans-Ruthenberg-Institute), Germany*

Phosphorous is the second most limiting mineral nutrient in agriculture. Being an integral part of photosynthesis, DNA, RNA, ATPs and the energy transfer chain, its deficiency causes reduction of growth, yield and most importantly quality of grains in cereals. It is acknowledged that large amounts of P are fixed in soils. Hence, there is need to search for sustainable options to make these fixed P resources accessible for plant uptake. A diverse group of microorganisms has the ability to release P from the fixed pool of soil-borne P. Consequently, the present study was conducted to evaluate the potential phosphate solubilising mechanism adopted by plant-associated bacteria. For this purpose, growth behaviour of four bacterial strains (*Bacillus subtilis* ZE15 and ZR3and *Bacillus megaterium* ZE32 and ZR19) were observed in P amended and unamended Pikovskaya's broth culture. In unamended media, most of the bacterial strains reached at stationary phase at 3rd day after incubation, while they behave differently in P amended media. Inoculation of strains ZE15, ZE32, ZR3 and ZR19 in P amended media has maximum increase up to 47 %, 62 %, 33 % and 31 % in bacterial population over unamended P media, respectively. Strain ZE15 has been recorded for maximum P solubilisation up to 130 μg mL^{-1} *in vitro*. A possible mechanism of P solubilisation might be the secretion of organic acids in P amended media, because a depletion of pH was recorded. Strains ZE32 and ZR3 were able to grow at a wide pH range (~ 3.0 – 10). Maximum phosphate esterase activity (insoluble vs soluble P, e.g. 3.65 and 1.08 nmol g^{-1} dry matter) was recorded for strain ZE15, while maximum ß-D glucosidase activity (insoluble vs soluble P, e.g. 2.81 and 0.81 nmol g^{-1} dry matter) was observed for strain ZE32. The phosphate solubilisation ability of *Bacillus* sp. and increase in soil exoenzyme activities in the rhizosphere supported P uptake through improved solubilisation and mineralisation. The tested strains proved their ability to promote growth and might thus represent candidates for effective biofertilisation in P deficient soils after inclusive assessment under field conditions.

Keywords: *Bacillus* sp., mechanism, mineralisation, phosphate esterase

Contact Address: Zafar Iqbal, The Islamia University of Bahawalpur, Department of Soil Science, 63100 Bahawalpur, Pakistan, e-mail: zaferiqbal31205@gmail.com

Prediction of Soil Erosion under Three Tillage Systems Using RUSLE and GUEST Models

G. Esaie Kpadonou[1], Samuel Kwasi Godfried Adiku[1], Dilys Sefakor Maccarthy[2]

[1]*University of Ghana, Dept. of Soil Science, Ghana*

[2]*University of Ghana, Soil and Irrigation Research Centre, Ghana*

Soil erosion is a major challenge to sustained agricultural production in the tropics. The severity of erosion is linked with the type of tillage. Yet, data and estimates of erosion under different tillage and land management systems are lacking. In this study, soil erosion under three tillage practices: namely (i) flat tillage, TF, (ii) ridge and furrow tillage, RF, and (iii) tied-ridge tillage, TR, were investigated using lysimeters. Maize (*Zea mays*) was cultivated on the lysimeters and data on water balance components and soil loss were collected over two growing seasons with varied rainfall amounts. In order to extrapolate findings to other situations, the Revised Universal Soil Loss Equation (RUSLE), one of the most widely used soil erosion model, and the Griffith University Erosion Simulation Template (GUEST) were used to simulate soil loss under the three tillage systems. The results showed that runoff to rainfall ratio were in the range of 41.3–32.3 %, 34.9–24.0 % and 15.0–0.0 % for RF, FT and TR, respectively, during the first (major) and the second (minor) season. Hence, the runoff order was RF > FT > TR. Similarly, soil loss followed the same order as runoff. TR significantly reduced soil loss by 43 % compared to RF during the major season. In the minor season TR recorded zero soil loss. In general, soil loss during the major season was 25–75 % higher than that of the minor season. Both models predicted the soil erosion well and captured the differing tillage effects. Comparisons between observed and simulated erosion were generally good. We conclude that modelling can enable the prediction of the impact of tillage type on soil erosion under varying soil management conditions and this could support agricultural planning for different agro-ecological zones.

Keywords: Conservation tillage, erosion, modelling, tillage

Contact Address: G. Esaie Kpadonou, University of Ghana, Dept. of Soil Science, Legon Campus University of Ghana, Accra, Ghana, e-mail: e.kpadonou@yahoo.com

Effects of Transhumance and Vegetation Type on Soil Quality of Rangelands in Northern Benin

Rodrigue V. Cao Diogo[1], Luc Hippolyte Dossa[2], Pénéloppe G. T. Gnavo[1], Eva Schlecht[3], Andreas Buerkert[4]

[1]*University of Parakou, Dep. of Sci. and Techn. of Animal Prod. and Fisheries, Benin*

[2]*University of Abomey-Calavi, Fac. of Agricultural Sciences, School of Science and Technics of Animal Production, Benin*

[3]*University of Kassel / University of Goettingen, Animal Husbandry in the Tropics and Subtropics, Germany*

[4]*University of Kassel, Organic Plant Production and Agroecosystems Research in the Tropics and Subtropics, Germany*

Increasing use of natural rangelands by animal herds transhumant from the Sahel to coastal countries may affect the ecology of rangeland soils and food production in these countries. To elucidate this further, we evaluated the effects of transhumance and vegetation type on soil quality of rangelands in two municipalities of northern Benin (Sinende and Tchaourou). Two zones with different intensities of transhumance (Strong: ST and weak: WT) and three vegetation types (VT) namely open forest/woodland savannah, wooded savannah/shrubland savannahs and crop field mosaic were studied. Soil samples (1 kg of pooled soil samples) were collected from 90 plots of 100 m^2 (60 on ST and 30 on WT zones) in 5 different spots of 1 m^2 using an auger. The composite soil samples were analysed for texture, pH and organic matter contents using standard methods. In addition, soil compaction was monitored using a penetrometer at 5 cm depth.

The results indicated that, irrespective of transhumance zone and vegetation type, soils encountered in Northern Benin rangelands have loamy-sandy and loamy textures. Also, the organic matter content was similar across zones ($p = 0.6571$) and varied across VT ($p = 0.0004$). However, soil compaction was significantly affected by VT ($p = 0.0386$) and transhumance zone ($p = 0.0026$) and was stronger in the ST than WT zone. In contrast, soil texture and pH were neither affected by transhumance ($p = 0.1083$) nor by VT ($p = 0.9995$). Moving herds strategically from one pasture to the other, may help avoid rangeland soil degradation, reduce overexploitation of pasture resources, and improve rangeland quality and productivity.

Keywords: Livestock, natural pasture, pastoral mobility, rangeland soils

Contact Address: Rodrigue V. Cao Diogo, University of Parakou, Dep. of Sci. and Techn. of Animal Prod. and Fisheries, 123 Parakou, Benin, e-mail: dcao_bj@yahoo.fr

Dietary Tannins Reduce Soil Respiration after Goat Manure Application on an Irrigated Sandy Soil in Oman

Mariko Ingold[1], Sibylle Faust[2], Philipp Holz[1], Aboud Fayad[1], Andreas Buerkert[1]

[1]*University of Kassel, Organic Plant Production and Agroecosystems Research in the Tropics and Subtropics, Germany*

[2]*University of Kassel, Soil Biology and Plant Nutrition, Germany*

The combined hot and moist conditions of irrigated agriculture on sandy soils in northern Oman lead to high gaseous carbon (C) and nitrogen (N) losses and impedes the accumulation of soil organic matter due to its high microbial turnover. In the millennia old oasis in Oman, goat manure is traditionally used as fertiliser and soil conditioner, which is, however, often of poor quality. By altering the goat's diets using condensed tannins, organic matter can be stabilised and N retention of soils improved. To investigate the effect of dietary tannins on soil respiration after application of manure to soil, a two-year field experiment was conducted on a sandy alluvial soil in the Al-Batinah Plain/northern Oman with the following treatments: mineral fertiliser (MIN); goat manure (GM); tannin-enriched goat manure (TM). GM was obtained from goats fed a basal diet of 50 % hay, 47 % maize and 3 % soybean and TM from goats additionally fed 3.4 % Quebracho tannins to the basal diet and applied to radish at 2.6 and 2.8 t C ha^{-1} $year^{-1}$ in 2011 and 2012, respectively. CO_2 emissions were analysed on field using a portable multi-gas monitor (Innova 1312, LumaSense Technologies A/S) at 1–7 day intervals starting with manure application in February until harvest end of March (n=453).

The CO_2-C emissions were highest during the first eight days after manure application in the GM and TM treatments. Generally, CO_2-C emissions were more than twice as high in the afternoon (at hottest day temperature) than early in the morning (coolest day temperature). In February 2011, daily mean flux rates under radish cultivation were 127 (se: 12) $<$ 263 (se: 21) $<$ 286 (se: 21) CO_2-C mg m^{-2} h^{-1} in MIN, TM, and GM, respectively, whereas in March mean flux rates were 103 (se: 8) $<$ 196 (se: 13) $<$ 211 (se: 13) CO_2-C mg m^{-2} h^{-1} in MIN, TM, and GM, respectively.

The results show that soil respiration was reduced in TM compared with GM, supporting previous results on organic matter decomposition, and C and nutrient release from manure applied to soil. Analysis of soil respiration during the second year are under way.

Keywords: Goat manure, radish cultivation, soil respiration, subtropics, tannins

Contact Address: Mariko Ingold, University of Kassel, Organic Plant Production and Agroecosystems Research in the Tropics and Subtropics, Steinstrasse 19, 37213 Witzenhausen, Germany, e-mail: ingold@uni-kassel.de

The Potential of Cultivating Fruit on Ex-Mined Soil in Indonesia

TEDI YUNANTO[1], FARISATUL AMANAH[2], DODDY HERIKA[3]

[1]*Ministry of Energy and Mineral Resources, Bandung Polytechnic of Energy and Mining, Indonesia*
[2]*Ministry of Energy and Mineral Resources, Directorate General for Mineral and Coal, Indonesia*
[3]*PT Berau Coal, Post-mining Division, Indonesia*

For centuries, forests have been food resources, particularly for local people. However, mining operations in forests result in a substantial reduction and even the loss of flora and fauna such as fruit plants. To restore the forest function as a food resource, fruit plants can be cultivated during mine reclamation. In terms of fruit production, fruit plants require macro and micro elements from the soil to achieve maximum yields and quality. Planting fruit plants in ex-mined areas proves to be challenging due to the poor soil condition such as low pH, inadequate organic matter and barrenness. This study, therefore, aims to review the potential of cultivating rambutan (*Nephelium lappaceum* L.) on ex-mined soil in comparison with the guideline on land suitability assessment provided by the Indonesian Ministry of Agriculture. The study also evaluates *N. lappaceum* production on Binungan Site, PT Berau Coal, East Kalimantan Province, Indonesia. The plantation of *N. lappaceum* requires 1,786 ha with 1,016 trees in three cultivars: Binjai, Rapiah and Garuda. The plantation of *N. lappaceum* was opened in 2005 and first harvested in 2010. The crop yield records in 2020 were over 1,345 kg. The soil analysis results show that Binungan Site had a very low value of pH (3.7–4.3); organic carbon of 0.31–1.34 %; nitrogen total of 0.06–0.19 %; cation exchange capacity (CEC) of 8.8-22.5 cmol kg^{-1}; P_2O_5 of 79–234 ppm; and K_2O of 80–214 ppm. Based on the guideline, this ex-mined soil is classified as "S2" which means the soil is quite adequate for plantation although with limiting factors. Despite lower soil quality in the ex-mined area, *N. lappaceum* can still grow and produce fruit. To improve the soil quality and fruit production, the intensification process, such as fertilisation and proper cultivar plantation, has to be conducted. The evaluation reveals that the ex-mined area can be cultivated as a food and new economy source for the local community around the mining area.

Keywords: Ex-mined soil, food source, fruit production, mine reclamation, soil suitability

Contact Address: Tedi Yunanto, Ministry of Energy and Mineral Resources, Bandung Polytechnic of Energy and Mining, Jl. Jenderal Sudirman No. 623, 40211 Bandung, Indonesia, e-mail: genom.tedi@gmail.com

Using the Best Land Preparation Leads to an Increase in Crop Productivity in Sudan

MAJDALDIN RAHAMTALLAH ABUALGASIM MOHAMMED[1], HANADI MOHAMED SHAWGI GAMAL[2]

[1]*National Center for Research, Natural Resources and Desertification Inst., Sudan*
[2]*University of Khartoum, Fac. of Forestry; Forest Products and Industries, Sudan*

The main purpose of tillage operations is to prepare the appropriate cradle for seeds by dismantling and loosening the soil and uprooting the grass plants which grow there. Success or failure of crop production depends on a good preparation of the land. In addition, tillage changes the physical properties of the soil. This study investigated the best tillage methods under desertified environments, in order to reduce costs of traditional tillage so as to increase crop production of the land. To achieve the objectives of this study, two tractors with different drag force were used: the first tractor was used as a tester and the second one as an auxiliary. Two primary ploughs (disc plough) and a chisel plough, and two secondary ploughs (disc harrow plough and chisel plough) were used and compared to an animal-drawn plough. Tillage parameters have been investigated. In a first test, the land has been prepared by using chisel plough to a depth of 30 cm, then opening furrows using Ridger plough. The second test used a disc plough to a depth of 20 cm, and then furrows have been made using a Ridger plough. In a the third test the harrow plough was used to a depth of 25 cm then furrows were made using a Ridger plough. The fourth test opened the furrows using a Ridger plough to a depth of 30 cm. The fifth test opened the furrows using the animal-drawn plough to a depth of 15 cm. The results recorded that a field efficiency for the chisel plough of 90.5, for the disc plough of 85.5, harrow plough 70.5, Ridger plough 50.5, and the animal-drawn plough showed an efficiency of 15.5. The results further showed that the fuel consumption in litter per ha recorded were 6.50, 3.30, 2.60 and 2.10, for the chisel plough, the disc plough, the harrow plough, and the Ridger plough, respectively. The study recommended that the most suitable practice is the Ridger plough which recorded the highest field efficiency and less fuel consumption in order to enhance soil moisture and increase crop productivity.

Keywords: Desertified land, land preparation, Sudan, tillage

Contact Address: Majdaldin Rahamtallah Abualgasim Mohammed, National Center for Research, Natural Resources and Desertification Inst., Khartoum, Sudan, e-mail: majdaldin_1974@yahoo.com

Organic amendments

Combining Organic/Mineral Fertilisers as a Climate-Smart Integrated Soil Fertility Management Practice in Sub-Saharan Africa: Meta-Analysis

Gil Gram[1], Dries Roobroeck[2], Pieter Pypers[2], Johan Six[3], Roel Merckx[4], Bernard Vanlauwe[2]

[1]*International Institute of Tropical Agriculture (IITA), Climate Change, Agriculture, and Food Security (CCAFS), Belgium*

[2]*International Institute of Tropical Agriculture (IITA), Kenya*

[3]*ETH Zurich, Dept. of Environmental System Science, Switzerland*

[4]*KU Leuven, Department of Earth and Environmental Sciences, Belgium*

Low productivity and climate change require climate-smart agriculture (CSA) for sub-Saharan Africa (SSA), through (i) sustainably increasing crop productivity, (ii) enhancing the resilience of agricultural systems, and (iii) offsetting greenhouse gas emissions. We conducted a meta-analysis on experimental data to evaluate the contributions of combining organic and mineral nitrogen (N) applications to the three pillars of CSA for maize (*Zea mays*). Linear mixed effect modeling was carried out for; (i) grain productivity and agronomic efficiency of N (AE) inputs, (ii) inter-seasonal yield variability, and (iii) changes in soil organic carbon (SOC) content, while accounting for the quality of organic amendments and total N rates. Results showed that combined application of mineral and organic fertilisers leads to greater responses in productivity and AE as compared to sole applications when more than 100 kg N ha^{-1} is used with high-quality organic matter. For yield variability and SOC, no significant interactions were found when combining mineral and organic fertilisers. The variability of maize yields in soils amended with high-quality organic matter, except manure, was equal or smaller than for sole mineral fertiliser. Increases of SOC were only significant for organic inputs, and more pronounced for high-quality resources. For example, at a total N rate of 150 kg N ha^{-1} season^{-1}, combining mineral fertiliser with the highest quality organic resources (50:50) increased AE by 20 % and reduced SOC losses by 18 % over 7 growing seasons as compared to sole mineral fertiliser. We conclude that combining organic and mineral N fertilisers can have significant positive effects on productivity and AE, but only improves the other two CSA pillars yield variability and SOC depending on organic resource input and quality. The findings of our meta-analysis help to tailor a climate smart integrated soil fertility management in SSA.

Keywords: Climate-smart agriculture, linear mixed effect modelling, maize productivity, soil organic carbon, yield variability

---

Contact Address: Gil Gram, International Institute of Tropical Agriculture (IITA), Climate Change, Agriculture, and Food Security (CCAFS), Michel-Zwaabstraat 12/3, 1080 Sint-Jans-Molenbeek, Belgium, e-mail: g.gram@cgiar.org

Importance of Livestock Manure in Crop Production in Tanzania

LUKE KORIR, NILS TEUFEL, HENRY KIARA

International Livestock Research Institute (ILRI), Kenya

Soil nitrogen depletion is considered one of the important constraints to improving the productivity of crop production in sub-Sahara Africa. It has been found that farmers cannot afford to apply recommended levels of mineral fertiliser to cropland, leading to nutrient mining. With the increasing importance of livestock in the livelihoods of poor farmers, livestock manure or leguminous forages can offer a potential solution to soil fertility loss in mixed farming-systems. Studies have found positive effects of manure application on crop yield, but the effects depend on the management practices when handling the manure. The objective of this study is to investigate how small-scale farmers in mixed systems in Tanzania manage soil fertility and manure and how their use of manure affects crop productivity. Data for this study is derived from a sample of 994 livestock keepers interviewed in 2017.
Results indicate that only about 23 % of the interviewed farmers applied mineral fertilisers on their farms while about 25 % of the farmers planted legumes with the intention of soil improvement. Although a majority (78 %) of the farmers applied livestock manure to their crops, less than 5 % of the farmers kept the manure enclosed before spreading it on crop-land. Mineral fertiliser is applied on average to only one crop, which is mostly maize, while manure is applied to a wide range of crops, on average two per farm, including vegetables and perennial crops, such as bananas. A comparison of farms using livestock manure versus those that do not, indicates that farms using livestock manure have a statistically significant higher value of crop production per acre than those that do not. This implies that building farmers capacity on effective livestock manure handling and application has the potential to considerably improve crop productivity, especially where soil fertility is threatened and access to mineral fertiliser is constrained.

Keywords: Crops, livestock manure, productivity, soil fertility

Contact Address: Luke Korir, International Livestock Research Institute (ILRI), Policies, Institutions & Livelihoods (PIL), P.O. Box 2840, 30100 Eldoret, Kenya, e-mail: lukorir@gmail.com

Rapid Early Growth Outranks Weed Competition in Horticultural Systems in Semi-Arid Bolivia

LAURA KUONEN[1], NOEMI STADLER KAULICH[2], LINDSEY NORGROVE[3]

[1]*Bern University of Applied Sciences, School of Agricultural, Forest and Food Sciences, Switzerland*

[2]*Mollesnejta - Centre for Andean Agroforestry, Cochabamba, Bolivia*

[3]*Bern University of Applied Sciences, School of Agricultural, Forest and Food Sciences, Switzerland*

Onion (*Allium cepa*) is an important crop in Bolivia, particularly in the semi-arid region of Cochabamba,. Here, water availability, as well as nitrogen (N), limit growth. We hypothesised that using locally available mulch might increase growth and yield more than the equivalent N in the form of urea as nutrient release may be better synchronised with crop demand. Mulch may also suppress weeds, thus improving relative yield.

We tested three mulches *Dodonaea viscosa, Melinis repens, Chamaecytus proliferus* (4 t DM ha^{-1}, equivalent to 44, 26 and 39 kg N ha^{-1}) versus two urea treatments (40 and 80 kg N ha^{-1}) in an onion monoculture, planted in January 2019. We also included an unfertilised, unmulched control and used a randomised complete block design (n=4). Growth parameters (plant height, circumference, number of green and of newly produced leaves) were evaluated every two weeks, weeded at 8 weeks after planting (WAP) and harvested at 19 WAP.

In the two urea treatments, onions grew faster with maximum circumference obtained by 8 WAP, and plants were significantly taller with more green leaves than all other treatments, despite having significantly higher weed biomass. Growth of onions in mulch treatments was not different from the control. At harvest, mulching treatments had higher aboveground biomass compared with the control and with the urea treatments. The urea-80 treatment had significantly higher onion yield than all other treatments. *Melinis repens* (26 kg N ha^{-1}) was the best-performing mulch treatment, with yields not different from the urea-40 treatment and the number of harvested onion bulbs being higher than the urea-80 treatment.

We conclude that yields were N-limited. There was no evidence of better nutrient synchrony with mulches; on the contrary, rapid early growth promoted later biomass allocation below ground and thus higher onion yield, so fertiliser can be recommended. Weed competition did not play an important role. Where farmers have limited purchasing power, the best-bet mulch was *Melinis repens*.

Keywords: *Allium cepa*, horticulture, mulch, nitrogen efficiency, semi-arid

Contact Address: Laura Kuonen, Bern University of Applied Sciences, School of Agricultural, Forest and Food Sciences, LÃ¤nggasse 85, 3052 Zollikofen, Switzerland, e-mail: laura.kuonen@gmail.com

Nutrient Recycling with Sugar Cane Ash in Urban Agriculture – Results for Lettuce and Cucumber

Raul Lopez[1], Luis Ricardo Saiz Rosales[2], Yans Guardia-Puebla[3], Bettina Eichler-Löbermann[4], Quirino Arias[3]

[1]*University of Granma, Plant Biotechnology Study Center, Cuba*
[2]*University of Granma, Agronomy, Cuba*
[3]*University of Granma, Chemistry, Cuba*
[4]*Rostock University, Agricultural and Environmental Faculty, Germany*

The reutilisation of sugar cane ash in agriculture is an important issue to create nutrient cycles and to save fertilisers in Cuba. Nutrient cycling is also an important issue in the urban garden system. in Cuba (so called organopónicos) which based on compost substrates. However, the use of biomass ashes in such gardens is rarely investigated until now. Therefore, agronomic impacts of sugar cane ash were investigated in combination with cucumber (*Cucumis sativus*) and lettuce (*Lactuca sativa*) in two experiments with compost substrates. In the first experiment pots were filled with 1 kg substrate and amended with two levels of sugar cane ash (9.6 and 14.4 g per pot). These treatments were compared with a commercial PK fertiliser and a control without mineral nutrient addition under controlled conditions. A second experiment was established in an urban garden where a treatment with ash application of 8.5 t ha^{-1} was compared with a control without ash application. Morphological plant characteristics, yields, and uptakes of plant nutrients as well as soil nutrients were included in the investigation program. In the pot experiment the fertiliser effect of sugar cane ash was comparable with the mineral PK fertilisation for both crops without any significant differences in shoot height, root length and yields. In the field experiments, the application of sugar cane ash resulted in a yield increase of cucumber fruits (22–34 %) and lettuce leaves (25–32 %). Despite the higher yields and following higher nutrient removals, increased levels of P and K were still found in the soil after plant harvest. Although the gardens based on nutrient-rich compost substrate, the results have shown that adding biomass ashes can further increase vegetable yields here. The ashes also can replace mineral fertilisers apart from nitrogen, which is not present in the ashes.

Keywords: Biomass ashes, compost, crops, organic fertiliser

Contact Address: Bettina Eichler-Löbermann, Rostock University, Agricultural and Environmental Faculty, Justus-von-Liebig-Weg 6, 18059 Rostock, Germany, e-mail: bettina.eichler@uni-rostock.de

Integrated Use of NPK Fertiliser, Cattle Manure and Frequency of Leaf Harvest for Improving Nutritional Quality of Butternut

Elisha Otieno Gogo[1], Martin Maluki[1], Lenard Mounde[1], James Mwololo[2]
[1]*Pwani University, Crop Sciences, Kenya*
[2]*International Crops Research Institute for the Semi-Arid Tropics (ICRISAT), Malawi*

There is limited knowledge on frequency of leaf harvest and soil nutrient replenishment effects on nutritional quality of butternut (*Cucurbita moschata* Duch. ex Poir). An experiment was conducted at the Agricultural Training Centre Matuga-Mkongani, Kenya, to determine effects of combined use of NPK fertiliser (17:17:17) and cattle manure, and frequency of leaf harvest on nutritional quality of butternut. A randomized complete block design with three replications was used. NPK was applied at 0, 250, 500, or 750 kg ha^{-1}. Cattle manure was applied at 0, 2,500, 5,000, or 7,500 kg ha^{-1}. Leaves were harvested at 2, 4, 6 or 8 week interval, starting 8 weeks after planting. Data was collected on carotenoids, vitamin C, protein, carbohydrates, and dietary fiber. The data were subjected to analysis of variance and means separated with Tukey's test, at $p \leq 0.05$. Combined use of NPK at 250 kg ha^{-1} and cattle manure at 7,500 kg ha^{-1} at 2-week leaf harvesting interval improved carotenoids, vitamin C, protein, carbohydrates, and dietary fiber on both leaves and fruit, similar to NPK applied at 750 kg ha^{-1} and cattle manure at 7,500 kg ha^{-1} and 8-week leaf harvesting interval. The findings demonstrate that combined application of NPK and cattle manure at lower rates permits more frequent leaf harvest, for consumption without compromising on nutritional quality of butternut fruit.

Keywords: Butternut squash, *Cucurbita* spp., leaf harvesting, plant nutrition, plant quality

Contact Address: Elisha Otieno Gogo, Pwani University, Crop Sciences, P.O. Box 195, 80108 Kilifi, Kenya, e-mail: e.gogo@pu.ac.ke

The Effects of Nutrient Enriched Biochar Placed in the Root Zone Varied with two Vegetable Species in Sandy Soil

Ghulam Haider[1], Muhammad Asif[2], Claudia I. Kammann[3]

[1] *National University of Sciences and Technology (NUST), Dept. of Plant Biotechnology, Pakistan*

[2] *MNS University of Agriculture Multan, Agronomy, Pakistan*

[3] *Hochschule Geisenheim University, Dept. of Applied Ecology, Germany*

Global climate is rapidly progressing, leading to unusual extreme weather conditions around the world. The major contributing factors identified are increasing greenhouse gas emissions and thereby global warming. On the other hand, increasing population puts growing pressure on existing farming practices to ensure global food security. Unfortunately, it often leads to non-judicious use of agrochemicals (more input-more output) like nitrogen fertiliser. Thus, mineral nitrogen losses as nitrate leaching or gaseous emission (nitrous oxide, a 300-fold potent greenhouse gas than carbon dioxide) are ever-increasing challenges for environmental sustainability since the late '70s. Biochar, a stable carbon product obtained by the pyrolysis of organic residues has demonstrated significant potential for carbon sequestration, soil health improvement, and crop yield. Based on our group's previous research, we observed that freshly produced biochar causes nitrogen uptake limitations in crops. Thus, in the present study, we aimed to improve our understanding of soil-plant nutrient interactions by addressing the following questions; i) Can fertilisers loaded onto biochar improve yields in vegetable crop plants, compared to the same fertiliser applied pure? and ii) Does the way of placement matter, i.e. homogenous mixing into the soil versus root zone placement? Two vegetable crops (red radish and celery) were grown in potted sandy soil in open field-controlled beds at Geisenheim University, Germany. Two nitrogen sources (analytical grade mineral nitrogen or cow-urine derived nitrogen) along with other nutrients were applied by either thorough mixing of nitrogen-enriched biochar into the topsoil layer or by more concentrated root-zone placement. Interestingly, the two vegetable species responded different to the treatments. The fresh and dry matter yield of radish was not ($p=0.184$) influenced by any application or placement method, or source of nitrogen. However, thorough mixing of urine nitrogen/mineral nitrogen loaded biochar (38.87 % and 5.31 %) and rootzone application significantly increased (39.36 % and 43.15 %, respectively) celery fresh mass over that of the controls without biochar. The celery dry matter was also increased in the range of 2.64 % to 39.15 %, with thorough mixing/root-zone placement of urine/mineral N enriched biochar over control. Thus, the results suggest that in some N-demanding vegetable crops, biochar may have the potential to increase N uptake by increasing crop yields, reducing nitrogen losses, and thereby improving crop productivity.

Keywords: Carbon sequestration, organic farming, waste management

Contact Address: Ghulam Haider, National University of Sciences and Technology (NUST), Dept. of Plant Biotechnology, H-12 Sector Nust, 44000 Islamabad, Pakistan, e-mail: ghulam.haider@asab.nust.edu.pk

Impact of Co-Compost Pellets on Growth and Yield of *Ipomoea batatas* and *Eleusine coracana*

A.W. Sagara Pushpakumara[1], Jayantha Weerakkody[1], Bandula Ranaweera[2], Felix Thiel[3]

[1]*Wayamba University of Sri Lanka, Dept. of Plantation Management, Sri Lanka*
[2]*Wayamba University of Sri Lanka, Dept. of Horticulture and Landscape Gardening, Sri Lanka*
[3]*Ruhr University Bochum / International Water Management Institute, Inst. of Geography / Rural Urban Linkages, Germany*

Dewatered Fecal Sludge (DFS) and organic Municipal Solid Waste (MSW) co-compost pellets have a high potential as an agricultural resource in Sri Lanka while elevating deficient MSW management. Three studies had been conducted to explore the importance of formulation and placement depth of these pellets. Using *Ipomoea batatas* one field study was carried out with biochar derived from empty fruit bunches of oil palm as additive to co-compost pellets to amend sandy loam soil and to evaluate its performance on the plant growth and yield Two studies were conducted to investigate the response of growth and yield of Eleusine coracana to different dosages and application depths of co-compost pellets under greenhouse conditions. The experiment of *Ipomoea batatas* tested seven pelletized treatments, namely MSW-DFS 30 % available nitrogen (N), MSW-DFS 100 % available nitrogen, MSW-DFS-biochar 30 % available nitrogen, MSW-DFS-biochar 100% available nitrogen, MSW-DFS-N enriched, MSW-FS-biochar N enriched, and a control with mineral fertiliser recommended by Department of Agriculture (DOA), Sri Lanka were used. The results revealed a significantly higher yield with MSW-DFS pellets (30 % available nitrogen) and MSW-DFS-biochar pellets (30 % available nitrogen) treatments against the recommended mineral fertiliser. It could be concluded that, harvest of 15 to 19.5 tons ha^{-1} could be achieved by amending soil with 16.8 tons ha^{-1} of MSW-DFS-Biochar pellets with 30% available Nitrogen. In *Eleusine coracana* trials, plots were laid according to Latin Square Design in a greenhouse. Eight different dosages of pellets based on available Nitrogen (10 %, 30 %, 50 %, 70 %, 90 %, 110 %, 130 % and 150 %) were applied as treatments with DOA recommended mineral fertiliser as the control in the dosage response trial. In the application depth trial, co-compost powder and pellets were applied at four different soil depths. Depth of application did not have any effect on growth and yield of *Eleusine coracana*. However, application of DFS-MSW co-compost recorded significantly higher growth and yield performance (63 % higher) compared to inorganic fertiliser. These results concluded that increased yield can be achieved with surface application of co-compost pellets with volumes that just cover 30 % of the mineral nitrogen demand.

Keywords: Biochar, co-compost, *Eleusine coracana*, *Ipomoea batatas*, pellets

Contact Address: A.W. Sagara Pushpakumara, Wayamba University of Sri Lanka, Dept. of Plantation Management, Kurunegala, Sri Lanka, e-mail: awspk.digit5@gmail.com

Does Mulching Increase Maize Yields in the Tropics? A Systematic Review

LAURA KUONEN, LINDSEY NORGROVE

Bern University of Applied Sciences, School of Agricultural, Forest and Food Sciences, Switzerland

Maize is one of the most important staples in the Tropics. The use of mulch in maize cropping is widespread and a plethora of trials have been conducted since the 1980s. Yet few studies have attempted to synthesise data to identify general patterns and research gaps. We hypothesised that mulching increases grain yield, depending on rainfall, the growing period of the maize, timing of mulch application, and that also applying mineral fertiliser might be synergistic. We conducted a systematic review in web of science. Inclusion criteria included that trials were monocultures, had a non-mulch control and were from the tropics. We found 54 papers fulfiling our criteria and from these we extracted grain yield data, as the proportion between relative grain yield and control, arcsine square root transformed them and analysed them in R v. 3.6.1.

The most commonly tested mulch was *Leucaena leucocephala* ($n = 13$). Treatments often involved additional factors, for example tillage, irrigation and fertiliser, whereas mulching was the only factor in 15 papers. Mulching improved maize grain yield (+ 30 %), yet the % increase was significantly higher in areas where precipitation > 1500 mm p.a. (+ 48 %) than where precipitation < 800 mm p.a. (9 %). Where mulch was applied after sowing, the highest % increase was obtained (+ 44 %), compared with mulch applied at (+ 29 %) or before sowing (+ 18 %). The shorter the maize growing season, the more pronounced the improvement with mulching. Where combined with mineral fertiliser, the improvement was accentuated, suggesting synergies from combining different inputs. Using fresh rather than dry material resulted in a greater % yield increase. Improvement with mulching over the control was greater in long-term trials compared to short-term trials. In conclusion, mulching increased grain yield, particularly for short season maize, optimally applied post-sowing, and showed synergisms with, rather than replacing fertiliser. Surprisingly, effects were most pronounced in higher rainfall areas, suggesting a role as a nutrient input rather than just a soil water conserving strategy.

Keywords: Mulch, systematic review, tropics, yield, *Zea mays*

Contact Address: Laura Kuonen, Bern University of Applied Sciences, School of Agricultural, Forest and Food Sciences, LÃ¤nggasse 85, 3052 Zollikofen, Switzerland, e-mail: laura.kuonen@gmail.com

Climate change

Climate change - Mitigation and adaptation

Oral Presentations

Posters

Embracing Uncertainty in Complex Systems – Assessing Alternatives to Face Climate Change Impacts in Mediterranean Climate Orchards

Gonzalo Rojas[1], Eduardo Fernandez[2], Cory Whitney[2], Eike Luedeling[2], Italo Cuneo[1]

[1] *Pontificia Universidad Católica de Valparaíso, Dept. of Agriculture and Food Sciences, Chile*

[2] *University of Bonn, Inst. Crop Sci. and Res. Conserv. (INRES) - Horticultural Science, Germany*

Subtropical areas such as the major fruit production zone of Chile are among the most vulnerable environments to climate change. Increasing frequencies of unusual weather events, e.g. excessive rainfall or hail and frost episodes, during the growing season can cause severe fruit losses. This can threaten the viability of farm. and have implications for local and global food security. When deciding on which strategies to adopt in the face of these climatic events farmers are faced with major uncertainties about costs, benefits and risks. They find it difficult to predict the outcomes of implementing new management practices and technologies. Decision Analysis approaches offer methods that can embrace uncertainty and offer forecasts to support such complex decisions. We applied these tools for interventions to protect sweet cherry orchards in two production systems (i.e. orchards with and without covers) under two climatic conditions (i.e. northern-central and southern Chile). We gathered experts to help us build a causal impact pathway model which we implemented as a Monte Carlo simulation to estimate the Net Present Value (NPV). The model was parameterised with inputs provided by the experts. We computed the Expected Value of Perfect Information (EVPI) for each variable to determine where more information might be helpful in choosing the best management option. Our model suggests no major differences for decision options in northern-central Chile, whereas in southern Chile use of the covers increases the probability of achieving positive NPV outcomes. Across sites, EVPI analysis revealed that additional information on market price and yield could improve the decision support offered by model. Our model suggests that orchards in northern-central Chile are more likely to obtain positive outcomes in the long term, independent of the cover system. Farmers in the south risk major losses when growing sweet cherries without protection. We expect this work to inspire farmers in complex agricultural systems to adopt Decision Analysis approaches for choosing between climate change adaptation strategies.

Keywords: Cover systems, decision analysis, fruit cracking, orchard protection, preharvest rain, spring frost risk

Contact Address: Cory Whitney, University of Bonn, Inst. Crop Sci. and Res. Conserv. (INRES), Auf Dem Hügel 6, 53111 Bonn, Germany, e-mail: whitney.cory@gmail.com

Agroforestry and Reforestation with the Gold Standard – Decision Analysis of a Voluntary Carbon Offset Label

Leonie Netter, Eike Luedeling, Cory Whitney

University of Bonn, Inst. Crop Sci. and Res. Conserv. (INRES), Germany

Projects that sequester carbon can receive an economic reward for mitigating climate change by selling these emission reductions in form of carbon certificates. Voluntary carbon offset standards were developed in order to standardise the quality of projects eligible for carbon offsetting. However, project managers are faced with uncertainty when deciding on investments in certification. The positive social, ecological, and economic benefits of certifying a carbon sequestration project do not always outweigh the costs and risks involved in establishing and maintaining the certification. Decision analysis provides a helpful set of tools that can support such complex decisions by forecasting outcomes under different scenarios. We generated a model for the certification of an additional site of a partially certified reforestation project and a new certification for an agroforestry project in Costa Rica with the voluntary carbon offset label Gold Standard.

The parameters considered important for the certification process were identified in interviews with decision-makers and stakeholders of the projects and translated into a model. After a calibration training, the experts provided probability distributions for the input parameters in form of estimated confidence intervals. The final decision model was run in a Monte Carlo simulation to project a plausible range of decision outcomes, expressed as Net Present Values and annual cash flows. We identified critical uncertainties by using the Expected Value of Perfect Information. Although the annual cash flow of the agroforestry project would most likely be negative in the early years due to high initial investment costs, certification with the Gold Standard would most likely have an overall positive impact on the profitability of the project. The results indicate that certification of the agroforestry area as well as the reforestation project would result in a positive Net Present Value. The partially low return on investment of the certification, however, underlines the importance of a thorough evaluation and the development of customized strategies for projects before applying for participation in a voluntary carbon offset scheme. The Decision Analysis approach we applied can help improve the process of decision making under uncertainty and should be widely adopted for evaluating the potential impacts of certification.

Keywords: Agroforestry, carbon credits, decision analysis, gold standard

Contact Address: Leonie Netter, University of Bonn, Inst. Crop Sci. and Res. Conserv. (INRES), Im Flürchen 83b, 66133 Saarbrücken, Germany, e-mail: leonie.netter@hotmail.de

Generating a Conceptual Framework of Agro-Climate Information Interventions in Vietnam

Thi Thu Giang Luu, Cory Whitney, Eike Luedeling
University of Bonn, Inst. Crop Sci. and Res. Conserv. (INRES), Germany

Farmers' agricultural practices in Vietnam are highly sensitive to variable weather and vulnerable to the impacts of climate change. The lack of timely and actionable weather forecasts and agricultural advisories can lead to significant yield loss and unprofitable investments in agricultural inputs, which can have severe consequences for resource-constrained farming communities, including many women and ethnic minorities. Development organisations in Vietnam have provided agro-climate information services (ACIS) in the form of climate-informed agricultural advisories to farmers. However, so far these interventions have been very limited in scale. Development organisations are advocating the government to consider upscaling the provision of climate advisory services, but a large-scale roll-out could strain their financial and human resources. We developed a conceptual model of costs, benefits and risks involved in such an intervention aiming to provide effective support to the government's decision-making process. The model was developed from expert knowledge gathered in workshops, which also served to generate data and offer space for learning and reflection. We used collaborative tools to explore the implications of the government's decision on investing in ACIS interventions in the complex and uncertain context of agricultural production in Vietnam.

We held a series of model building workshops in 2019 with a team of technical staff from government and NGOs with experience in implementing ACIS and expertise in meteorology, climate change, crop production, animal husbandry, communications, gender and poverty. The workshops were designed to generate and validate a conceptual framework of the decision. The resulting model includes: (1) costs for generating forecasts and translating them into advisories for farmers, including transfer, support, monitoring and evaluation; (2) benefits, including increased yields, reduced losses of agricultural inputs, reduced poverty and gender inequality and improved health; (3) risks including the possibility of inaccurate weather forecasts leading to misguided advice causing yield losses and inefficient use of costly agricultural inputs.

The framework offers a foundation for the design, implementation and evaluation of ACIS. It provides pragmatic guidance on ACIS interventions, and can be adjusted to fit specific circumstances on the ground.

Keywords: Agro-climate information service, conceptual framework, cost-benefit

Contact Address: Thi Thu Giang Luu, University of Bonn, Inst. Crop Sci. and Res. Conserv. (INRES) - Horticultural Science, Auf Dem Hügel 6, D-53121 Bonn, Germany, e-mail: s7thluuu@uni-bonn.de

Leaf Gas Exchange of Lowland Rice in Response to Nitrogen Source and Vapor Pressure Deficit

Duy Hoang Vu[1], Alejandro Pieters[2], Sabine Stürz[2], Folkard Asch[2]

[1]*Vietnam National University of Agriculture, Dept. of Cultivation Sci., Vietnam*
[2]*University of Hohenheim, Inst. of Agric. Sci. in the Tropics (Hans-Ruthenberg-Institute), Germany*

In anaerobic lowland rice fields, ammonium (NH_4^+) is the dominant form of nitrogen (N) in the soil, whereas nitrate (NO_3) is the main form of N in aerobic fields. During drained periods in water-saving irrigation practices, nitrification is favoured and thus, plants take up a higher share of N as NO_3. Therefore, the progressive implementation of water-saving irrigation practices arouses increased interest in physiological responses of rice to different N sources. Since assimilation and translocation of the two N forms differ in their demand of photosynthates, leaf gas exchange may be subject to adjustments in response to N source. Two experiments were carried out to study gas exchange of four lowland rice varieties (IR64, BT7, KD18, Jasmine 85) in response to N source (sole $NH_4^+$and sole NO_3). Since assimilation rate (A) strongly depends on stomatal conductance (gs), which is highly sensitive to vapour pressure deficit (VPD), the experiments were conducted at varying VPD. Plants were grown in the greenhouse in standard Yoshida nutrient solution during pre-cultivation. At the plant age of 3 weeks, nutrient solution with sole NH_4^+ or NO_3 was provided for another 2 weeks. At the end of this period, leaf gas exchange was measured, and after, morphological parameters were determined and starch concentration in the leaves was analyzed. Both, after 2 weeks at low VPD and after transferring the plants from low to high VPD, rice varieties did not show significant differences in A and gs between the two N sources. However, after 2 weeks at high VPD, NO_3 nutrition led to higher A and gs in IR64 and BT7, whereas no significant difference was found in KD18 and Jasmine 85. Increased root/shoot ratio and specific leaf area of plants in NO_3 solution was observed after 2 weeks at high VPD. At the same time, starch concentration in the leaves was not altered, indicating an increased allocation of photosynthetic products to the roots. Therefore, we conclude that under long-term high evaporative demand, NO_3 nutrition can increase assimilation rates of rice via an increased stomatal conductance in some, but not all varieties.

Keywords: Ammonium, nitrate, photosynthesis, stomatal conductance, VPD

Contact Address: Duy Hoang Vu, Vietnam National University of Agriculture (VNUA), Dept. of Cultivation Sci., Ngo Xuan Quang - Gialam, 131000 Hanoi, Vietnam, e-mail: vdhoang87@gmail.com

Calibration and Validation of the LandscapeDNDC Model for Simulating Biomass Production in West African Savannah Ecosystems

Jaber Rahimi[1], Rüdiger Grote[1], David Kraus[1], Edwin Haas[1], Clemens Scheer[1], Sina Berger[1], Expedit Evariste Ago[2], Håkan Torbern Tagesson[3], Olivier Roupsard[4], Bernard Cappelaere[4], Amit Kumar Srivastava[5], Ulrike Falk[6], Pierre Hiernaux[7], Augustine Ayantunde[8], Klaus Butterbach-Bahl[1]

[1]*Karlsruhe Institute of Technology (KIT), Inst. of Meteorology and Climate Research, Atmospheric Environmental Research (IMK-IFU), Germany*

[2]*Université d'Abomey-Calavi (UAC), Lab. d'Hydraulique et de Maîtrise de l'Eau, Benin*

[3]*University of Copenhagen, Department of Geography and Geology, Denmark*

[4]*University of Montpellier, CIRAD, France*

[5]*University of Bonn, Inst. Crop Sci. and Res. Conserv. (INRES), Germany*

[6]*Deutscher Wetterdienst (DWD), Germany*

[7]*Géosciences Environnement Toulouse (GET), France*

[8]*International Livestock Research Institute (ILRI), Burkina Faso*

Savannah ecosystems, as a valuable component of the West Africa's vegetation types, cover more than 80 % of its area and provide a wide range of ecosystem services from both ecological (e.g. regulating climate) and economical (e.g. producing biomass for humans and animals) perspectives. According to the recent surveys, West African savannahs are experiencing rapid land cover and climatic changes and these changes will at the end affect their biodiversity and productivity. However, from modelling perspectives, there is still a need for parameterisation of process based biogeochemical models for modelling the behaviour of these highly dynamic ecosystems. Here, we report simultaneous calibration of LandscapeDNDC model, which is a process-based ecosystem model that simulates C, N and water cycling, using multiple observation types from flux towers and satellite images and for various typical land cover types (i.e. grasslands, shrublands, arable lands, and woodlands) in West African Sudanian and Sahelian zones. Model performance of the newly parameterised vegetation module of LandscapeDNDC was assessed on basis of several *in situ* above-ground biomass production measurements gathered from different sources across the Sudanian and Sahelian agro-ecological zones. Our results indicate that the LandscapeDNDC model is able to simulate biomass growth as well net ecosystem CO_2 exchange (NEE) and Leaf Area Index (LAI) realistically over periods of several years. In addition, potential applications of the LandscapeDNDC model for managed and natural ecosystems in semi-arid environments, such as impacts of climate change on seasonal biomass production and changes of regional C exchange due to land use change will be discussed.

Keywords: Calibration, landscapeDNDC model, savannah, West Africa

---

Contact Address: Jaber Rahimi, Karlsruhe Institute of Technology (KIT), Inst. of Meteorology and Climate Research, Atmospheric Environmental Research (IMK-IFU), 82467 Garmisch-Partenkirchen, Germany, e-mail: jaber.rahimi@kit.edu

Quantifying On-farm Greenhouse Gas Emissions in Smallholder Livestock Systems in Western Kenya: Life Cycle Assessment

Phyllis Ndung'u[1], Taro Takahashi[2], Linde Du Toit[1], Melanie Robertson-Dean[3], Lutz Merbold[4], Svenja Marquardt[4], John Goopy[4]

[1] *University of Pretoria, Dept. of Animal and Wildlife Sciences, South Africa*
[2] *Rothamsted Research, United Kingdom*
[3] *University of New England, School of Science and Technology, Australia*
[4] *International Livestock Research Institute (ILRI), Mazingira Centre, Kenya*

Low productivity combined with increasing livestock populations is resulting in negative environmental effects such as greenhouse gas (GHG) emissions (methane (CH_4) and nitrous oxide (N_2O)) among African livestock systems. This study aims at quantifying the GHG emissions intensities (EIs) of cattle in smallholder farms in Western Kenya using primary field data in order to show the distribution of farm EIs across counties and assess the total crude protein (CP) output. Emission intensities were quantified using the life cycle assessment (LCA) technique, a "cradle to farm gate" systems boundary was adopted, and with milk and meat being the primary outputs. Crude protein was chosen as the functional unit to harmonise milk and meat outputs. Enteric CH_4 emissions were estimated with a metabolisable energy requirement (MER) approach. Simultaneously, three manure management systems were identified i.e. manure left on pasture, cattle enclosures such as bomas, and manure piles and subsequent emissions of CH_4 and N_2O were estimated using accurate and appropriate yield conversation factors for CH_4-C and N_2O-N. The median farm emission intensity for the counties Nandi, Bomet and Nyando were 67, 66 and 135 kg CO_2-eq/kg CP respectively. Despite the fact that all investigated smallholder farms are low-input systems, farm EIs varied substantially with Nyando having higher variation than Nandi and Bomet counties caused by a wider range of production level. Milk CP contributed 80–85 % and meat CP 15–20 % to the total farm output. Enteric CH_4 contributed the largest proportion (up to 95 %) of all GHG emission sources. Quantile regression analysis revealed that specific management features such as herd size, proportion of females in the herd and average parity were highly influential to EI at the farm level, irrespective of agro-ecology. Both meat and milk were significant drivers of EI across all quantiles but mean milk yield per cow, (rather than milk production per farm) was the most important driver of EI. Pursuing these characteristics as focused management objectives have the potential to move low input smallholder farms towards high emissions efficiency operations, reducing GHG emissions per unit of milk and meat produced and potentially lowering GHG emissions from ruminant production.

Keywords: African livestock system, crude protein, emission intensity, GHG emission, primary field data, quantile regression analysis

Contact Address: Phyllis Ndung'u, Univ. of Pretoria, Dept. of Animal and Wildlife Sci., Lynnwood road, Hatfield, 0002 Pretoria, South Africa, e-mail: wanjuguphyllis@ymail.com

Greenhouse Gas Emissions and Fertiliser Quality from Cattle Manure Heaps in Kenya

Sonja Leitner[1], Donal Ring[2], George Wanyama[1], Daniel Korir[1], David Pelster[3], John Goopy[1], Lutz Merbold[1]

[1]*International Livestock Research Institute (ILRI), Mazingira Centre, Kenya*
[2]*Trinity College Dublin, Dept. of Botany, Ireland*
[3]*Agriculture and Agri-Food Canada, Canada*

Livestock farming is essential for smallholder farmers in sub-Saharan Africa (SSA) whose livelihoods depend on it. In addition, manure is an important nitrogen source for croplands, because mineral fertiliser is often unaffordable for smallholders. But manure emits 10–25 % of agricultural greenhouse gases (GHG), and due to poor feeding and manure management manure fertiliser quality is low. There are few *in situ* measurements of manure GHG emissions from SSA, and agricultural GHG budgets from African nations rely largely on IPCC default values, which might not represent smallholder farms well. To address this knowledge gap, we conducted two manure incubation experiments in Kenya, using manure from local Boran (*Bos indicus*) cattle fed with local feeds. Manure was collected daily and piled in uncovered heaps, representing the most common manure storage in Kenyan smallholder systems. CH_4 and N_2O emissions were measured over 140 days. In the first trial, cattle were either fed at 120 % maintenance-energy requirement (i.e. receiving enough food to support their metabolism), or at sub-maintenance energy levels to simulate feed scarcity, common particularly during the dry season. Manure N_2O emissions from hungry cows were lower than from cattle fed at maintenance energy levels because of lower manure-N concentration and a wider C:N ratio, indicating lower fertiliser quality. Furthermore, in sub-maintenance cows excreted N shifted from urine-N (mostly inorganic) to faecal-N (mostly organic), indicating higher resistance to decomposition and conversion to N_2O. Across all diets, manure N_2O and CH_4 emissions were lower than the IPCC default emission factors for solid storage in tropical regions. In the second trial, Boran cattle were fed with three tropical forage: Napier (*Pennisetum purpureum*), Rhodes (*Gloris gayana*), and *Brachiaria* (*Brachiaria brizantha*). Manure from the Rhodes grass diet had the lowest N concentration and the lowest CH_4 emissions, whereas manure from the *Brachiaria* diet had a slightly better fertiliser quality. N_2O emissions did not differ between diets. Again, CH_4 and N_2O emissions were lower than IPCC default factors. These results help reducing uncertainties in agricultural GHG emissions in SSA. If African nations use IPCC default values for GHG reporting, emissions are likely to be overestimated.

Keywords: Boran cattle, fertiliser quality, methane, nitrous oxide, smallholder farms

Contact Address: Sonja Leitner, International Livestock Research Institute (ILRI), Mazingira Centre, Kabete Campus, Old Naivasha Rd., 00100 Nairobi, Kenya, e-mail: s.leitner@cgiar.org

Enteric Methane Production from Cattle Fed on Three Tropical Grasses in East Africa

Daniel Korir[1], Alan Sanchez[2], Uta Dickhoefer[2], Svenja Marquardt[1], Lutz Merbold[1], John Goopy[1], Richard Eckard[3]

[1]*International Livestock Research Institute (ILRI), Kenya*
[2]*University of Hohenheim, Inst. of Agric. Sci. in the Tropics (Hans-Ruthenberg-Institute), Germany*
[3]*University of Melbourne, Fac. of Agriculture and Veterinary Sciences, Australia*

Tropical planted grasses are an important feed resource in many East African mixed crop-livestock systems. There is, however, inconclusive information on the nutritive value of these grasses and the consequences of feeding these grasses on enteric methane emission from cattle, primarily caused by a lack of in-situ measurements. A feeding experiment was conducted with growing Boran steers (n=18, live weight (LW): 216 $\pm$ 6 kg) fed on either *Pennisetum purpureum* (Napier), *Chloris gayana* (Rhodes) or *Brachiaria brizantha* (Brachiaria) for 65 days on ad libitum basis. Voluntary nutrient intake, apparent total tract digestibility, LW gain and enteric methane production (using cattle respiration chambers) were measured in a 3×3 crossover Latin square design. Crude protein (CP) content (g kg^{-1} dry matter (DM)) of the fodder fed to the animals were 89 $\pm$ 16.2, 77 $\pm$ 12.6 and 83 $\pm$ 9.9 (Mean $\pm$ SD) for Napier, Rhodes and *Brachiaria* grass respectively, while NDF (g kg^{-1} DM) content ranged from 648 $\pm$ 22.7 (Napier) to 686 $\pm$ 24.9 (Rhodes). There were no differences ($p > 0.05$) in voluntary DM intake and average daily LW gain (441 $\pm$ 107.5 g) across the three treatments. Animals fed on Napier grass had higher ($p < 0.05$) total tract DM and organic matter (OM), CP and fibre digestibility compared to animals fed on Rhodes, but not different from animals fed on *Brachiaria*. Napier grass (35 $\pm$ [2.5] g kg^{-1}) and *Brachiaria* (32 $\pm$ 2.2 g kg^{-1}) had higher ($p < 0.01$) methane yield per OM intake than Rhodes grass (31 $\pm$ 3.6 g kg^{-1}) but the differences were not observed when expressed per digestible OM intake ($p = 0.59$).
We conclude that utilisation and enteric methane yield per OM intake of grasses grown and fed to cattle in East Africa, at animal level, differ from grass to grass. Accounting for the differences in methane yields for the grasses grown in the different agro-ecological zones in the region can improve the Accuracy of the emission factors used for East African countries. More data on a wider range of grasses is needed to validate the findings of the present study.

Keywords: Digestibility, weight gain, yield

Contact Address: John Goopy, International Livestock Research Institute (ILRI), Nairobi, Kenya, e-mail: j.goopy@cgiar.org

Recycling of Coffee By-Products by Composting Regarding Climate Relevant Emissions and Products

Macarena San Martin Ruiz, Martin Reiser, Martin Kranert

University of Stuttgart, Inst. for Sanitary Engineering, Water Quality and Solid Waste Management, Germany

Throughout the world, the agriculture, sanitation and waste management sectors are mainly carried out in isolation, resulting in permanent nutrient drainage and large amounts of greenhouse gas emissions due to inadequate or excessive use of fertilisers. The purpose of this study is to develop an innovative experimental methodology for the sustainable recycling and improved treatment of coffee by-products to produce organic compost, which can be used in agricultural crops including coffee plantations. The methodology will be implemented in a Mill in Costa Rica, due to the current waste management and its potential to reduce GHG emissions during its production cycle. This study investigated the performance of aerobic windrow systems by using coffee by-products and green waste to reduce gaseous emissions. Thereafter a comparison with the current treatment and gaseous emissions at a Coffee Mill in Costa Rica was made. Different composting studies were performed in Germany and in Costa Rica. Temperature, water content, and pH were the key parameters controlled over 35 days and 60 days. Moreover, CH_4 emission rates were quantified by a FTIR and by a portable gas detector device where the emissions were reduced up to 100 times than using the current method at the Mill. It was found that CH_4 emissions could be avoided, following the key parameter for composting and a proper mixture. With this, the reduction of CH_4 emissions at the Mill in Costa Rica could be achieved in the future, providing a circular economy and towards to a reduction in the carbon footprint during the management of the coffee-by-products in the sector.

Keywords: greenhouse gases, coffee by-products, coffee pulp, composting, emission rates, methane

Contact Address: Macarena San Martin Ruiz, University of Stuttgart, Inst. for Sanitary Engineering, Water Quality and Solid Waste Management; Chair of Waste Management and Emissions, Bandtaele 2, 70569 Stuttgart, Germany, e-mail: Macarena.sanmartin@iswa.uni-stuttgart.de

Environmental Impact Assessment of Rice Cultivation in Da Nang, Vietnam – Options for Sustainable Production Systems

Sebastian Awiszus, Ziba Barati, Sasinee Pansailom, Joachim Müller
University of Hohenheim, Inst. of Agric. Sci. in the Tropics (Hans-Ruthenberg-Institute), Germany

Rice is the main staple food in Vietnam. which occupies the biggest share of the agricultural areas. This also applies to the environs of the city of Da Nang, located in the central region of Vietnam. The overuse of inputs, in terms of fertilisers, pesticides, and water in the rice production systems resulted in environmental issues and also health problems for the population. In this study, the environmental impacts of the traditional rice-farming system in Da Nang was assessed following ISO 14040 and ISO 14044. Potential alternatives were modelled and evaluated based on the functional unit (FU) of 1 kg harvested rice. Accordingly, the traditional cultivation (baseline) was modified regarding the direct incorporation of harvesting residues, irrigation management and the on-field burning as well as the composting of rice straw. The environmental impacts of the scenarios were evaluated regarding the global warming potential (GWP), the primary energy demand (PED), the eutrophication potential (EP) and the water consumption (WC). The GWP ranged from 0.8 – 1.76 kg CO_2-eq FU^{-1} and showed advantages when applying alternate wetting and drying (AWD) techniques compared to the baseline (1.07 kg CO_2-eq FU^{-1}). The incorporation of residues resulted in a higher GWP in the pairwise comparison with the baseline and the AWD scenarios. Burning (1.12 kg CO_2-eq FU^{-1}) and composting (1.76 kg CO_2-eq FU^{-1}) of rice straw resulted in the highest GWP. The EP ranged from 1.97×10^{-4} – 2.36×10^{-4} kg P-eq FU^{-1} indicating that an incorporation of residues or application of composted straw have positive effects. This also applies for the PED, ranging from 7.63 – 9.23 MJ FU^{-1} and the WC varying from 11.64–14.28 kg FU^{-1}. Furthermore, results show complex interlinkages and tradeoffs in both, within and between the chosen impact categories. Applying alternative rice cultivation systems can result in a lower GWP, but causing a higher WC, PED or EP instead, highly dependent on the local conditions. Therefore, further studies should clarify best practice solutions based on site specific-data and should also take social aspects into consideration.

Keywords: Global warming potential, life cycle assessment

Contact Address: Sebastian Awiszus, University of Hohenheim, Inst. of Agric. Sci. in the Tropics (Hans-Ruthenberg-Institute), Garbenstr. 9, 70599 Stuttgart, Germany, e-mail: sebastian.awiszus@uni-hohenheim.de

Cassava Yield Gap and Variability: A Simulation Study in Nigeria

Amit Kumar Srivastava, Thomas Gaiser, Martina Pesch, Frank Ewert
University of Bonn, Inst. Crop Sci. and Res. Conserv. (INRES), Germany

An increase in crop productivity is needed to ensure sufficient food supply for the continuously growing world population. Thereby, Cassava production is of great importance for food security in Africa and a major source of human caloric intake for the local population. Nigeria is the largest producer of Cassava, but there is still a need to increase the production of Cassava for export purpose. The climate is an important factor for the Nigerian agriculture, as temperature, radiation, humidity and water are main factors for crop growth and yield. Climate variability is a determining factor for Cassava yields. For this study, a crop model was used to assess the yield and yield gaps of Cassava based on different climatic variables such as temperature, radiation and precipitation in 10 states of Nigeria over 16 years from 1995 to 2010. The crop model LINTUL5, embedded into a general modelling framework SIMPLACE (Scientific Impact Assessment and Modelling Platform for Advanced Crop and Ecosystem Management), was used to estimate potential yield (Yp), water-limited yield (Yw), water- and nitrogen-limited yield and water- and nutrient-limited yield. Multiple regression shows the correlation between the observed yields and simulated yields, as well as yield gaps with the climate variables. The yield gaps in Cassava production are high and show variability across the different regions. The potential yield gap ranges from 6 t ha^{-1} to 9 t ha^{-1} depending on the region as well as over the 16 years. Spatially, the potential yield gap correlates negatively with mean temperature and precipitation and shows a positive correlation with radiation. To close the high Cassava yield gaps in Nigeria, an improvement in farm management, including soil and crop management, is indispensable.

Keywords: Cassava, LINTUL5, Nigeria, yield gap

Contact Address: Amit Kumar Srivastava, University of Bonn, Inst. Crop Sci. and Res. Conserv. (INRES), Katzenburgweg 5, 53115 Bonn, Germany, e-mail: amit@uni-bonn.de

Analysis of the Impact of Climate Change on the Streamflow at Chaghasrai Watershed in Afghanistan

Fazal e Azeem Khilgee[1,2], Bernhard Tischbein[2], Fazlullah Akhtar[2]

[1]*University of Bonn, Agricultural Sciences and Resource Management in the Tropics and Subtropics (ARTS), Germany*

[2]*University of Bonn, Center for Development Research (ZEF), Dept. of Ecology and Natural Resources Management, Germany*

Afghanistan is a water-scarce country and is host to only 12 % of its area as arable which is not enough to meet the food security demands of the country. Around 11 % of the total population of the country has been strongly hit by hunger and the situation might further escalate the already fragile conditions of the agriculture and water resources sector. Beside political instability and population pressure, climate change is another factor that is projected to melt down the permanent snow lying at the peaks of the watershed upstream. which may alter the flow regime of the rivers flowing downstream. In order to manage and mitigate the adverse impacts of climate change, it is vital to estimate the present and project the future flow of the rivers. Therefore, we used a semi-distributed physical based hydrologic model, SWAT (Soil Water Assessment Tool) to evaluate the impacts of climate change on the river hydrology of the Chaghasrai watershed in Afghanistan. The SWAT-model was calibrated and validated for streamflow observed at Chaghasrai gauging station during the prediction period. The performance of SWAT model was evaluated using the Nash–Sutcliffe Efficiency (NSE) and the Coefficient of determination (R^2). The future climate changes scenarios of RCP 4.5 and RCP 8.5 were extracted from the Coordinated Regional Downscaling Experiment (CORDEX). Upon calibration of the model, the streamflow was simulated for a period of 2014–2030 while using the selected climate change scenarios. The analyses show that the project site is expected to experience a temperature increase of 1.8 °C while mean annual precipitation is expected to increase by 3 % (20 mm) under both scenarios during 2014–2030. The overall streamflow is expected to be decreased by 20 % due to increase in temperature which triggers the rate of evapotranspiration during the study period. This study provides a base for estimation of the water supply (availability) in the watershed but future studies are required to consider water demand side (irrigation requirements.

Keywords: Climate change, Kabul river basin, precipitation, streamflow

Contact Address: Fazal e Azeem Khilgee, University of Bonn, Agricultural Sciences and Resource Management in the Tropics and Subtropics (ARTS), 53115 Bonn, Germany, e-mail: fazaliazeem52@gmail.com

Remotely Sensed Yield Modelling of Household Fields to Monitor Child Undernutrition and Climate Change Impacts

Isabel Karst[1], Isabel Mank[2], Issouf Traoré[3], Raissa Sorgho[2], Kim-Jana Stückemann[4], Séraphin Simboro[3], Ali Sié[3], Jonas Franke[1], Rainer Sauerborn[2]

[1]*Remote Sensing Solutions Gmbh, Germany*

[2]*Heidelberg University, Heidelberg Institute of Global Health (HIGH), Germany*

[3]*Institut National de Santé Publique (INSP), Centre de Recherche En Santé de Nouna (CRSN), Burkina Faso*

[4]*University of Osnabrück, Institute of Computer Science, Germany*

Climate change has a large impact on agricultural fields, resulting in food instability, particularly among rural smallholder farmers in sub-Saharan Africa. Based on satellite remote sensing and extensive field observations from 2018, we conducted an interdisciplinary study to monitor and predict crop yields of household fields (median 1.4 ha) to support research and policies on food security and child undernutrition. The study was conducted in the Nouna Health and Demographic Surveillance (HDSS) area, in the Sudano-Sahelian region of North-Western Burkina Faso. The area is dominated by subsistence agriculture, which is characterised by manual labour and rainfall dependent farming practices. This makes the farmers extremely vulnerable to weather unpredictability from climate change. Seasonal food insecurity is high in the Nouna HDSS, where, as of September 2018, stunting and wasting of children below five years of age reached 26 % and 7 %, respectively.

We established a crop yield estimation and prediction regression model based on monthly metrics of vegetation indices derived from multi-temporal Sent-inel-2 data. In-situ harvest measurements taken from five crops typical for the region (maize, beans, sorghum, millet, peanuts) served as model inputs. The models produced adjusted R^2 between 0.32 and 0.5 and, e.g. enabled maize yield predictions up to two months before harvest. The spatial application resulted in yield maps, in which each field can be associated and spatially linked to an individual household. These quantitative yield estimates can then be translated to crucial socio-economic information, such as crop-specific calorie intake per household. The quantitative yield estimation is, therefore, a predictor and direct determinant for potential food insecurity.

Our findings provide a scientifically and politically relevant earth observation approach since it allows monitoring and predicting yields of individual households at 10 m spatial resolution. These results can consequently be used to identify local yield losses, which is crucial for implementing early-onset prevention measures aimed at minimising food insecurity and subsequently, child undernutrition. Moreover, they can serve as a basis to evaluate the impact of climatic factors, agricultural interventions or management on households' economic and child nutritional status.

Keywords: Burkina Faso, household yield, subsistence farming, vegetation index

Contact Address: Isabel Karst, Remote Sensing Solutions Gmbh, Dingolfingerstr. 9, 81673 München, Germany, e-mail: karst@rssgmbh.de

Zonal and Seasonal Methane Emissions from Rice Production in the Vietnamese Mekong Delta

Thi Bach Thuong Vo[1], Folkard Asch[1], Reiner Wassmann[2], Bjoern Ole Sander[3]

[1]*University of Hohenheim, Inst. of Agric. Sci. in the Tropics (Hans-Ruthenberg-Institute), Germany*
[2]*Karlsruhe Institute of Technology, Germany*
[3]*International Rice Research Institute (IRRI), Vietnam Office, Vietnam*

In Vietnam rice is produced on 7.7 million ha making Vietnam the world's 6th largest rice producer. In the Mekong River Delta (MRD) comprises lowland rice with MRD providing 55 % of all Vietnamese rice production. Lowland rice production is a source of greenhouse gases (GHG) due to emissions of methane (CH_4) and – to a lesser extent nitrous oxide (N_2O). Rice production in the MRD can be classified according to seasons and zones: early year (October to June), mid-year (May to November), and late year season (December to April) as well as saline zone (close to the sea), alluvial zone (mid-delta) and the flood-prone zone (upper delta, close to the river). So far, these seasonal and environmental differences have not been reflected in the generic IPCC guidelines that provide default emission factors at the sub-continental scale. We present here a database derived from field measurements conducted with the closed chamber method and standardised crop management at 12 sites with 24 cropping seasons. Reflecting a large variation in total, emissions in the saline zone (31 to 357 kg CH_4 ha^{-1} $season^{-1}$) were lower than in the alluvial zone (150 to 440 kg CH_4 ha^{-1} $season^{-1}$) and the flooded zone (80 to 914 kg CH_4 ha^{-1} $season^{-1}$). While saline conditions are known to inhibit microbial methane production, it was surprising that the differences were not more pronounced. This can be attributed to adjusted cropping calendars in the saline zone to avoid a standing crop in the critical time window (February to April). At the peak of salinity levels, rice production is limited to locations with improved infrastructure to control salt intrusion into the canals. In those areas with persistent salt intrusion, the dominant land use is shrimp farming instead of rice. Thus, the actual rice seasons in this zone are characterised by similar conditions for microbial methane production as in other zones – even though the name of the zone suggests otherwise. Finally, the newly generated data is set into the context of GHG estimates for the Mekong Delta using the IPCC guidelines.

Keywords: Emission factor, greenhouse gas, IPCC Tier 2, methane, salinity zone

---

Contact Address: Thi Bach Thuong Vo, University of Hohenheim, Inst. of Agric. Sci. in the Tropics (Hans-Ruthenberg-Institute), Garbenstr. 13, 70599 Stuttgart, Germany, e-mail: thibachthuong.vo@uni-hohenheim.de

Varietal Effects of Five Contrasting Rice Varieties on Diurnal Methane Emissions at Different Development Stages

Theresa Detering[1], Thi Bach Thuong Vo[1], Reiner Wassmann[2], Bjoern Ole Sander[3], Folkard Asch[1]

[1]*University of Hohenheim, Inst. of Agric. Sci. in the Tropics (Hans-Ruthenberg-Institute), Germany*
[2]*Karlsruhe Institute of Technology, Germany*
[3]*International Rice Research Institute (IRRI), Vietnam Office, Vietnam*

Currently, concentrations of atmospheric methane (CH_4) are rising faster than at any time in the last two decades. In order to combat climate change, it is essential to quantify and mitigate methane emissions. Rice production is one of the main anthropogenic sources of methane. While CH_4 emissions from paddy fields are subject to diurnal variations, these are often neglected in similar field studies, leading to an over- or underestimation of total daily or seasonal methane fluxes. In collaboration with the International Rice Research Institute (IRRI), a field experiment was conducted during the dry season 2019/20 in the An Giang province of the Vietnam Mekong Delta (VMD) to evaluate diurnal variations of CH_4 fluxes at tillering, panicle initiation, and flowering stage for five contrasting rice varieties.CH_4 samples were collected at 0, 15 and 30 min after chamber closure with a 60-ml syringe using the manual closed chamber method and stored in a 30 ml evacuated glass vial. Analyses were performed with an SRI 8610C gas chromatograph (GC) at the IRRI laboratory in Los Banos, Philippines. Preliminary results show a distinct diurnal pattern that differed strongly among varieties, with maximum emissions in the early afternoon (12:00 to 15:00), followed by a decline to a minimum around midnight. Data obtained in this study will be used to correct regional emission factors to reflect the effects of diurnal variations and improve the accuracy of CH_4 extrapolations. Varietal differences in methane emissions in combination with appropriate water management would allow mitigating methane emissions in rice production systems.

Keywords: Diurnal variation, greenhouse gas mitigation, manual closed chamber, *Oryza sativa*, plant growth stages

Contact Address: Theresa Detering, University of Hohenheim, Inst. of Agric. Sci. in the Tropics (Hans-Ruthenberg-Institute), Eugenstraße 22, 70182 Stuttgart, Germany, e-mail: theresa.detering@gmail.com

Gaseous N and C Losses During Sun-Drying of Goat Manure – Effects of Drying Conditions and Feed Additives

Mariko Ingold[1], Herbert Dietz[2], Eva Schlecht[3], Andreas Buerkert[1]

[1]*University of Kassel, Organic Plant Production and Agroecosystems Research in the Tropics and Subtropics, Germany*

[2]*Royal Court Affairs Muscat, Royal Gardens and Farms,*

[3]*University of Kassel / University of Goettingen, Animal Husbandry in the Tropics and Subtropics, Germany*

Animal manure is a key resource in farming systems of arid and semi-arid regions. Its quality is often low due to inappropriate storage. Thus, it is important to assess the effects of drying and storage conditions on nutrient losses during typical sun-drying of manure, and how these losses can possibly be mitigated by changes in manure properties. Charcoal and tannins are known to stabilise organic matter and increase nitrogen (N) retention in soils and compost. Therefore, they were used as feed additives in order to stabilise carbon (C) and N in manure.

During sun-drying at three times during the cropping season, differing in temperatures and resulting drying rates, NH_3-N, N_2O-N, and CO_2-C losses from goat manure were measured. Manure was obtained from goats fed a diet of 50 % hay, 47 % maize and 3 % soybean with or without the addition of either 2.6 % activated charcoal (AC) or 3.4 % Quebracho tannin extract (QT). The cumulative N and C losses reached up to 1.7 % and 0.3 % of the initial N and C contents, respectively. During the drying process, control manure lost 2.5 to 6.1 mg N g^{-1} of the initial N content, AC lost 3.3 to 4.8, and QT 2.3 to 3.9 mg g^{-1} of the initial N content via NH_3 volatilisation and N_2O emissions. Carbon losses during drying ranged from 0 to 2.3, 0 to 1.9, and 0.1 to 1.3 mg g^{-1} of the initial C concentration for Co, AC, and QT manure, respectively. Slow drying (up to 84 h) favoured CO_2 emissions and reduced NH_3 volatilisation due to higher microbial activity and an immobilisation of mineral N in manure. In comparison, two times more NH_3-N was lost during quick drying of manure (4 h) than during slow drying, even after manure reached constant weight. In contrast to dietary charcoal, tannins reduced N and C losses by up to 60 % during manure drying and storage, increasing its quality as organic fertiliser and improving nutrient recycling.

Keywords: Activated charcoal, ammonia, carbon dioxide, goat manure, nitrous oxide, tannin

---

Contact Address: Mariko Ingold, University of Kassel, Organic Plant Production and Agroecosystems Research in the Tropics and Subtropics, Steinstrasse 19, 37213 Witzenhausen, Germany, e-mail: ingold@uni-kassel.de

Strategies to Achieve the GHG Mitigation Goals of the Livestock Sector in Latin America

Alejandro Ruden[1], Jacobo Arango[1], Deissy Martinez-Baron[1], Ana Maria Loboguerrero[1], Alexandre Berndt[2], Mauricio Chacon[3], Carlos Felipe Torres[4], Walter Oyhantcabal[5], Juan A. Gomez[6], Patricia Ricci[7], Juan Ku-Vera[8], Stefan Burkart[1], Jon M Moorby[9], Ngonidzashe Chirinda[1]

[1]*International Center for Tropical Agriculture (CIAT), Colombia*

[2]*Empresa Brasileira de Pesquisa Agropecuária (EMBRAPA), Research and Development, Brazil*

[3]*Ministry of Agriculture and Livestock, Costa Rica*

[4]*Clima Soluciones S.A.S, Colombia*

[5]*Ministry of Livestock, Agriculture and Fishing, Uruguay*

[6]*Universidad Nacional Agraria La Molina, Peru*

[7]*Instituto Nacional de Tecnología Agropecuaria INTA, Animal Production, Argentina*

[8]*University of Yucatan, Faculty of Veterinary Medicine and Animal Science, Mexico*

[9]*Aberystwyth University, Institute of Biological, Environmental and Rural Sciences, United Kingdom*

Livestock production is a fundamental source of income and greenhouse gas (GHG) emissions in Latin American countries. 20 percent of the region's emissions come from agriculture, 70 percent of which comes from livestock. There are several management and technology options with enteric methane mitigation potential that have been evaluated and their scale is expected to contribute to achieving the GHG emission reduction targets under the Paris Agreement. These technologies include management of the animal diet, reproductive control, administration of supplements, and reduction of the age at slaughter, among others. However, widespread adoption of promising mitigation options remains limited, raising questions about whether the planned emission reduction targets are achievable. Using the results of local studies, we have explored the mitigation potentials of currently proposed management technologies and practices to mitigate enteric methane emissions from livestock production systems in Latin American countries with the highest emissions. We then discuss the barriers to adopting innovations that significantly reduce enteric methane emissions from livestock and the main changes in policies and practices necessary to raise national ambitions in high-emission countries. Drawing on today's latest science and thought, we take our perspective to an inclusive approach and reimagine how the academic, research, business and public policy sectors can support and incentivize the

Contact Address: Alejandro Ruden, International Center for Tropical Agriculture (CIAT), Tropical Forages Program, Km17 recta Cali - Palmira, Cali, Colombia, e-mail: d.ruden@cgiar.org

changes necessary to raise the level of ambition and achieve goals of sustainable development, taking into account actions from the farm to the national scale. Some improvements identified and that need to be made are improving access to information through effective technology transfer plans, access to financial products by small producers, and establishing seed multiplication systems for fodder materials.

Keywords: Enteric methane, Latin America, NDC, Paris agreement (COP 21), SDG targets

(Agro)forestry and (agro)biodiversity

Trees and agriculture

Evaluation of Local and International Cacao Cultivars in Monoculture and Agroforestry Systems

MARCO PICUCCI[1], MONIKA SCHNEIDER[1], JON KEHLET HANSEN[2], JOACHIM MILZ[3], LAURA ARMENGOT[1]

[1]*Research Institute of Organic Agriculture (FiBL), Switzerland*

[2]*University of Copenhagen, Department of Geosciences and Natural Resource Management, Denmark*

[3]*ECOTOP, Bolivia*

Locally selected cultivars of cacao (*Theobroma cacao*) are supposedly better adapted to local environmental conditions compared with commercially selected cultivars. In this context, the objective of this study was to compare production and disease incidence of local cultivars with international cultivars and to test for interactions between cacao production systems and cultivars.

In the 1990s, the Bolivian cacao farmers' cooperative El Ceibo carried out a selection programme in the Bolivian Alto Beni region, by collecting germplasm from well-performing cacao trees, which originated from a governmental programme that distributed hybrid seeds to cacao farmers between the 1960s and 1980s.

From the El Ceibo program, four cultivars were selected and tested together with four commercial cultivars and four full-sib families encompassing five different production systems: two monocultures and two agroforestry systems under organic and conventional management, and one successional agroforestry system without external inputs. The long-term field trial was established in 2009 by the research institute FiBL and local partners in the Alto Beni region, Bolivia. Data on cacao yield and fungal disease incidence was recorded for each tree every 15 days between 2015 and 2019.

Across all years, the two monocultures were the most productive systems with an average production of 4.8 kg tree^{-1} (fresh beans with fruit pulp) under conventional and 4.3 kg tree^{-1} under organic management. Conventional and organic agroforestry systems obtained an average production of 2.7 kg tree^{-1}, while the successional agroforestry system had an average production of 2.1 kg tree^{-1}.

The local cultivars showed significantly higher yield in the five production systems across all years. The two best performing cultivars showed an average production of 6.6 and 6.4 kg tree^{-1} (fresh beans), respectively. The international cultivars had an average production of 3.6 kg tree^{-1}. The full-sib families performed very poorly with an average production of 1.3 kg tree^{-1}. There was a significant interaction between production systems and cultivars. Nevertheless, the rank between cultivars across production systems did not really change. The incidence of fungal diseases was low in all systems and slightly lower in local cultivars compared with the international ones. Our results highlight the relevance of selecting local genetic material.

Keywords: Agroforestry, breeding, cacao cultivars, fungal diseases, germplasm, long-term system comparison, organic farming, *Theobroma cacao*

Contact Address: Marco Picucci, Research Institute of Organic Agriculture (FiBL), Rehhagweg 15, 4147 Aesch, Switzerland, e-mail: marcoflaviopicucci@gmail.com

Microenvironment and Leaf Traits Regulate Transpiration of Cocoa (*Theobroma cacao* L.) under Different Cultivation Systems

Francisco Saavedra[1], Ernesto Jordan[1], Monika Schneider[2], Kasuya Naoki[1]

[1] *San Andres University, Biology, Bolivia*
[2] *Research Institute of Organic Agriculture (FiBL), Dept. of International Cooperation, Switzerland*

The response of plant species to environmental conditions influences changes in functional traits associated with the process that determines biological fitness and ecosystem processes. Thus, the variation of functional traits will be associated with the gradient of variation in environmental conditions. However, documenting these responses remain largely elusive in agroecological cultivation systems. We analyzed, how do cultivation systems influence changes in environmental variables and leaf traits? and how do leaf traits and environmental variables influence the transpiration rate of cocoa trees under different cultivation systems? Fieldwork was carried out at the Sara Ana experimental station in Alto Beni, La Paz, Bolivia. We sampled four trees in each of eight plots; four plots for each cultivation system (organic monoculture vs. organic agroforestry). From each tree, two mature, sunlit and healthy leaves were collected to make measurements of leaf traits (i.e., leaf area, specific leaf area, leaf relative water content, stomatal density and stomata size), environmental variables (i.e., canopy cover, temperature, and absolute air humidity) and transpiration rate. We found that canopy cover was higher in the agroforestry than monoculture systems. Temperature and absolute air humidity were similar between cultivation systems. The specific leaf area was greater in agroforestry systems but the stomata size and transpiration rate were both significantly higher in monoculture systems. The leaf relative water content was slightly higher in agroforestry systems and no differences were found between cultivation systems for stomata density. Temperature had a positive relationship with transpiration rate in both cultivation systems, whereas canopy cover and specific leaf area had a negative relationship in the agroforestry system. Our results suggest that cultivation system caused changes in microenvironmental conditions and on the expression of morphological and physiological leaf traits that regulate water flow through the plant. Cocoa plants have reduced the transpiration rate in agroforestry systems due to the mutual effects of canopy cover, larger leaves and smaller stomatal size. Consequently, agroforestry systems could be used as an adaptative strategy to minimise the negative effects of higher temperatures and less humidity in the context of climatic change.

Keywords: Agroforestry, canopy cover, specific leaf area, stomata size, temperature

Contact Address: Francisco Saavedra, San Andres University, Biology, University Campus Cota Cota Av. Andres Vello Calle 27 S/n, 591 La Paz, Bolivia, e-mail: fsaavedraa@fcpn.edu.bo

Improved Food Security via On-Farm Wood Production and Consumption. Evidence from Subsistence Farmers in Tanzania

Johannes Michael Hafner[1], Götz Uckert[1], Frieder Graef[1], Harry Hoffmann[1], Stefan Sieber[1], Todd Rosenstock[2], Anthony Kimaro[3]

[1]*Leibniz Centre for Agric. Landscape Res. (ZALF), Sustainable Land Use in Developing Countries (SusLAND), Germany*
[2]*World Agroforestry Centre (ICRAF), Land Health Decisions, DR Congo*
[3]*World Agroforestry Centre (ICRAF), Tanzania Country Programme, Tanzania*

Food security is a persistent challenge for many small-scale farmers in Tanzania. Fuelwood availability can affect food insecurity at household level by determining the amount of cooking that can be done and the degree meals can be cooked. Growing fuelwood (e.g. intercropping shrubs/ trees) and using on-farm fuels during cooking as well as the utilisation may therefore indirectly improve food security while having positive outcomes for forest degradation and soil erosion but also provide on-farm organic fertiliser.
We assessed the foliage production potential from intercropped shrubs of the species *Gliricidia sepium* with a spacing of 4 m by 4 m in a complete randomised block design. The tested intercropping systems were (1) maize and *G. sepium* and (2) maize, pigeonpea and *G. sepium*. In addition, we conducted a controlled cooking test to assess consumption patterns of on-farm fuels *G. sepium* and pigeon pea and compared it to the off-farm species (*Mimusops obtusifolia*) with regard to cooking time and firewood consumption per meal. In total 46 cooking tasks using traditional three-stone-fire stoves with the meal "rice and vegetables" were conducted. We found that integrating agroforestry species on farm can increase available biomass and increase efficiency cooking time. Biomass production, after 3 years of establishment, for *G. sepium* was 7.1 tons ha^{-1} ha in year 4 and 5.6 tons ha^{-1} in year 5 under when intercropped with maize, respectively 5.0 tons ha^{-1} in year 4 and 4.1 tons ha^{-1} in year 5 under when intercropped with maize and pigeonpea. With regard to the cooking time, *G. sepium* (-24.7 %) and pigeon pea (-34.6 %) used significantly ($p < 0.05$) less time compared to the off-farm species to conduct the cooking task. The controlled cooking tested showed that less *G. sepium* (-20.1 %) and significantly less pigeonpea (-31.8 %) fuel was required for completing the cooking task. The results show that *G. sepium* adds substantial amount of green manure towards the soil improving its properties. As shown, on-farm fuels enhance the amount of fuel at household level and might have knock-on effects such as better-quality diets, reduced greenhouse gas emissions during cooking as well as positive impacts on biodiversity and wildlife.

Keywords: Agroforestry, biodiversity, controlled cooking test, environmental degradation, intercropping

Contact Address: Johannes Michael Hafner, Leibniz Centre for Agric. Landscape Res. (ZALF), Sustainable Land Use in Developing Countries (SusLAND), Müncheberg, Germany, e-mail: johanneshafner@gmx.net

Extent and Rate of Deforestation in West Bugwe Central Forest Reserve, Uganda

Fatuma Mutesi[1], David Mfitumukiza[1], Cory Whitney[2], John Robert Stephen Tabuti[1]

[1]*Makerere University, College of Agricultural and Environmental Sciences (CAES), Uganda*

[2]*University of Bonn, Inst. Crop Sci. and Res. Conserv. (INRES), Germany*

Forests play an important role in supporting social wellbeing and livelihoods. Forests and forest cover over most of sub-Saharan Africa is being cut down (deforestation) or converted into other vegetation or other land cover forms (degradation). Knowledge on the status of forest cover and how it has been changing is vital for developing the appropriate restoration and other management strategies. Detailed assessments of land cover change at local scales, and assessments of the factors that lead to the change, are lacking for much of Uganda. We showcase methods to fill this gap by offering a case study combining GIS and key informant surveys to assess local trends in West Bugwe Central Forest Reserve (WBCFR). Our findings show that the forest in WBCFR has been extensively and rapidly deforested and degraded by humans due to a number of poverty related drivers. We used Remote sensing and Geographic Information Science (GIS) techniques to map and quantify the LCC in WBCFR for the period 1986 to 2016. We obtained Landsat images (MSS 31/03/1986, TM 02/04/1995, ETM +06/03/2006 and OLIS/TIRS 11/04/2016) from USGS Earth explorer. We used supervised classification in ERDAS IMAGINE 2014 with support of ground truth data and the land change modeler in TerrSet 18.2 to analyse the pattern of land cover change. We also used household survey, key informant interview, and focus group discussions to determine the drivers of land cover change. Our results indicate that the forest in WBCFR has declined by over 82 % from 1,682 ha to 311 ha since 1986. Forest cover has decreased at an average annual rate of 1.2 % and shrubland has increased 1.5 % per year. The wetland has also been extensively degraded and drained, as with the forest much of the former wetland is now shrub land. Key informants indicated that the key drivers that have led to this change are poverty, population growth and associated harvesting of woody products for subsistence and income generation. Re-afforestation is needed to reverse forest loss. Alternative sources of fuel should be encouraged to reduce pressure on WBCFR.

Keywords: Drivers, land cover change, land-use change

Contact Address: Cory Whitney, University of Bonn, Inst. Crop Sci. and Res. Conserv. (INRES), Auf Dem Hügel 6, 53111 Bonn, Germany, e-mail: whitney.cory@gmail.com

Herbaceous Species Cacao Production Systems: Biotic Homogenisation and Dynamics Over Time in a Long-Term Trial in Bolivia

Luis Marconi Ripa[1], Renate Seidel[1], Laura Armengot[2]

[1]*San Andres University, Inst. of Ecology, Bolivia*

[2]*Research Institute of Organic Agriculture (FiBL), Switzerland*

The different types of cocoa production have an impact on the spontaneous diversity of the cropping systems. Plants, and specifically herbs, are one of the most susceptible groups to the transformation of forests into cropping land. We study the plant species of the herbaceous stratum in an experimental trial in Bolivia, where five production systems representing a gradient of management intensity were compared: two monocultures and two agroforestry systems under conventional and organic farming containing a planted cover crop layer, and a complex successional agroforestry system with no external inputs.

In a first study we explored the role of potential role of agroforestry systems and management intensity in diversity conservation and against biotic homogenisation. We did not find significant differences in species richness between production systems, but higher number of species was found in the successional agroforestry system. However, community composition did change following the management intensity gradient. In addition, we found that widely distributed species, including some exotic species, were associated to intensive management, i.e. monocultures and conventional systems with high solar exposure levels and/or glyphosate application. Conversely, successional agroforestry and organic systems harbored species with a geographical distribution range restricted to the Neotropics or South America. Accordingly, cocoa organic and agroforestry systems, could contribute to both biodiversity conservation and the minimisation of biotic homogenisation.

In a second study based on Braun-Blanquet samplings of herbaceous strata over seven years, we found that the differences in community composition were established at a very early stage and time had a minor role compared with the selective pressure of the production system. In the systems with more available resources (light, space) we registered higher number of new species, but the pool from which they come from depended on the production system.

So far we have found 171 different herb species in the trial. We have identified some species that could be used as cover crops if kept in the system under proper management, which could reduce the weeding efforts. We have also identified species selected and promoted by the use of glyphosate in the conventional systems.

Keywords: Agroforestry, biodiversity conservation, community composition, full-sun monocultures, organic farming

Contact Address: Laura Armengot, Research Institute of Organic Agriculture (FiBL), Ackerstrasse 113, 5070 Frick, Switzerland, e-mail: laura.armengot@fibl.org

Cracking the Brazil Nut Puzzle : Can Nut Gathering and Timber Harvesting Coexist ?

Jacques Kohli, Juliana Frizzo, Jürgen Blaser, Lindsey Norgrove

Bern university of applied sciences, School of Agricultural, Forest and Food Sciences, Switzerland

Non-timber forest products (NTFPs) can provide a livelihood for forest communities in the tropics. Yet, forests are becoming more accessible to logging companies. Thus, it is vital to develop forest management practices that do not jeopardise NTFPs. Brazil nut, *Bertholletia excelsa*, is an important species, due to its high market value, and the need for the near intact forest for production. We aimed to evaluate which forest management practices might promote brazil nut production, which might lead to declines and thus to develop best practice guidelines. We conducted a systematic review in 2019 to capture publications detailing factors that influence brazil nut yield per tree and assess how forestry practices could improve timber production without compromising current nut yield levels. We also sought to assess how the social and policy contexts influence practices. Both endogenous and exogenous factors impacted brazil nut yield per tree. The main yield influencing factors such as size, crown position, liana load, yearly variation, genetics, climate, soil, and pollination, were well researched. Tree size was the most important endogenous factor with trees exceeding 40 cm diameter at breast height (DBH) factor being productive in 98 % of cases whereas smaller trees were rarely productive. However, at DBH exceeding 100 cm, crown-size was the more important trait determining productivity. An emergent rather than a suppressed crown was an essential criterion for productivity. Higher productivity was achieved on soils with higher cation exchange capacities. Liana cutting greatly increased yield. Shifting cultivation improved yield compared to in logging gaps, given increased light and nutrient availability. Although coexistence appears feasible in theory, in practice institutional barriers such as overlapping land tenure and lack of transparency have prevented the sensible integration of the two livelihood strategies on the same plot to date.

Keywords: Brazil nut, multiple forest uses, NTFP, Para nut, reduced-impact logging

Contact Address: Jacques Kohli, Bern university of applied sciences, School of Agricultural, Forest and Food Sciences, Route de Payerne 62, 1773 Léchelles, Switzerland, e-mail: jacques.kohli@students.bfh.ch

Analysis of Dynamic Agroforestry Systems and Cocoa Plant Health in in Western Ghana

Elisa Bossi[1], Ingrid Fromm[1], Christian Andres[2], Monika Schneider[3]

[1]*Bern University of Applied Sciences, School of Agricultural, Forest and Food Sciences, Switzerland*

[2]*Swiss Federal Institute of Technology (ETH), Dept. of Environmental Systems Science, Switzerland*

[3]*Research Institute of Organic Agriculture (FiBL), Dept. of International Cooperation, Switzerland*

Agroforestry can help mitigate the numerous negative ecological effects caused by cocoa production with full-sun monocultures providing at the same time the diversification of livelihoods through the selling of by-products. Because the high vulnerability of young cocoa trees represents one of the major constraint for farmers, this study in Western Ghana compared cocoa health and field management quality in monoculture and in agroforestry systems during the field establishment phase. Cocoa mortality rate, vigour, growth rate and field management quality were assessed on 20 Dynamic Agroforestry (DAF) and on 9 monoculture fields aged one, two or three years. The results showed that cocoa mortality rate is lower and the growth rate is higher in DAF than in monocultures. Cocoa vigour and field management quality do not particularly differ between the two cultivation systems. However, management quality strong influences the three plant health parameters. With better field management, cocoa plant health is higher. Many factors not necessarily dependent on cultivation system were found to influence cocoa health such as field planting scheme and accuracy during weeding practices. On the other hand, shade percentage on the field was not a significant factor for higher cocoa health in DAFS. The findings highlighted the potential of DAFS and of good management practices to improve cocoa health during the vulnerable field establishment phase. For this reason, it is important to encourage farmers to diversify their livelihood with DAFS but also to sensitise farmers to improve the management quality to enhance cocoa health during the field establishment phase.

Keywords: Cocoa production, dynamic agroforestry, field establishment of cocoa, field management, plant health

Contact Address: Ingrid Fromm, Bern University of Applied Sciences, School of Agricultural, Forest and Food Sciences, Laenggasse 85, 3052 Zollikofen, Switzerland, e-mail: ingrid.fromm@bfh.ch

Agroforestry and Anti-Erosion Practices of Soils Conservation in Soudanian Zone (Benin)

Clement Soloum Teteli, Antoine Elie Padonou
Nationale University of Agriculture, Tropical Forestry School, Benin

The commune of Ouaké is part of the Sudanese zone of Benin where we are witnessing advanced soil degradation. In order to contribute to the restoration of these soils, agroforestry and anti-erosion soil conservation practices (PAACS) were evaluated in this commune. Surveys and data collection were carried out from individual and group interviews with 215 farmers in 22 villages, covering all the districts of the commune, as well as at the level of agricultural and forestry development structures. A diagnosis was carried out on each PAACS by identifying their strengths, weaknesses, opportunities and threats followed by their prioritisation. A total of eight agroforestry practices (with a dominance of agroforestry parks in *Parkia biglobosa* (83.94 %) and *Vitellaria paradoxa* (74.16 %)) and five anti-erosion practices (with a predominance of the practice of furrows in a direction perpendicular to the slope (80.63 %)) have been identified. The adoption rate of these PAACS differs according to the villages and districts, socioeconomic groups, ethnicities, sex, age, level of education and main activity. Agroforestry parks in *Parkia biglobosa* and *Vitellaria paradoxa* have a high priority followed by agro-pastoralism. Compared with anti-erosion practices, anti-erosion hedges and stony cords have high priority. The development of PAACS must be based on these priority practices. This study, based on agroforestry and anti-erosion practices for soil conservation in the commune of Ouaké, will no doubt have a positive impact on safeguarding soil quality, improving the level of production of agricultural crops and also improving ecological functions. Agroforestry as a sustainable management system will ensure effective protection of available natural forest resources, conserve biodiversity, maintain an ecological balance in order to promote a productive and resilient environment. The implementation of good agroforestry and anti-erosion practices adapted to farmers' needs could therefore encourage them to put these systems into practice, thereby increasing crop yields. This increase in yield could improve the food and socioeconomic situation. They will thus have food security and well-being in this commune.

Keywords: Agroforestry systems, commune of Ouaké, erosion, soil conservation

Contact Address: Clement Soloum Teteli, Nationale University of Agriculture, Tropical Forestry School, 1721 Abomey-Calavi, Benin, e-mail: cteteli29@gmail.com

Permaculture as Efficient and Environmentally Sound Agro-Forstry Approach for Smallholders in Sierra Leone

Brigitte Amara-Dokubo[1], Edwin Sam Mbomah[2], Falko Feldmann[3]

[1]*Löwe für Löwe E.v. Germany, Germany*

[2]*Njala University, Inst. of Environm. Management and Quality Control, Sierra Leone*

[3]*Julius-Kühn-institut, German Federal Research Centre for Cultivated Plants, Inst. for Plant Protection in Horticulture and Forests, Germany*

More than 50 % of the Sierra Leonean population is concerned with agriculture. Nevertheless, a considerable amount of people is food insecure. Efficient, environmentally adapted strategies for food production are sought to overcome food deficiency. At the same time, the quality of the food has to reduce nutrient deficiency ("hidden hunger") of children significantly.

Therefore, we decided to combine intensive horticultural sustainable food production, with agro-forestry, namely with the integration of native fruit trees and exotic *Moringa oleifera* trees in a so called permaculture approach.

Permaculture integrates design principles reflecting whole system thinking, simulating, or is directly utilising resilient features observed in natural ecosystems. Permaculture includes integrated water harvesting methods and resources management that develops regenerative and self-maintained habitat and agro-florestal systems.

The twelve principles of permaculture include Observe and Interact, Catch and Store Energy, Obtain a Yield, Apply Self Regulation and Accept Feedback, Use and Value Renewable Resources and Services, Produce No Waste, Design From Patterns to Details, Integrate Rather Than Segregate, Use Small and Slow Solutions, Use and Value Diversity, Use Edges and Value the Marginal, and Creatively Use and Respond to Change.

To be successful with such an complex approach we started to build up the *Moringa* Start-up and Innovation Centre (MOST) in the peri-urban area of Waterloo, Sierra Leone. Together with local people, a plantation of the under-utilised multi-purpose tree *Moringa oleifera* is currently installed, followed by annual and biannual crops in two and three dimensional rotation systems. As a first step, the aim is a high crop diversity for subsistence and local markets. The first *Moringa* biomass will be used for Hugelculture systems and at the same time for the starting of water harvesting and nutrient recycling systems what finally will result in a production widely independent on agro-chemicals. The second step, production of processed *Moringa* products is already anticipated and a factory built.

During the installation of MOST concurrent multi-level discussion of observations with all stakeholders and teaching of smallholders of the neighbourhood will take place to create awareness for the huge potential of permaculture.

Keywords: Food security, Moringa, permaculture, Sierra Leone

Contact Address: Falko Feldmann, Julius-Kühn-institut, German Federal Research Centre for Cultivated Plants, Inst. for Plant Protection in Horticulture and Forests, Messeweg 11-12, 38104 Braunschweig, Germany, e-mail: falko.feldmann@julius-kuehn.de

Effect of Shade Trees in Cocoa Agroforestry Systems on Water and Light Availability in Dry Seasons

Dennis Kyereh

Czech University of Life Sciences Prague, Department of Crop Sciences and Agroforestry, Czech Republic

The objective of this paper was to assess the influence of single standing shade trees in cocoa agroforestry systems on soil moisture and light availability for cocoa in the dry seasons and how these environmental factors affect potential pod yields of cocoa. The research was conducted in a moist semi-deciduous forest zone of Ghana. Seven different shade trees that were commonly found in cocoa systems were selected. An effect ratio was used to compare tree sub-canopy effects to the open area effects.The averages of the individual tree sub-canopy and open area readings were calculated to represent their respective replicate values.Data were analysed as one-way analysis of variance (ANOVA) using the R Statistical Package. For each variable, normal distribution was tested using the Shapiro-Wilk normality test for homogeneity of variances. Significant ANOVAs were subsequently assessed using Tukey's Honestly Significant Difference (HSD) test and probability was set at 0.05 for the statistical tests. *Morinda lucida* (0.19), *Spathodea campanulata* (0.16) and *Ficus capensis* (0.13) showed favourable soil moisture conditions, however *Citrus sinensis* (-0.28) revealed a lower soil moisture content in the sub-canopy. *Entandrophragma angolense* and *Terminalia superba* had the highest transmitted percentage light of 69.2 % and 67.1 %, respectively and the lowest being *Mangifera indica* (3 %). The potential pod yields of cocoa were higher under *Morinda lucida* (0.40), *Terminalia superba* (0.40) and *Entandrophragma angolense* (0.34) but lowest under *Mangifera indica* (-0.55). *Morinda lucida, Spathodea campanulata, Entandrophragma angolense* and *Terminalia superba* in cocoa agroforestry systems potentially ensure higher soil moisture content and light availability in the sub-canopy, especially during the dry seasons, which could translate into higher cocoa pod yields.

Keywords: Cocoa agroforestry, cocoa yields, dry seasons, light, shade trees, soil moisture

Contact Address: Dennis Kyereh, Czech University of Life Sciences Prague, Department of Crop Sciences and Agroforestry, Kamycka 129, 165 21 Prague, Czech Republic, e-mail: kyereh@ftz.czu.cz

The Maya Breadnut Tree: Providing Sustainable Nutrition and Forestry from the Ancient Maya Until Today

Scott Forsythe[1], Sebastian De La Hoz[2], Andrew Puente[3]

[1]*Forest Foresight, Beekeeping / Reforestation, Guatemala*

[2]*Green Balam Forests, Agroforestry Management, Guatemala*

[3]*Journeys in Conservation, Conservation Management, Guatemala*

Brosimum alicastrum, a keystone rainforest tree species ranging naturally between Mexico and the Amazon region, known variously within the Yucatan region as ramón, breadnut, Maya nut or yaxox, has been a staple of lowland Maya nutrition since pre-columbian times. Its resistance to drought, its high germination and survival rates, its ubiquity, productivity, and seed storability are all reasons for its importance in Maya nutrition. Stands of ramón are commonly associated with Mayan archaeological sites. It figures in the concept of the Maya forest garden, in Maya forest management, and provides forage for a wide variety of forest wildlife. Its leaves provide preferred forage for domestic animals. Today, there is international demand for its prolific seeds when processed into flour or into a substitute coffee drink, thus providing income for local collecting and processing cooperatives, usually organised by women. Considering this knowledge, we are employing *B. alicastrum* as the predominant species in reforestation of post-deforestation cattle pastures in the heavily deforested, multiple-use zone of the Maya Biosphere Reserve in northern Guatemala. In this reforestation initiative, the ca. 25 species planted are strictly native, yet economically or nutritionally useful. In our presentation, we intend to illustrate the exceptional nutritional, productivity and drought resistance aspects of this traditionally significant species and its potential to stimulate nutrition and agroforestry initiatives, especially in the Petén area of Guatemala, where only a few decades ago existed the largest unbroken tract of rainforest in Central America. Cattle ranches surrounding and inside the Maya Biosphere Reserve are the source of extensive fires which encroach into the forested core area. Both deforestation and climate change combine to lower water tables, which in turn is limiting the profitability and/or feasibility of cattle ranching. Thanks to government forestry incentives, and with proper selection of native species, we can provide a working model demonstrating the long-term, economically attractive option of reforestation over cattle ranching.

Keywords: Agroforestry, breadnut, *Brosimum alicastrum*, cattle pasture, cattle ranching, cooperatives, Guatemala, Maya Biosphere Reserve, women

Contact Address: Scott Forsythe, Forest Foresight, Beekeeping/Reforestation, El Remate, Guatemala, e-mail: forsythe_scott@hotmail.com

The Pacuare Reserve Landscape: Land Cover Change and Implications for Biodiversity Conservation in Costa Rica

José Andrés Rodriguez Zumbado[1,2], Carlos A. Muñoz Robles[1], Claudia Raedig[2], Bernal Herrera-Fernández[3]

[1]*Autonomous University of San Luis Potosí, Desert Areas Research Institute, Mexico*

[2]*TH Köln, Inst. for Technology and Resources Management in the Tropics and Subtropics, Germany*

[3]*European Union & Ministry of Environment and Energy, Post2020 Biodiversity Framework and Nature-Based Solutions, Costa Rica*

Habitat loss due to land use change (LUC) has been identified as the main cause of global environmental change, responsible for biodiversity decline and the deterioration of the ecological processes. In tropical regions, habitat loss and fragmentation have been driven by processes of LUC such as deforestation, agricultural expansion and intensification, urbanisation, and globalisation. The objective of this research was to determine the effects land use change on the process of habitat loss and the patterns of fragmentation in the surrounding landscape of the Pacuare Reserve (PR) in the Caribbean lowlands of Costa Rica. The PR is a private protected area of 800 ha surrounded by an agricultural landscape with a history of over 150 years of Musaceae monocultures (bananas and plantains) and pasture lands. Landsat satellite images from 1978 to 2020 were used to conduct a temporal analysis of LUC around the PR (years: 1978, 1986, 1992, 1997, 2001, 2015 and 2020). Patterns of change were explored using fragmentation metrics and new connectivity routes were drawn from the PR to other protected areas using the least cost path method. Overall, forest cover decreased during the study period at a rate of -4.8 % per year during the period of 1992–1997. In the year 2001 it reached its lowest cover and then increased at a mean annual rate of 1.6 %. A mean overall accuracy of 93 % was obtained for the land classification process. A clear fragmentation pattern was observed, as shown by a decreased in forest mean patch area and largest patch index and by the increase in patch density. Although forest cover increased in the last decade, fragmentation metrics suggest this recover happened in a spatially scattered manner, due to agricultural land abandonment. Connectivity maps showed the importance the already established coastal biological corridor has on the movement of species to and from the PR, however it also evidenced the lack of connectivity of this particular forest remnant to other inland protected areas and the need to promote restoration projects in the agricultural matrix.

Keywords: Biological corridors, connectivity, land use change

Contact Address: José Andrés Rodriguez Zumbado, TH Köln, Inst. for Technology and Resources Management in the Tropics and Subtropics, Betzdorfer Strasse 2, 50996 Köln, Germany, e-mail: jose_andres.rodriguez_zumbado@smail.th-koeln.de

Workers Map Social Landscape Values in a Colombian Oil Palm Plantation

Adriana Gomez, Stephanie Domptail

Justus Liebig University Giessen, Inst. of Agricultural Policy and Market Research, Germany

Humans are participants in landscape-thinking, feeling and acting giving meaning to landscapes and places. Additionally, humans attach value to landscapes, even to highly modified landscapes, and possibly to an "already-created" landscape such as the oil palm plantation. Workers interact daily with the oil palm plantation and they could confer meaning and value to the natural habitats and the oil palm vegetation. For the present study, the landscape units refers to land covers present in the farm. The natural habitats (e.g. gallery forests, natural savannahs) are embedded within the Macondo oil palm plantation. With this in mind, the present study evaluates whether workers perceive, value and use social landscape values in the plantation. To collect data, first, we conducted three focus group discussions to identify the social landscape values. Second, we applied 35 structured participatory mapping interviews to workers. Each worker registered on a map of the plantation the perceived social landscape values attached to the landscape units on a map of the plantation. This article shows that workers perceive, identify and use social landscape values in the oil palm plantation. During the focus groups discussions, we identified the social landscape values to construct the participatory mapping interview guide. The 35 respondents mentioned 511 times social landscape values within the landscape units such as shade, food, fresh water, biological control, fauna, flora, beauty, soil, and leisure. The respondents mentioned more frequently shade (97), flora (79), water (66) and contemplation of fauna (60) within all landscape units. Furthermore, workers also perceived occupation risks such as fall accidents and injuries due to obstacles on the ground (e.g. leaves or trunks) or the inadequate use of machinery. In general, the use of the participatory mapping proved to be a tool to identify and localise the values that the workers perceive, use and identify in the oil palm plantation. The participants highlighted the presence of natural habitats within the oil palm plantation, to observe flora, fauna, find fresh water and shade. These results demonstrate that the plantation provide several values and benefits for the workers beyond the economic advantage.

Keywords: Participatory mapping, social landscape values

Contact Address: Adriana Gomez, Justus Liebig University Giessen, Inst. of Agricultural Policy and Market Research, 35396 Giessen, Germany, e-mail: adriana.gomez@agrar.uni-giessen.de

Linking Physiological Response to Shade with Growth and Yield in Different Coffee Agroforestry Systems in Ecuadorian Amazonia

KEVIN PIATO[1], CRISTIAN SUBÍA GARCÍA[2], DARÍO CALDERÓN[2], FRANÇOIS LEFORT[3], LINDSEY NORGROVE[1]

[1]*Bern University of Applied Sciences, School of Agricultural, Forest and Food Sciences, Switzerland*

[2]*National Agricultural Research Institute, Cocoa and Coffee Research Program, Ecuador*

[3]*Institute Land Nature Environment, HES-SO University of Applied Sciences and Arts Western Switzerland, Life Sciences, Switzerland*

We assessed how agroforestry shade cover, shade types and farming practices affected *Coffea canephora* (robusta) growth, physiology and yield at 5 years old, in Ecuadorian Amazonia. We hypothesised that shade would increase chlorophyll concentrations, further enhanced by nitrogen-fixing trees and would consequently increase yields. The experiment was planted in 2015 and the five treatments were 1) full sun; 2) *Myroxylon balsamum*; or two N-fixing trees 3) *Inga edulis*; 4) *Erythrina* spp. or 5) *Erythrina* spp. plus *M. balsamum*. Four farming practices assessed were: conventional farming at either 1) moderate or 2) intensified input and organic farming at 3) low or 4) intensified input. The experiment was a RCBD with 20 treatment combinations, replicated three times. Each plot was zoned according to distance from the nearest shade tree and measurements were weighted according to the prevalence of each zone in order to compute plot means. Shade cover above coffee was assessed with an MP-200 pyranometer on 4 coffee plants for each zone defined in 2018. Two chlorophyll measurements were made per leaf and four leaves per branch in 2 pairs of leaves at the middle third of the branch with an MP-100 chlorophyll meter. One branch was measured per plant, which resulted in 8 measures per plant on 18 plants per plot. Berry yields are being measured and will be completed in June 2020.
Coffee trees were, on average, taller under the N-fixers, *Inga edulis* and *Erythrina*, than in either the full sun control or under other treatments. Chlorophyll concentrations in coffee leaves ranged from 410 to 581 μmol m^{-2}, showing a strong, positive relationship with shade level (r^2=0.61). They were highest in the *Inga edulis* and *Erythrina* treatments and lowest in the full sun. Initial results suggest that shade, particularly with N-fixing trees, leads to higher leaf chlorophyll content. Early partial coffee berry yields were 22 % higher under full sun than in the agroforestry treatments, regardless of shade tree species. No correlation was found between early yield and leaf chlorophyll content but final yields will only be available in June 2020.

Keywords: Chlorophyll concentration meter, Ecuadorian Amazon region, growth, pyranometer, robusta agroforestry system, shade, yield

Contact Address: Kevin Piato, Bern University of Applied Sciences, School of Agricultural, Forest and Food Sciences, Route de Bertigny 28, 1700 Fribourg, Switzerland, e-mail: kevin.piato@students.bfh.ch

Developing Agroforestry Around Myanmar's Inle Lake, Supporting Small-Scale Farmers and the Local Ecosystem

Theresa Dunkel, Christoph Studer, Alessandra Giuliani

School of Agricultural, Forest and Food Sciences (HAFL), Bern University of Applied Sciences (BFH), Switzerland

Myanmar's Inle Lake is located in the heart of the Shan Plateau and described as a magical place. The lake matters profoundly to the livelihood of the local communities, by being a vital source of income. However, the combined effects of unsustainable use of resources, increasing population, climate change and rapid tourism development have led to environmental degradation of the lake and its watershed. Reversing degradation and conserving the lake's ecosystem have become a main concern. A local community-based organisation called "PweHla Environment Conservation And Development" (PHECAD) continuescca the UNDP Lake Conservation and Rehabilitation Project, launched in 2015 with the overall goal to reduce environmental degradation and uplift the livelihoods of the local communities. As tree coverage has a direct impact on the lake downstream, PHECAD focuses on reforestation and biodiversity conservation, by developing on-farm tree cultivation with small-scale farmers. A master thesis project has been engaged to support the local farmers in adopting agroforestry practices in the Northern watershed area of the Inle Lake. This study investigates whether and how agroforestry practices can benefit small-scale farmers and their households as well as the environment in the Inle Lake watershed area. The project focuses on three main outcomes: (i) participatory identification of agroforestry options that suit the socio-economic and environmental situation in the area; (ii) piloting of agroforestry practices together with local smallholder farmers; and (iii) identification of the best way to support them in the implementation of agroforestry practices. The first part of the research work has been accomplished during a field visit between November 2019 and January 2020. Existing farming systems and the socio-economic situation of the small-scale farmers were analysed through individual and key informant interviews. In focus group discussions farmers identified agroforestry practices that are attractive to them and organised the implementation of pilot plantings. The use of a participatory approach assures that selected options fit well the needs and expectations of the local population. Piloting and implementation of agroforestry practices are planned for the upcoming rainy season, combined with in-field trainings and field schools, and possibly a second field visit.

Keywords: Agroforestry, environmental conservation, Inle Lake, livelihood, Myanmar, small-scale farmers

Contact Address: Theresa Dunkel, School of Agricultural, Forest and Food Sciences (HAFL), Bern University of Applied Sciences (BFH), Länggasse 85, 3052 Zollikofen, Switzerland, e-mail: theresa.dunkel@students.bfh.ch

(Agro-)biodiversity

Oral Presentations

Posters

Homegarden Structure Dynamics in Communities of a Wetland Conservation Area in the Pacific Coast of Guatemala

Pedro Daniel Pardo Villegas, Claudia Burgos, Pablo Lee

Universidad de San Carlos de Guatemala, Biology School/ Dept. of Ecology, Guatemala

An ethnobotanical study centred in homegardens was performed in five communities of a wetland conservation area in the Pacific coast of Guatemala. The purpose of the study was to record the ethnoflora and the traditional knowledge associated with it. In 2011, through 101 semi-structured interviews and the collection of plant samples, a total of 181 species were registered. In terms of the use given to the reported plants, 40 are edible, 91 medicinal, and 50 species are edible and medicinal. Between 15 to 30 edible species and 3 to 25 medicinal species were registered growing or planted all together at the homegardens. Of the registered plants, 60 % are native, found growing in remnants of natural vegetation or in homegardens. Considering that homegardens function as important reservoirs of useful plants of the area, in 2018 a follow-up assessment was carried out, which focused in 31 homegardens. This information allowed to identify important changes in homegarden structure between 2011 and 2018. Most of the changes are related to size reduction and species diversity loss in homegardens. In general terms, a simplification in homegarden structure in the area is evident. This can have a long-term consequence in terms of food security and family economy, since some of the fruits produced complement the family income. Taking this into account, a demonstration garden was designed at the reserve's visitor centre, which has approximately 40 useful species. This, as a means to educate the new generations on the role of the homegardens, in addition to contributing to the in-situ conservation of the ethnobotanical diversity of the area.

Keywords: Agrobiodiversity, agroforestry, edible plants, ethnoflora, mangrove reserve, medicinal plants

Contact Address: Pedro Daniel Pardo Villegas, Universidad de San Carlos de Guatemala, Biology School/ Dept. of Ecology, Edificio T-10, segundo nivel, Ciudad Universitaria zona 12, 01012 Guatemala, Guatemala, e-mail: pardo.pedro@usac.edu.gt

Variation in Baobab (*Adansonia digitata* L.) Root Tuber Development, Yield Potential and Leaf Number among Different Growth Conditions for Five Provenances in Malawi

LENNART JANSEN[1], DIETRICH DARR[1], NELE HANSOHM[1], JENS GEBAUER[1], KATHRIN MEINHOLD[1], CHIMULEKE R. Y. MUNTHALI[2], FLORIAN WICHERN[1]

[1]*Rhine-Waal University of Applied Sciences, Fac. of Life Sciences, Germany*

[2]*Mzuzu University, Department of Forestry, Malawi*

The baobab tree is an underutilised indigenous fruit tree in sub-Saharan Africa which, at the same time is vulnerable to overexploitation in areas close to centres of demand, as currently baobab use is limited to wild baobab trees. Baobab seedlings are known to form root tubers, but little is known about their growth characteristics and its yield potential. This study aims to investigate the root tuber and leaf development of baobab seedlings grown from seeds of five provenances, sown at three different planting distances in two nursery trials at climatically distinct locations in Malawi, namely Mzuzu and Mangochi. The observed yield data was fed into preliminary farm-gate profitability analyses for three different scenarios that differed by planting distance. Results indicate increased growth rates for root dry mass and number of developed leaves with increasing planting distance. However, we did not find a significant effect of seedling provenance on any of the measured plant growth parameters. Seedlings invested mainly into root development during the growth period, with root tubers reaching an average fresh weight of 41 ±,39 g and an average length of 24 ±,11.9 cm at 138 days after sowing. Profitability analyses showed a potential total net benefit of 12.78 USD per harvest cycle of 16 weeks and per 100 m^2 of land cultivated with baobab root tubers, which was better than an alternative scenario of maize cropping on the same area that showed with a negative total net benefit when cost of family labour was included. However, the heterogeneity of root tuber development as affected by abiotic and biotic factors like soil fertility and water availability, as well as genetic origin warrant further investigations.

Keywords: Baobab, indigenous fruit tree, Malawi, root tubers, underutilised plants

---

Contact Address: Lennart Jansen, Rhine-Waal University of Applied Sciences, Fac. of Life Sciences, Kleve, Germany, e-mail: lennart.jansen@hsrw.org

Is the Forest Half-Empty or Half-Full? The Role of Conucos in Providing Bushmeat in the Gran Sabana, Venezuela

Izabela Stachowicz[1], José Rafael Ferrer Paris[2], Ada Sánchez-Mercado[2]

[1]*University of Lodz, Department of Biodiversity Studies and Bioeducation, Poland*

[2]*University of New South Wales, School of Biological, Earth and Environmental Sciences, Australia*

Overexploitation of bushmeat in tropical forests has increased in recent years, creating debate about the sustainability of current hunting rates. The Empty Forest hypothesis predicts that current hunting rates in tropical forests can lead to a widespread loss of biodiversity and a reduction in vertebrate abundance. Alternatively, the Garden Hunting hypothesis states that heterogeneous agroforestry landscapes can maintain similar species richness as pristine forests, but with species composition dominated by savannah related species. Here, we combined cameras trap surveys and interviews to Pemón indigenous communities in the Gran Sabana to examine the generality of Empty Forest hypothesis. We fitted occupancy models and MANOVA to assess how important are human activities (indigenous farming and hunting activity), and landscape characteristics (forest cover and fragmentation) to explain wildlife occupancy, and changes of species composition across landscape. Consistent with Garden Hunting hypothesis predictions, we found higher occurrence of savannah related herbivores in habitat with medium disturbance than in unperturbed habitats. Evidence for decreasing predator's occurrence in perturbed habitats was mixed, with some species being attracted and other repealed by the human presence and agricultural activity. Although overhunting reduces population density of targeted game species, the current scheme of resource use does not seem to produce a generalised pattern of defaunation. Mammal diversity seems to respond to amount and distribution of remaining forest cover, suggesting that deforestation has a larger impact than hunting. We discuss opportunities and challenges of implement land sharing approaches for landscape and wildlife management in complex landscape of high cultural and biological diversity.

Keywords: Camera traps, empty forest, garden hunting hypothesis, indigenous, shifting agriculture

Contact Address: Izabela Stachowicz, University of Lodz, Department of Biodiversity Studies and Bioeducation, Banacha 1/3, 90-237 Lodz, Poland, e-mail: stachowicz.izabela@gmail.com

Sustainability Performance of Monoculture and Agroforestry Palm Oil Production Systems of Smallholders in Sarawak (Malaysia)

Lukas Kliewe[1], Rainer Weisshaidinger[2], Lukas Baumgart[2], Alexander Follmann[1]

[1]*University of Cologne, Inst. of Geography, Germany*

[2]*Research Institute of Organic Agriculture (FiBL), Austria*

Palm oil is the world's most important vegetable oil, and Malaysia and Indonesia dominate global export trade in crude palm oil with 85 % of the world´s total. The oil palm's incomparably high yield (3.5–4.0 $tha^{-1}y^{-1}$) and the multiple use in food products, soap and industrial purposes, including biofuels, make it a highly valuable commodity. Sustainability in the palm oil industry is debated with growing concern due to ecological and social trade-offs, e.g. destruction of tropical forests, greenhouse gas emissions, land disputes and forced labour.

Using the SMART Farm Tool, we have comprehensively assessed the sustainability performance of 16 non-certified palm oil smallholdings in Sarawak (Malaysia). SMART is an indicator-based multi-criteria tool measuring the 58 sustainability (subtheme-)goals defined in the FAO-SAFA framework. The farmers were selected according to a snowball approach in two regions (1st and 4th division). The dominant cultivation system of the oil palm farms studied is monoculture. However, two of the 16 farms – one in each region – practice agroforestry (AF), which is seen as a strategy to improve environmental sustainability in palm oil production.

As for monoculture systems, the goal achievements per SAFA-goal show that sustainability performance is largely lacking, and usually ranges between 20 % and 60 % goal achievement. Good Governance and Social Well-Being are dimensions which in particular show low scores at all observed farms. One AF-farm performs equally in all dimensions compared to monocrop farms. This is due a relatively small proportion and a low diversity of the AF-system, as well as the use of mineral fertiliser and pesticides. In contrast, the second AF-farm, which does not use chemical inputs, performs better in all ecological and interestingly in most economic subthemes ranging from 60 % to 89 % and 23 % to 97 %.

These findings underscore the need to improve the sustainability of mono-cropping palm oil production not only to reduce ecological and social trade-offs, but also to enhance the economic resilience of smallholders. Although controversially discussed in scientific literature, agroforestry appears to be a potentially more sustainable strategy, especially for smallholders, but that depends on the setting.

Keywords: Agro-forestry, palm oil, Sarawak/Malaysia, smallholder, SMART-Farm Tool, sustainability, sustainability assessment

---

Contact Address: Rainer Weisshaidinger, Research Institute of Organic Agriculture (FiBL), Sustainabilty Assessment & International Cooperation, Doblhoffgasse 7/10, 1010 Vienna, Austria, e-mail: rainer.weisshaidinger@fibl.org

The Role of the Albanian Gene Bank in Addressing and Improving the Livelihoods of the Poorest Farmers. Case Study - The Northern Alps of Albania

Anika Totojani, Belul Gixhari

The Agricultural University of Tirana, Institute of Plant Genetic Resources (IPGR), Albania

The livelihoods of many farmers in the northern Albanian Alps depend on agricultural production. This study assessed the impact of the Albanian Gene Bank on the poorest farmers through providing improved seeds of maize, beans and vegetables.
In a case study 55 surveys and semi-structured interviews were held. The livelihood framework was used as a methodological framework to analyse the food security aspect and the well-being of farmers.
The results revealed that farmers are totally dependent on agricultural production for their livelihood. Only a few households have diversified their rural income via agro-tourism. Men are exclusively head of the households and the ones who contribute to the rural economy.
Farmers cultivate small parcels of land in the range of 0.1- 1 ha. The fragmentation of the parcels, is one of the main reasons for the low yields encountered. The autochthone varieties of maize, beans and vegetables (garlic, pumpkin, squash, egg-plant, and peper) are good adapted to the local conditions. Farmers trust these local varieties, which are conserved for over 80-150 years. Farmers produce for self-subsistence of the households. Although agricultural productivity is low, farmers continue and prefer the use of the local varieties rather than the proposed hybrid varieties. The Albanian Gene Bank interferes in the livelihood framework by providing hybrid seeds and improving the techniques of local seeds preservation. Recently, some of the local production is consumed in the households for agro-tourism purposes. In so doing, there is a diversification of activities, and farming is supporting the rural economy even more.
Therefore, preserving the agro-biodiversity of the local Albanian seeds and combining it with adapted high yielding varieties is enhancing and fostering the rural development. The Gene Bank can be one of the crucial actors to improve the livelihood of smallholder farmers. A combination of autochthone seeds and well-adopted hybrid seeds for certain crops can be a very good option. There is a need to deepen this topic in future studies. In the near future, the local government should work closely with the Gene Bank to improve the rural economy.

Keywords: Adopted local varieties, Albanian gene bank, autochthone seeds, farmers, livelihoods framework

Contact Address: Anika Totojani, The Agricultural University of Tirana, Institute of Plant Genetic Resources (IPGR), Siri Kodra Street 132/1, 132/1 Tirana, Albania, e-mail: anikatotojani@yahoo.com

ICTs in Agriculture: State of the Art Tools for Broader Access to Tropical Forage Knowledge

José Luís Urrea Benítez[1], Michael Peters[2], Stefan Burkart[1]

[1]*International Center for Tropical Agriculture (CIAT), Tropical Forages, Colombia*
[2]*International Center for Tropical Agriculture (CIAT), Tropical Forages, Kenya*

The development of information technologies and internet connectivity strongly increased access to scientific knowledge. However, this also comes along with e.g. quality issues, affordability (restricted access to publications, download payments) and increasing numbers of "predatory" publishers. To provide stakeholders with high-quality tropical forage knowledge, various research institutes (CSIRO, QDPIF, CIAT, ILRI), with funds from ACIAR, BMZ/GIZ, DFID and CATAS, developed two tools: 1) Tropical Grasslands-Forrajes Tropicales (TGFT), a bilingual peer-reviewed open-access journal, indexed in the most recognised databases/journal directories (www.tropicalgrasslands.info). Since its inception in 2013, TGFT has shown sustained growth, reaching in 2019 >228,000 visits, and >492,000 abstract and >696,000 PDF eBook downloads, respectively. Its main metrics are JCR Impact Factor (0.441), CiteScore (0.80), Journal Rank (0.28) and i10-index (38). It is a RoMEO Green Journal with Gold Open Access status. TGFT provides access to all papers published in the former journals Tropical Grasslands (1967−2010) and Pasturas Tropicales (1979−2007). 2) Tropical Forages, a tool for selecting forage species for local conditions, launched in 2005 (www.tropicalforages.info). It is among the most widely used (250–480k annual visits) and cited (450 citations) tropical forages databases and provides information on >170 forage species with potential for use in animal production, identified and characterised by leading tropical forages researchers, including e.g. information on morphology, agronomic management, nutritional value, productive potential or promising accessions. A set of 17 variables allows users filtering through the species and refining a shortlist for their specific local conditions. Seed samples can be requested from the linked CGIAR genebanks. It is a valuable information source for e.g. researchers, extension services or farmers seeking to improve animal productivity and sustainability, which is evidenced by constantly increasing pages visits (2018–2019: from 798k to 1,414k). Both tools have promising outlooks: TGFT's goal is to become a global benchmark in forage research, supporting the publication of results from the global tropics by following rigorous scientific standards. Tropical Forages is in the process of finalizing its first major update and will be relaunched in 2020, with content updates and notable technical improvements, such as a revamped interface responsive to multiple devices, a mobile application and automatic translation.

Keywords: Digital tools, forage database, ITCs, open access, scientific publications, tropical forages

Contact Address: Stefan Burkart, International Center for Tropical Agriculture (CIAT), Trop. Forages Program, Km 17 Recta Cali-Palmira, Cali, Colombia, e-mail: s.burkart@cgiar.org

Effectiveness of Czech NGOs and Public Pressure on the Protection of Rainforests in Developing Countries

Tereza Lysakova

Czech University of Life Sciences Prague, Fac. of Tropical AgriSciences; Dept. of Economics and Development, Czech Republic

The civil pressure and NGOs initiatives against deforestation represent a powerful movement capable to influence both the public and the government towards rainforest protection. Therefore, the purpose of this research was to study the activities of NGOs in the Czech Republic concerning globalisation induced deforestation. Mainly related to high world demand for commodities such as palm oil originating from Indonesia or soya beans and beef meat originating from Brazil.

An in-depth face to face interviews were conducted with representatives of local NGOs. The interviews were recorded, transcribed and analysed using the content analysis technique. Even though the respondents claim the consumers related campaigns on the palm oil production affecting the Indonesian rain forests were relatively successful in the Czech Republic, the consumer campaigns in support of Brazilian rain forests, specifically related to the livestock farming and meat consumption, has not been resonating so loudly. The consumers in the Czech Republic are generally aware about negative issues related to the palm oil production, including deforestation, loss of biodiversity, pollution or land degradation. Although the problems related to the Brazilian forests are not that much known, as the problem is mainly indirect, through imports of animal feed to the European Union or by consumption of beef meat. Most Czech cattle and livestock depend on the feed containing soybean coming from the areas connected with deforestation.

Therefore, the attention now needs to be directed towards the politicians and producers. Change in policies of the key players, such as the European Union, represents an opportunity to influence adverse production practices, for instance via the Renewable Energy Directive. As the EU market power represents a key lever on the importing countries which need to comply with the rules if they want to sell their products. There is also a strong need for thorough adherence of the palm oil certified by the Roundtable on Sustainable Palm Oil. The campaign for protection of Brazilian rain forests shall be intensified among the consumers as it is crucial to alter the patterns of meat consumption, to promote a more plant-based diet and to reduce overconsumption and food waste.

Keywords: Beef, consumers, deforestation, environmental campaign, environmental protection, NGOs, palm oil, producers, public pressure, soybean

Contact Address: Tereza Lysakova, Czech University of Life Sciences Prague, Fac. of Tropical AgriSciences; Dept. of Economics and Development, Kamýcká 129, 165 00 Praha 6 - Suchdol, Czech Republic, e-mail: lysakova@ftz.czu.cz

Genetic Diversity of Aguaje (*Mauritia flexuosa*) in Peruvian and Ecuadorian Amazon

Dita Mervartová[1], Bohdan Lojka[1], Marie Kalousová[1], Jorge Manuel Revilla Chávez[2], Goldis Perry Davila[3]

[1]*Czech University of Life Sciences Prague, Fac. of Tropical AgriSciences, Dept. of Crop Sciences and Agroforestry, Czech Republic*

[2]*Peruvian Amazon Research Institute (IIAP), Directorate of Research in Amazonian Terrestrial Biological (DBIO), Peru*

[3]*National University of Ucayali (UNU), Fac. of Agricultural Sciences, Dept. of Agricultural Sciences, Peru*

Mauritia flexuosa (aguaje) is a dioecious indigenous species from the family Arecaceae important for local people and also for Amazon forest as a whole. The specific environments along the banks of large rivers or inside the lowland jungle forest, where water is retained and where nutrients are slowly decomposed, are called aguajales. Particularly these swampy areas, where *M. flexuosa* occurs in abundance, provide a refuge for different kinds of fauna and flora. People from the local communities living in the Amazon depend on its nutritive fruits. The fact that it is a dioecious species presents a number of difficulties in cultivation. Therefore, a large part of the production is still obtained from wild populations, usually by cutting the female trees. Unfortunately, this bad management of the fruit collection leads to degradation of aguajales and to biodiversity loss.

The main objective of this research was to assess the intra-population and inter-population genetic diversity of *M. flexuosa* in Peruvian and Ecuadorian Amazon by SSR markers. Totally, 145 trees from 15 populations were sampled and successfully analysed using seven polymorphic microsatellite loci. The populations were characterised by high values of genetic diversity and very low levels of inbreeding. The high molecular variance was determined especially among individuals. According to STRUCTURE analysis, no different genetic composition was indicated in the populations. The UPGMA diagram divided populations into two main groups corresponding to the region's isolation; (1) Ecuador and (2) Peru. A very low correlation between geographic and genetic distances was determined using a Mantel test.

The results can be explained by the high level of gene flow along the Amazonian rain forest, where the rivers are used as bio corridors. Also, the human activity through the history of the Amazonian probably has had a significant impact on the *M. flexuosa* distribution. Based on the high genetic diversity, the selection of superior individuals for further breeding can be explored.

Keywords: Arecaceae, biodiversity, genetic variability, microsatellites, population genetics

---

Contact Address: Dita Mervartová, Czech University of Life Sciences Prague, Fac. of Tropical AgriSciences, Dept. of Crop Sciences and Agroforestry, Prague 6 - Suchdol, Czech Republic, e-mail: mervartova.d@centrum.cz

Advances in Testing Multi-Species Pastures for Productivity and Environmental Benefits: Influence on Pollinators

Juan Andrés Cardoso[1], Luis Miguel Hernandez[1], Paula Espitía[1], Jacobo Arango[2], Michael Peters[3], Mauricio Sotelo[2]

[1]*Alliance Bioversity-CIAT, Tropical Forages, Colombia*
[2]*International Center for Tropical Agriculture (CIAT), Colombia*
[3]*International Center for Tropical Agriculture (CIAT), Tropical Forages, Kenya*

In the American Tropics, livestock production is mostly based on pasture grazing systems. Such pastures tend to be dominated by a single forage species (mostly grasses), and if combined with legume(s) (herbaceous, shrubs or trees), the diversity of planted species within such pasture systems is still low (~three species in total). There is a growing interest in increasing diversity of forage species within pastures as a method of enhancing agricultural production while providing ecosystem services. Recently, the Tropical Forages Program of the Alliance Bioversity-CIAT (ABC), in collaboration with Grow Colombia Project (growcolombia.org), is testing the suitability of diverse forage mixes for agricultural productivity while providing positive environmental impacts. The project is investigating stress resilience, forage quality, yield, nutritional value from multi-species mixtures (up to six species of grasses, legumes and herbs) and their effects upon soil health compared to that of a grass-legume (1 grass, 1 legume) system under field conditions. Field trials were established in November 2019 and located at the regional office of ABC in the Americas. Furthermore, the trials are used to undertake pollinator surveys. Preliminary results show that even within the limited period since trials were established, there was over a two-fold increase in richness and diversity of insects, including pollinators, in multi-species pastures. The relevance of the preliminary results is high as there has been a steady decline of pollinators worldwide. Our preliminary results suggest that establishment of multi-species pastures can rapidly provide an environment friendly to pollinators and thereby mitigate their reductions as shown elsewhere.

Keywords: Biodiversity, entomofauna, multi-species swards

Contact Address: Juan Andrés Cardoso, Alliance Bioversity-CIAT, Tropical Forages, Km 17 Recta Cali Palmira, Palmira, Colombia, e-mail: j.a.cardoso@cgiar.org

Diversity of Tree and their Use in Cocoa Agroforestry Systems in Alta Verapaz, Guatemala

Carlos Villanueva Gonzalez[1,2], Bohdan Lojka[2], Carlos Ernesto Archila[1], José Alejandro Ruiz-Chután[2]

[1] *Rafael Landivar University, Fac. of Agricultural and Environmental Sciences, Guatemala*

[2] *Czech University of Life Sciences Prague, Fac. of Tropical AgriSciences, Dept. of Crop Sciences and Agroforestry, Czech Republic*

The agroforestry systems (AFS) provide a bunch of opportunities for the biodiversity conservation and livelihoods development for local people. We evaluated the tree diversity and their local use in cacao AFS (*Theobroma cacao*), randomly distributed in four Municipalities of Alta Verapaz department in Guatemala: Lanquín, Cahabón, Panzos y Cobán. Fieldwork was carried out over a 10-month period, during April 2019 and February 2020. We measured and evaluated the typology and diversity of trees in 40 cocoa AFS (sampling plots of 50 x 50 m). To valuate the food security and the contribution to well-being, we used classical ethnobotanical methods among the local population, such as structured interviews, focus groups, transect tours in productive plots and carried out the tree inventory. In each AFS, all individuals with DAB (diameter at brest heigh) > 10 cm were identified and measured. In order to inventory and record the common and Latin name of the species were identified in each sampling plot. In total we found 1,559 trees in an area of 10 hectares, belonging to 63 species. The AFS of the municipality of Lanquin presented the highest levels of diversity, compared to the rest of the AFS evaluated. Cocoa producers benefit directly from the diversity of trees through the provision of firewood, building materials, medicine, food and fodder. These findings reinforce the potential of AFS for biodiversity conservation in an area, but in addition, they are able to reduce food insecurity, especially among highly vulnerable local populations, due to the species diversity within the system and the variability of current and potential uses they possess.

Keywords: Agroforestry systems, biodiversity, conservation, livelihoods

Contact Address: Carlos Villanueva Gonzalez, Rafael Landivar University, Fac. of Agricultural and Environmental Sciences, 12av.1-71, 16001 Cobán, Guatemala, e-mail: villanueva@ftz.czu.cz

Bamboo for Landscape Restoration

Bhoke Masisi, Astrid Zabel, Jürgen Blaser
Bern University of Applied Sciences (BFH), School of Agriculture, Forest, and Food Science, Switzerland

Responding to the African Forest Landscape Restoration Initiative (AFR100), 30 countries have pledged to bring more than 100 million hectares of land across Africa into restoration by 2030. Nature-based solutions that can help achieve this ambitious target to the benefit of the rural poor are much sought-after. In many policy circles, there is a nascent discussion on the role that bamboo could play as one component of landscape restoration strategies. Major advantages are that it is a fast-growing versatile woody grass, that can prosper on degraded land.

So far, landscape restoration strategies have been driven by climate policy debates, where bamboo has largely been neglected. This circumstance is attributable to its taxonomic classification as grass rather than tree. Moreover, most empirical research on bamboo was conducted in Asia, leaving open questions on the transferability of the results to the African context.

In this paper, we contribute to the debate by investigating to what extent bamboo can contribute to agroecological systems with the aim of sustaining livelihoods of poor rural societies, especially in degraded ecosystems. We structure our analysis around the IPBES ecosystem services framework and lay a special focus on bamboo's carbon sequestration potential. Taking Tanzania as a case study, we present empirical results on the variation of bamboo's carbon content (i) along an altitudinal gradient, (ii) between indigenous and exotic species, and (iii) between intensive and extensively managed ecosystems.

Our results contribute to developing a more nuanced picture of the advantages and disadvantages of incorporating bamboo into landscape restoration efforts. The novel findings on variation in bamboo's carbon sequestration potential may be a first step toward unlocking future climate finance.

Keywords: Bamboo, carbon, ecosystem services, landscape restoration, Tanzania

---

Contact Address: Bhoke Masisi, Bern University of Applied Sciences (BFH), School of Agriculture, Forest, and Food Science, Laenggasse 85, 3052 Zollikofen, Switzerland, e-mail: bhoke.masisi@students.bfh.ch

Morphological and Wood Anatomical Study on *Acacia nilotica* (L.) Willd. Ex Delile, Grown in Sudan

HAYTHAM HASHIM GIBREEL[1], HANADI MOHAMED SHAWGI GAMAL[2]

[1]*University of Khartoum, Fac. of Forestry; Forest Silviculture, Sudan*

[2]*University of Khartoum, Fac. of Forestry; Forest Products and Industries, Sudan*

Acacia nilotica (L.) Willd. ex Del. is a multipurpose leguminous species in the family Fabaceae. The species has an extensive and diverse natural distribution in Sudan. It is valuable mainly for timbers and fuel wood production; other benefits include environmental, tannin production and it has medicinal attributes. The diversity of site and climatic conditions under which *Acacia nilotica* grows led to the evolution of an extremely variable species. Thus, the taxonomic history and nomenclature of the species was puzzled.

In Sudan, *Acacia nilotica* is represented by three subspecies with wide range of distribution, namely the subspecies *nilotica, tomentosa* and *adstringen*. Additionally, a new group of *Acacia nilotica* trees were identified with new morphological characteristics of pods, seeds, leaves. This new group is not mentioned in the taxonomical key. However, with no doubt there is not a complete and clear picture of the pattern of variation within and among *Acacia nilotica* with it's currently known subspecies.

The overall objective of the present study is to update the taxonomy of *Acacia nilotica* and understand the level of morphological and anatomical diversity among its subspecies in the Sudan. Therefore, the taxonomic relationships among the three identified subspecies and the new group were studied by analysing some of the quantitative and qualitative morphological characteristics of mature trees and of the seedlings at early germination stages as well as after one year after germination. The anatomical characteristics among them were also studied in wood samples of one year old seedlings, using the stereological account technique. In total, thirteen variable study sites were selected as sample sources, this across the agro-ecological zones of the Sudan where *Acacia nilotica* populations are growing.

The results revealed significant variation in all of the studied morphological and anatomical characteristics among the new group and the three identified subspecies in this study. The study concluded that, *Acacia nilotica* population in the Sudan included three main subspecies namely subsp. *nilotica, tomentosa* and *adstringens* beside one newly identified morphological group intermediate between subspecies *nilotica* and *tomentosa*.

Keywords: *Acacia nilotica*, anatomical variation, morphological, taxonomy

Contact Address: Haytham Hashim Gibreel, University of Khartoum, Fac. of Forestry; Forest Silviculture, Khartoum, Sudan, e-mail: haythamgibreel@yahoo.com

Validation of the X-ray Densitometry Method for Wood Density Determination of Acacia seyal Tree Species

HANADI MOHAMED SHAWGI GAMAL[1], HAYTHAM HASHIM GIBREEL[2]
[1]*University of Khartoum, Fac. of Forestry; Forest Products and Industries, Sudan*
[2]*University of Khartoum, Fac. of Forestry; Forest Silviculture, Sudan*

Wood density is a variable influencing many of the technological and quality properties of wood. It is its single most important physical property. Determining the wood density values is important for its end use. The X-ray technique, traditionally applied to softwood species to assess the wood quality properties, due to its simple and relatively uniform wood structure. In the other hand very limited information is available on the validation of using this technique for hardwood species. The suitability of using X-ray technique for the determination of hardwood density has a special significance in countries like Sudan where only a few timbers are well known. this will not only save the time consumed by using the traditional methods but it will also enhance the investigations of the great number of the lesser known species, the thing which will fill the huge gap on information of hardwood species growing in Sudan.

The current study aimed to evaluate the validation of using the X-ray densitometry technique to determine the wood density of *Acacia seyal* var. *seyal*. To achieve this aim, a total of thirty trees were collected randomly from four states in Sudan. The wood density was determined using the air dry gravimetric method as well as the X-ray densitometry method in order to assess the validation of X-ray technique in wood density determination.

The results reveal no significant differences between the values obtained by the two methods. These results confirmed the validation of using the X-ray technique for *Acacia seyal* var. *seyal* density radial trend determination. It also promotes the suitability of using this method for other hardwood species.

Keywords: *Acacia seyal* var. *seyal*, wood density, x-ray densitometry

Contact Address: Hanadi Mohamed Shawgi Gamal, University of Khartoum, Fac. of Forestry; Forest Products and Industries, Alemarat Street 61, Khartoum, Sudan, e-mail: hanadishawgi1979@yahoo.com

Variation in Wood Anatomical Properties of *Acacia seyal* var. *seyal* Tree Species Growing in Sudan

Hanadi Mohamed Shawgi Gamal[1], Ashraf Mohamed Ahmed Abdalla[1], Haytham Hashim Gibreel[2]

[1]*University of Khartoum, Fac. of Forestry; Forest Products and Industries, Sudan*
[2]*University of Khartoum, Fac. of Forestry; Forest Silviculture, Sudan*

Sudan is endowed by a great diversity of tree species, nevertheless the utilisation of wood resources has traditionally concentrated on a few number of species. With the great variation in the climatic zones of Sudan, great variations are expected in the anatomical properties between and within species. This variation need to be fully explored in order to suggest best uses for the species. Research on wood has substantiated that the climatic growth condition has significant effect on wood properties. Understanding the extent of variability of wood is important because the uses for each kind of wood are related to its characteristics. The present study demonstrates the effect of rainfall zones on some anatomical properties of *Acacia seyal* var. *seyal* growing in Sudan. For this purpose, twenty healthy trees were collected randomly from two zones. One zone with relatively low rainfall (273 mm annually) and the second with relatively high rainfall (701 mm annually). From each sampled tree, a stem disc (3 cm thick) was cut at 10% from stem height. One radius was obtained in central stem dices. Two representative samples were taken from each disc, (at 10% and 90% distance from pith to bark), in order to represent the juvenile and mature wood. The investigated anatomical properties were fibers length, fibers and vessels diameter, lumen diameter and wall thickness as well as cells proportions. The result of the current study reveals significant differences between zones in mature wood vessels diameter and wall thickness as well as juvenile wood vessels wall thickness. The higher values were detected in the drier zone. Significant differences were also observed in juvenile wood fiber length, diameter as well as wall thickness. Contrary to vessels diameter and wall thickness, the fiber length, diameter as well as wall thickness decreased in the drier zone. No significant differences have been detected in cells proportions of juvenile and mature wood. From these results *Acacia seyal* var. *seyal* seems to be well adapted with the change in rainfall and may survive in any rainfall zone.

Keywords: *Acacia seyal* var. *seyal*, anatomical properties, rainfall zones, variation

Contact Address: Hanadi Mohamed Shawgi Gamal, University of Khartoum, Fac. of Forestry; Forest Products and Industries, Alemarat Street 61, Khartoum, Sudan, e-mail: hanadishawgi1979@yahoo.com

Livestock

Monogastrics plus alternative livestock

Oral Presentations

Posters

Dietary Supplementation of Garlic, Propolis and Wakame Improves Recuperation in Cadmium Exposed Japanese Medaka Fish (*Oryzias latipes*)

HENRY UJEH, MASAAKI KURASAKI

Hokkaido Univeristy Japan, Environmental Science, Japan

Deleterious effects due to cadmium (Cd) exposure in fish can become protracted during recuperation, defying the protective capacity offered by endogenous antioxidants. The prospects of exogenous supply of antioxidants from plant based sources have been explored by numerous studies. Based on their immune stimulating and antioxidative merits, the present study was undertaken to assess the ameliorating potentials of dietary supplemented garlic (D1), propolis (D2) and wakame (D3) on Cd bioconcentration and toxicity during recuperation in juvenile medaka fish. Fish were exposed to 0.0 or 0.3 mg l^{-1} Cd for twenty one days, during which a control diet (C1) was fed. This was followed by a twenty one day 0.0 mg l^{-1} Cd exposure (depuration/recuperation), during which C1 or C2 (positive control diet), D1, D2, and D3 were fed. Results showed antioxidants superoxide dismutase (SOD) and total glutathione (GSH) levels in fish tissues (gill, liver, and muscle) increased significantly following Cd exposure. Values were not significantly restored in the C2 fed groups during recuperation. Furthermore, elevated tissue Cd burden, metallothionein (MT) expression, and lipid peroxidation (LPO) levels were not significantly restored in the C2 fed groups. On the other hand, D1 and D2 elicited significant restoration in elevated tissue Cd burden and LPO levels which supported greater detoxification and recuperation. Condition indices in D1 and D2 fed groups were also significantly higher than D3 and C2 fed groups. In conclusion, dietary supplementation significantly increased recuperation and tissue functions in juvenile medaka fish, with an observed trend of D1>D2>D3.

Keywords: Cadmium exposure, depuration, garlic, japanese medaka, *Oryzias latipes*, propolis, recuperation, wakame

Contact Address: Henry Ujeh, Hokkaido Univeristy Japan, Environmental Science, Hokkaido University. N10W5 Sapporo, Hokkaido, 060-0810 Sapporo, Japan, e-mail: ujehhenryo@yahoo.com

Meat Yields and Sensory Evaluation of two Commercial Hybrid Strain of Broiler Chicken in Nigeria

Jude T. Ogunnupebi[1], Rosemary Ozioma Igwe[1], Isaac Ikechukwu Osakwe[2], Solomon Olubiyi Olorunleke[1]

[1]*Ebonyi State University Abakaliki, Animal Science, Nigeria*
[2]*Alex-Ekwueme Federal University, Animal Science, Nigeria*

The research was conducted to evaluate the effect of sex and strain on the meat yields and sensory evaluation of an exotic and an improved local commercial hybrid strain of broiler chicken in Nigeria. It was conducted at the poultry unit of the teaching and research farms of Ebonyi State University. One hundred and ninety two (192) day old broilers comprising of 96 each of ross 308 and funaab alpha strains of broiler chicken were randomly allotted to four treatment groups in a 2 x 2 factorial experimental design. Each treatment was replicated 3 times with 16 birds per replicate. The birds were reared for 56 days. At the end of the rearing period, 12 birds (one per replicate) of average body weight were selected and slaughtered. Different meat yield parameters were recorded. Data collected were analysed using the general linear model of minitab 18 software. There were no significant differences in the effect of sex on the meat yields while there were significant differences in the effects of strain on the defeathered weight (1958 g, 1576.67 g), breast meat % (28.22, 9.44), thigh meat % (10.39, 5.90), drumstick meat % (9.17, 5.18), breast bone % (4.4, 2.88), thigh bone % (2.59, 1.75) with the ross 308 having the higher values. In the interaction effect of sex and strain on the meat yield, there were significant differences in the live weight (2400 g, 1866.67 g), defeathered weight (2166.67 g, 1533 g), breast meat % (26.15, 8.38), thigh meat % (10.87,5.31), drumstick meat % (9.35, 5.15), breast meat to bone ratio (6.39, 2.86) with the ross 308 male recording the highest values while the funaab alpha male recorded the lowest values. In the sensory evaluation, the funaab alpha had the better acceptability (7.83, 5.1). funaab alpha was observed to have a higher coefficient of variation in most of the parameters measured. Ross 308 had a higher meat yield while the funaab alpha was more acceptable and a higher variability in meat yield which suggests the need for more selection for meat yield in the funaab alpha.

Keywords: funaad alpha, hybrid strain, meat yield, ross 308, selection

Contact Address: Jude T. Ogunnupebi, Ebonyi State University Abakaliki, Animal Science, CAS Campus, Abakaliki, Nigeria, e-mail: jtogunnupebi@gmail.com

Locusts for Food, Feed and other Uses: A Potential Game-Changing Approach in Managing Locust Plagues

James Peter Egonyu, Sevgan Subramanian, Chrysantus Mbi Tanga, Thomas Dubois, Sunday Ekesi

International Centre of Insect Physiology and Ecology (ICIPE), Plant Health Theme, Kenya

Locusts comprise about twenty species of grasshoppers in fifteen genera in the Family Acrididae, Order Orthoptera. They breed in arid areas and can change from solitary to gregarious phases under high population pressure, forming massive swarms that migrate over very long distances. The locust plagues feed voraciously along their paths, massively destroying crops and pastures. For instance, an outbreak of the desert locust, *Schistocerca gregaria* in Africa in 2003–2005 covered 13 million ha in 22 countries. The locust plague caused 80–100 % crop losses which affected over 8 million people. Almost half a billion US dollars was spent to control the plague and provide food aid to the affected people. The locusts were largely controlled by spraying with over 13 million litres of broad-spectrum neurotoxic insecticides. Though this chemical control approach was successful, the hazardous effects of insecticides on humans and the environment are too costly. In many countries across the globe, locusts are gathered for subsistent use as food and/or ingredients in livestock feeds during outbreaks. Literature indicates that locusts are highly nutritious, in some cases superseding conventional livestock and crop sources of food or feed. However, mainstreaming the use of locusts for food and feed, faces several research challenges. These include, technologies for mass trapping/mass-rearing of locusts, systematic assessment of nutritive and anti-nutritive factors of locust meal for use as food and feed, post-harvest processing and storage technologies, socio-economic, market and policy assessments, and value-chain analysis. Here we review the state-of-the-art techniques and practices of harvesting and processing locusts into food, feed and other valuable products globally. Key knowledge gaps to be addressed and potential interventions that are necessary to foster this strategy as a game-changing approach to the management of locust plagues as an alternative to massive application of insecticides are discussed.

Keywords: Harvesting, impacts of locusts, nutrition, processing, sustainable locust management

---

Contact Address: James Peter Egonyu, International Centre of Insect Physiology and Ecology (ICIPE), Plant Health Theme, Nairobi, Kenya, e-mail: pegonyu@icipe.org

Effect of Melatonin and Lighting Regime on Physiological Responses and Reproductive Traits of Layers

Rosemary Ozioma Igwe[1], Udo Herbert[2], Isaac Ikechukwu Osakwe[3]

[1]*Ebonyi State University Abakaliki, Animal Science, Nigeria*

[2]*Michael Okpara University of Agriculture Umudike, Animal Breeding and Physiology, Nigeria*

[3]*Alex-Ekwueme Federal University, Animal Science, Nigeria*

Two experiments were conducted to evaluate the effect of melatonin and lighting on physiological responses and reproductive traits of two strains of laying birds using 324 laying birds. Each of the experiments consisted of 162 birds. Experiment I was with Nera Black strain while experiment II was with Isa Brown strain. Each of the experiments was grouped into 9 treatments which were further subdivided into three replicates of six birds each in a 2×3 factorial in a completely randomised design. Melatonin and lighting at three levels were administered to the birds four times weekly for 30 weeks. The three levels of melatonin were 0 mg, 5 mg and 10 mg while lighting were 12 hours, 15 hours and 18 hours daily. Melatonin was dissolved in 2 mls warm water and the birds were drenched while 100 watt bulbs were used to provide lighting. Results from the physiological responses showed that rectal temperature (RT), respiratory rate (RR) and heart rate (HT) were significantly ($p < 0.05$) influenced by both melatonin and lighting regime in the two experiments. Melatonin at 5 mg significantly ($p < 0.05$) reduced the RT, RR and HT with values of 41.55 °C, 142.56 and 327.11 respectively in experiment I and 40.55 °C, 139.44 and 320.22 respectively in experiment II. Performance characteristics were significantly ($p < 0.05$) influenced by both melatonin and lighting regime. The hen day egg production, feed conversion ratio and egg weight of 91.66 %, 1.73 and 61.52 g respectively for experiment II and 84.77 %, 1.61 and 70 g respectively for experiment I were recorded. The weight of the reproductive organs and the number of follicles were significantly ($p < 0.05$) improved by melatonin in both experiments. Large yellow follicles and small white follicles increased significantly with increasing level of melatonin in both experiments. Interaction between melatonin and lighting improved the overall performance of the birds as the groups on 5 mg of melatonin and 15 hrs lighting performed better than the other groups in both experiments. It is therefore concluded that 5mg melatonin and 15 hours lighting should be used to enhance egg production and improve the behavioural characteristics of laying birds during thermal stress.

Keywords: Follicles and performance, layers, lighting, melatonin, physiological, reproductive

Contact Address: Rosemary Ozioma Igwe, Ebonyi State University Abakaliki, Animal Science, Abakaliki, Nigeria, e-mail: alorosemaryozioma@gmail.com

Domestication of a Wild Silkworm *Borocera cajani* in Madagascar: An Alternative Source of Food and Protein

Razafindrakotomamonjy Andrianantenaina[1], Ravaomanarivo Lala Harivelo[2]

[1]*University of Antananarivo, Doctoral School Life Sciences and Environment, National Center for Applied Research for Rural Development (FOFIFA), Madagascar*
[2]*University of Antananarivo, Departement of Entomology,*

In Madagascar, 42 % of households have an unbalanced diet and in particular a deficiency in proteins and vitamins. Malagasy farmers have a poorly diversified diet and they consume mainly cereals and tubers and very little meat, fish, legumes, dairy products, fruits, vegetables, sugar and oil. Insects can be used as a source of food and protein. In the Amoron'i Mania Region (Central Highlands of Madagascar), the rate of food insecurity is 30 %. Insect consumption has been practised by the people of this region for a long time. In general, the population collects insects in the forest. The species *Uapaca bojeri* (Euphorbiaceae), commonly known as 'Tapia', is a plant endemic to this region, alone forming the wild forest of 'Tapia'. This forest is the natural habitat of the wild silkworm *Borocera cajani* (Lasiocampidea), a bivoltine species with 2 generations per year. The chrysalis of this insect are consumed by the local population but they are only available for a short period of the year during the rainy season, between January and March, the second generation hatches between the end of April and November. The bush fire perpetually threatens the *U. bojeri* forest and the massive gathering of pupae can lead to the disappearance of this species. The development of a mass rearing technique for *B. cajani* outside its natural ecological niche is being undertaken as part of this study in order to reduce the pressure on this species and to produce these pupae in quantity for the food. Laboratory tests have evaluated the potential of using guava leaves *Psidium guyava* (Myrtaceae), substituting those of Tapia, to feed the insect. The optimum temperature is 24 ° C $\pm$ 0, 5 and the relative humidity is 70 $\pm$ 5 % the rearing yield is 70.12 %, the duration of obtaining the chrysalis is 102.4 $\pm$ 2.0 days and the average weight is 1.77 grams per pupa.

Keywords: Animal proteins, entomophagy, mass rearing, *Psidium guyava*, wild silkworm

Contact Address: Razafindrakotomamonjy Andrianantenaina, University of Antananarivo, Doctoral School Life Sciences and Environment, National Center for Applied Research for Rural Development (FOFIFA), Antananarivo, Madagascar, e-mail: nantenaina73@gmail.com

Insect Farming for Food and Feed Security in Myanmar

Aye Aye Myint, Tin Htut
Yezin Agricultural University, Department of Animal Science, Burma

In Myanmar most edible insects are harvested in the wild, but a few insect species like silkworms and bees are domesticated because of their commercially valuable products as silk and honey. Moreover, by-products like processed pupae from silk worm and bee larvae from beekeeping are eaten as protein rich delicacy in Myanmar. However, edible insect rearing for human consumption is a new farming system for Myanmar creating a new opportunity for poverty stricken smallholder rural populace. Recently, through edible insect project "Production and Processing of Edible Insect for Improved Nutrition ProciNut", Yezin Agricultural University initiated and explored the opportunity for affordable small-scale cricket farming. Two species, Field cricket *Gryllus bimaculatus* (Degeer) and Giant cricket *Brachytrupes portentosus* (Lichtenstein) are produced. The rearing methods are as simple as backyard farming, needing no expensive materials. Day-old chick feed are fed for young stage and vegetable scraps from pumpkins, carrot and cabbage are fed for adults. However the lack of technical backstopping, it is felt that current production system has not been well developed. On the other hand, high-quality rearing techniques and food safety are essential for the widespread use of insects as human food. Another major challenge is rearing insects in large numbers, which may require understanding ecology of particular edible insect species, legal instruments, Environmental Impact Assessment (EIA) and Social Impact Assessment (SIA). Training on insect farming and processing methods are urgently needed for small scale producers. However, it is believed that edible insects as a new sources of food and feed for innovative and integrated food system in Myanmar.

Keywords: Cricket, feed, food, insect farming, processing, production

Contact Address: Aye Aye Myint, Yezin Agricultural University, Department of Animal Science, Yezin Agricultural University, 15013 Zayarthiri Township, Burma, e-mail: ayeayemyint2006@gmail.com

Dietary Sodium Diformate Improves Growth Performance and Nutrient Digestibility in Broilers against Negative and Positive Controls

Christian Lückstädt, Stevan Petrović

ADDCON, Germany

Broiler growth rate and feed efficiency are key to economic performance through to market. Dietary formic acid and its salts act against pathogens, helping to decrease pressure on the immune system and improving nutrient digestibility. Previous studies on the antimicrobial impact of organic acids and their salts, including sodium diformate (Formi NDF, ADDCON), placed less emphasis on the impact in the GI-tract of birds. This formed the impetus for this study, which assessed the impact of the additive on pH-levels at different locations of the GI-tract and digestibility parameters. In a trial conducted at a research farm in Iran, 0.1 % NDF was tested in a typical corn-soy diet, against both a negative (NC) and positive control (PC) containing an antibiotic growth promoter (500 mg Trimethoprim−Sulfadiazine per kg). 216 day-old broiler chicks (male Ross 308) were randomly selected into 3 treatment groups with 6 replicates of 12 birds each. Feed, in mash form, and water were available *ad libitum* throughout the 42-day trial period. The effects of dietary NDF on performance (body weight gain BWG, FCR, broiler index EBI), pH in the gizzard and ileum, Protein Efficiency Ratio (PER) and apparent ileal digestibility of protein and minerals were examined at the end of the trial. Data were analysed using the t-test and a confidence level of 95 % defined for these analyses.

Performance was boosted in the birds fed 0.1 % NDF. Treated birds had a significantly increased BWG against NC and PC respectively (2126 g vs. 2007 and 2006 g; $p < 0.05$), while the FCR tended ($p < 0.1$) to be improved (1.78 vs. 1.87 and 1.86). EBI was enhanced by almost 11 % against both NC and PC. Utilisation of nutrients was also significantly improved in the NDF-fed broilers, especially for crude protein, crude ash, calcium and phosphorus. Calculated as PER, the usage of NDF led to an increase of protein utilisation against both controls by more than 5 %.

This study demonstrates that including NDF in broiler diets is a sustainable tool for improved performance and nutrient utilisation, thereby saving nutrient resources, even compared to an antibiotic growth promoter.

Keywords: Broiler, growth performance, nutrient digestibility, sodium diformate

Contact Address: Christian Lückstädt, ADDCON, Parsevalstrasse 6, 06749 Bitterfeld-Wolfen, Germany, e-mail: christian.lueckstaedt@addcon.com

Neutrophil and Lymphocyte Counts in Broilers Administered Aqueous *Vernonia amygdylina* as Natural Growth Promoter

Darlington Godslove Ogan, Ruth Tariebi S. Ofongo
Niger Delta University, Dept. of Animal Science, Nigeria

Plant extracts (phytogenics) are a group of natural growth promoters (NGPs) used as feed additives. They are derived from medicinal plant (herbs) and spices and are commonly regarded as favourable alternatives to antibiotic growth promoters (AGPs) in livestock production. Currently, organic farming is taking place in many countries in the world with the ban of antibiotics as growth promoters. The effect of plant extracts *in vivo* cuts across improvement in growth performance, antibacterial, antioxidant and immune response modulatory effect with few discrepancies in results obtained from different studies. This experiment was designed to determine the effect of aqueous *V. amygdylina* extract on neutrophil and lymphocyte counts in four weeks old broiler chickens. A total of one hundred and fifty (150) Anak 2000 day-old chicks were randomly distributed to 3 treatments having five replicates of ten birds per replicate. Birds in treatment 1 (control group) were not administered antibiotic or plant extract. Birds in treatment 2 were administered antibiotics while birds in treatment 3 were administered 1.00 ml of aqueous *V. amygdylina* per bird on day 7, 14 and 21 as natural growth promoter. A standard starter diet was supplied ad-libitum for 4 weeks. The experiment was designed as a complete randomised design. Blood was collected into EDTA bottles from 2 birds per replicate to determine, neutrophils and lymphocytes counts. Values obtained were subjected to one-way analysis of variance (ANOVA) using Turkeys post hoc test and significant means ($p < 0.05$) were evaluated using Minitab v.17 software.

Neutrophil and lymphocyte counts were significantly ($p < 0.05$) affect by Aqueous *V. amygdylina* administration. Neutrophil count was 52.2±3.30 in birds administered *V. amygdylina* with a value of 43.7±2.98 in birds administered antibiotics and 34.2±5.27 in the control group. lymphocyte counts were significantly ($p < 0.05$) lower in birds administered aqueous *V. amygdylina* (40.7±4.15), compared to the control group (53.3±2.72). this value was numerically lower but not significantly different ($p > 0.05$) from that recorded for antibiotic administration (50.2±1.84).

It can be concluded that a phytogenic extract like Aqueous *V. amygdylina* positively impacts neutrophil and lymphocyte counts in broiler chickens.

Keywords: Broiler chicken, lymphocytes, neutrophil, phytogenics

Contact Address: Ruth Tariebi S. Ofongo, Niger Delta University, Dept. of Animal Science, Wilberforce Island, P.M.B. 071, Yenagoa, Nigeria, e-mail: tariruth@live.de

Opportunities and Hurdles Relating to Full Exploitation of Edible Caterpillars in Africa

Elizabeth Kusia[1], Sevgan Subramanian[1], Chrysantus Mbi Tanga[1], Christian Borgemeister[2]

[1]*International Centre of Insect Physiology and Ecology (ICIPE), Kenya*

[2]*University of Bonn, Center for Development Research (ZEF), Ecology and Natural Resources Management, Germany*

Edible caterpillars form an important part of traditional diets in sub-Saharan-Africa. The caterpillars consumed mainly belong to families Saturniidae, Notodontidae, Noctuidae and Sphingidae. They are a rich source of proteins, vitamins, fats and minerals. Additionally, edible caterpillars provide a source of income for rural communities mainly in western (e.g. Nigeria), central (e.g. Congo) and southern (e.g. Zimbabwe) Africa. However, this resource is underexploited owing to several challenges. We seek to unpack the opportunities, challenges and possible channels to maximise the utilisation of edible caterpillars in Africa. Caterpillars consumed in Africa are mainly harvested from the wild, which is unsustainable, often seasonal, and raises major conservation concerns due to overharvesting and habitat destruction. Harvesting from the wild, processing and packaging is mostly done by women which is tedious and time consuming. Production in captivity could be an effective way to curb this problem. We focus on rearing opportunities, possible challenges facing mass-rearing (for example entomopathogens, parasitoids, pupal and egg diapause and unsynchronised emergence of male/female moths) and potential channels to involve communities in their rearing. Poor storage, sanitation and handling also leads to bacteria and fungi contamination of the caterpillars raising food safety concerns. This calls for improved rearing, harvesting, processing, storage and handling methods to enhance the quality of the end product. Women and youths can be trained on more efficient methods of value addition and packaging of the product to fetch better prices. Finally, we discuss existing government policies supporting edible caterpillars and critical policy gaps for improvement.

Keywords: Edible caterpillars, edible insects, entomophagy, insect consumption, saturniidae

Contact Address: Elizabeth Kusia, International Centre of Insect Physiology and Ecology (ICIPE), P.O. Box 30772, 00100 Nairobi, Kenya, e-mail: ekusia@icipe.org

Feeding Restriction with Cassava Flour on Carcass Composition of Broilers

Youssouf Toukourou, Dassouki Issifou Sidi, Ibrahim Alkoiret Traore
University of Parakou, Fac. of Agronomy, Dept. of Sciences and Techniques of Animal and Fisheries Production, Benin

In order to promote poultry farming in resource-limited rural areas, the effects of feeding restriction with cassava flour on the carcass composition of broilers was studied. After three weeks on a restrictive diet (step 1), the broilers were re-fed during four weeks according to their physiological needs (step 2). In total, 75 four-weeks old chicks were randomly divided into three lots of 25 subjects. Lot I (control) is fed without cassava flour. The lots II and III are fed with diets containing respectively 10 and 30 % of cassava flour, with energetic and protein density of 85 and 70 % of the control. Eight broilers of each lot have been randomly selected and slaughtered at the end of each step. At the end of the restrictive step, the carcass yields and the weights of the digestive tracts are 67.1, 66.3, and 64.7 % and 178.5, 170.0, and 113.3 g respectively for the lots I, II, and III with a significant difference ($p \leq 0.05$) between lot I and III and then between lots II and III. After 4 weeks of re-feeding, the lots I, II, and III had respectively 69.9, 73.2, and 67.7 % of carcass yield as well as digestive tract weights of 178.3, 180.8, and 156.0 g. The carcass yield had been entirely made up ($p \geq 0.05$) to the broilers previously submitted on a restrictive diet. However, the weight of the empty cold carcass was not fully compensated ($p \leq 0.05$). It appears from the present study that cassava flour can partially replace maize corn under alternating restrictive feeding conditions in broilers.

Keywords: Broiler, carcass productivity, cassava flour, compensatory growth

Contact Address: Youssouf Toukourou, University of Parakou, Fac. of Agronomy, Dept. of Sciences and Techniques of Animal and Fisheries Production, Box 123, Parakou, Benin, e-mail: ytoukourou@gmail.com

Impact of Collective Marketing of Local Poultry Breeds on Smallholder Farm Incomes in Eastern Kenya

David Michael Ayieko[1], Eric Bett[2]

[1]*University of Nairobi, Agricultural Economics, Kenya*

[2]*Kenyatta University, Agribusiness Management and Trade, Kenya*

Agriculture is a viable exit strategy from the poverty cycle in Kenya, where it contributes to 23 % of the gross domestic product (GDP). Moreover, collective marketing in agriculture has the potential to reduce on transaction costs and mitigate on marketing risks, which reduce on farm incomes. However, less than 20 % of households in Eastern Kenya are currently participating in collective marketing of the farm produce. Therefore, the objective of this study was to determine the factors that determine participation in collective marketing of local poultry breeds and its impact on farm income and gross margins in Eastern Kenya. A random sample of 237 households was obtained from Wote, Kee and Kaiti divisions within Makueni County. First, we randomly picked three districts and a division from each of these districts. Subsequently, we collected data using pretested questionnaires and from focus group discussions and key informant interviews. Thereafter, we used a logit regression, propensity score matching (PSM) and average treatment effect on treated (ATT) to estimate effects of socioeconomic factors on collective marketing and its impact on gross margin. The results indicate that participation in collective marketing is significantly affected by education of household head (0.21)**, frequency of contact with extension worker (0.02)**, distance to nearest market(-0.11)**, distance to all weather roads (-0.17)**, source of labour(-0.15)** and marketing channel(-0.08)**. Moreover, participants in collective marketing improved gross margins by Ksh 2125.36*** (US$20), compared to the non- participants in collective action. Therefore, to enhance participation in collective action farmer field schools and open field days should be used to promote collective action among smallholder in Makueni County. Furthermore, innovative strategies such as subsidised extension services should be adopted. Basic infrastructure such as markets and all weather roads should be improved to improve participation in collective action. There is also need to invest in training of labour and promotion and adoption of efficient and reliable sources of labour. Finally, shorter and efficient marketing channels should be promoted among smallholder to improve on their participation in collective marketing. Therefore, Government policy, specifically on rural development and extension, should focus on consolidating smallholder farmers into commodity marketing groups.

Keywords: Breeds, collective, propensity, marketing, margins

Contact Address: David Michael Ayieko, University of Nairobi, Agricultural Economics, P.O. Box 29053-00625, 00200 Nairobi, Kenya, e-mail: ayiekodav@gmail.com

Assessment of Rabbit Production Value Chain in South-West Nigeria

Esther Okpakpor[1], Lydia Adeleke[2], Muyiwa Adegbenro[1], Gbenga Onibi[1]

[1]*Federal University of Agriculture Akure, Dept. of Animal Production and Health Technology, Nigeria*

[2]*Federal University of Technology Akure, Dept. of Fisheries and Aquaculture Technology, Nigeria*

Rabbit farming is an aspect of mini-livestock production that has the ability of increasing the income level, providing alternative livelihood and supplying animal protein for man. This study assess the value chain of rabbit production in South West, Nigeria. Both primary and secondary data were used. Interview schedule and structured questionnaire were administered on one hundred and five (105) respondents to get relevant information on feeding, medication and marketing of rabbits in the study area. Data collected were analysed descriptively using SPSS 16.0 package. The result from this study showed that 93 % of the respondents used forages, 56 % used concentrates while 63 % used kitchen waste. Ten percent (10 %) of the respondents used commercial rabbit feeds, 17 % formulated rabbit feeds on their own while 31 % of the respondents used poultry feed for rabbits. Most of the farmers (53 %) sell directly to the final consumers while 15 and 26 percent of the respondents reported activities of wholesalers and retailers respectively. About 72 % of the respondents sell live rabbit while only 21 % process rabbits into meat before sales. Other by-products sold by 8 % respondents include rabbit feaces and urine. The value chain description of rabbit production in the study, exhibit all the levels: primary, secondary and tertiary/points of distribution chains. This study therefore reveals the need to encourage livestock feed producers to invest in rabbit feed production to reduce the use of poultry feed in rabbit production. Likewise more work needs to be done with respect to tropical forage production and preservation for feeding rabbits especially in the dry season.

Keywords: Feed, forage, Nigeria, rabbit, value chain

Contact Address: Esther Okpakpor, Federal University of Agriculture Akure, Dept. of Animal Production and Health Technology, Akure, Nigeria, e-mail: estherugochi@gmail.com

Reproductive Performance of Rabbits Fed Diets Containing Varying Dietary Levels of two Varieties of Composite Sweet Potato Meal in A Palm Kernel Based Diet

Ibikunle Olaleru[1], Ahmed Abu[2]

[1]*National Root Crops Research Institute, Farming Systems Research Program, Nigeria*

[2]*University of Ibadan, Animal Science Department,*

Twenty-five rabbit does of mixed breeds (New Zealand White, California and Chinchilla) aged between 6-7 months were randomly allotted to five experimental diets; T1 as control, T2, T3, containing 25 and 50 % of orange flesh CSPM, and T4, and T5 containing 25 and 50 % white flesh CSPM in a completely randomised design to evaluate the reproductive performance of rabbits fed two varieties of composite sweet potato (*Ipomoea batatas*) meal (CSPM). The diets contains 10.6-12.6 % crude fiber, 16.4-17.6 % crude protein and 2610-2788 Kcal kg^{-1} metabolisable energy.

There was no significant difference in the total feed intake of does on the different treatments ($p > 0.05$). The T5 (50 % of white CSPM) does had the highest daily feed intake (103.75 g day^{-1}) followed by T1 (0 % of CSPM) does (103.45 g day^{-1}) and T4 (25 % of white CSPM) does (103.20 g day^{-1}), while T3 (50 % of orange CSPM)and T2 (25% of orange CSPM) does had the lowest values of (99.65 g day^{-1}) and (97.95 g day^{-1}) respectively. T4 and T5 does had the highest litter size at birth of 5.00 (though not significantly different) from the other treatments. There was no significant difference in the initial average body weight and gestation length of the does on the different treatments as well as in the litter weight at birth.

Litter size at weaning was highest in T4 and T3 with values of 3.00 and 2.80 respectively. Litter weight at weaning was highest in T1 and T3 groups recording a weight of 569.9 and 553.0 g respectively. The average daily weight gains per kit were 11.14, 8.56, 10.83, 10.84 and 10.28 g day^{-1} for T1, T2, T3, T4 and T5 treatments, respectively. There were no significant differences ($p > 0.05$) in average daily weight gain per kid and in milk yield across all the treatments. Milk yield was higher in T1, but not significantly different against T2, T3, T4 and T5 does. In conclusion composite sweet potato meal can replace maize without adverse effect on the reproductive performance of rabbits.

Keywords: Composite sweet potato meal, kits, rabbits doe, reproductive performance, survivability

Contact Address: Ibikunle Olaleru, National Root Crops Research Institute, Farming Systems Research Program, Umudike, 100265 Umuahia, Nigeria, e-mail: olaleru.ibikunle@gmail.com

Economics of Production of Weaner Rabbits Fed Graded Levels of two Varieties of Composite Sweetpotato (*Ipomoea batatas* [L.] Lam) Meal in A Palm Kernel Based Diet

IBIKUNLE OLALERU[1], AHMED ABU[2]

[1]*National Root Crops Research Institute, Farming Systems Research Program, Nigeria*

[2]*University of Ibadan, Animal Science Department, Nigeria*

Whole sweetpotato (*Ipomoea batatas* [L.] Lam) plant plays a significant role in crop-livestock farming systems in Africa. In this study the economic efficiency of inclusion of orange fresh (CIP 440293) and white flesh (TIS 87/0087) varieties of composite sweetpotato meal- CSPM (65 % root and 35 % leaf and vine) as replacement for maize and their economic implications for sustainable rabbit production in South western Nigeria.

Eighty-four young doe of mixed breeds aged between 6–8 weeks, weighing between 550–600 g were used to determine the effect of feeding levels of two varieties of composite sweetpotato (*Ipomoea batatas*) meal on the profitability of rabbit production. The rabbits were randomly allocated to seven treatments T1-T7. T1 as control (without sweetpotato), T2, T3, T4, containing 25, 50 and 75 % of CIP 440293 (orange flesh), respectively, and T5, T6 and T7 containing 25, 50 and 75 % TIS 87/0087 (white flesh), respectively CSPM with four replications per treatment in a complete randomised design (CRD). The diets contains 10.6–12.6 % crude fiber, 16.4–17.6 % crude protein and 2610 - 2788 Kcal Kg^{-1} metabolisable energy. The diets were formulated to meet the nutrient requirements of growing rabbits.

There was significant different across the dietary treatment. The research showed that the diet affected the economics of production such that feeding dietary treatments containing the CSPM compares well in terms of cost of feed to gain 1 kg weight in the rabbit for nine weeks i.e. T1 (₦380.00 kg^{-1}), T2(₦325.81 kg^{-1}), T3(₦335.82 kg^{-1}), T4(₦259.55 kg^{-1}), T5(₦266.56 kg^{-1}), T6 (₦326.55 kg^{-1}) and T7 (₦275.76 kg^{-1}). This research shows that both varieties of CSPM can serve as a substitute for maize in female rabbits diets up to 75 % and would the help save as much as 20 % cost of production, hence positively affects the economics of production parameters. The usage of the sweet potato composite meal will help to reduce cost, scarcity of rabbit feeds and also reduce dependency on conventional feed ingredients such as grains that bring competition between man and animals.

1$= ₦360

Keywords: Composite sweet potato meal, economic variables, profit margin

Contact Address: Ibikunle Olaleru, National Root Crops Research Institute, Farming Systems Research Program, Umudike, 100265 Umuahia, Nigeria, e-mail: olaleru.ibikunle@gmail.com

Ruminant systems

Oral Presentations

Posters

Differences in Carbon Footprint of Dairy Production Systems in the Rural-Urban Interface of Bengaluru, India

Marion Reichenbach[1], Anjumoni Mech[2], Ana Pinto[3], Pradeep K. Malik[2], Raghavendra Bhatta[2], Sven König[3], Eva Schlecht[1]

[1] *University of Kassel / Georg-August-Universität Göttingen, Animal Husbandry in the Tropics and Subtropics, Germany*

[2] *National Institute of Animal Nutrition and Physiology, Bengaluru, India*

[3] *Justus-Liebig-University Giessen, Inst. of Animal Breeding and Genetics, Germany*

The global environmental impact of dairy production systems (DPS), as measured by their greenhouse gas emission, varies according to productivity level and feeding strategy. In the (sub-)Tropics, the feeding strategy of dairy producers is affected by shifts in resource availability due to an urbanizing environment. India, where milk is a major source of animal protein, hosts the largest dairy herd worldwide and is rapidly urbanizing. Therefore, the aim of this study was to quantify emission intensity of four DPS (ubiquitous extensive DPS-1, rural (semi-)intensive DPS-2, DPS-3 and DPS-4; intensification level based on feeding (use of pasture, forages self-cultivation) and breeding strategy (cattle flows, specialised dairy genotypes) coexisting within the rural-urban interface of the Indian megacity of Bengaluru. Feedstuff offer and offtake of energy corrected milk (ECM) were quantified on 24 dairy production units (DPU), six per DPS, at 6 week intervals for one year (8 rounds). Following the 2006 IPCC Guidelines (https://www.ipcc-nggip.iges.or.jp/public/-2006gl/) the calculated carbon footprint (CF) included methane and nitrous oxide emissions due to enteric fermentation and manure management system. It was expressed in carbon dioxide equivalents (CO_2-eq) per DPU and round. Results show that the CF of Bengaluru's DPS differs with quality of diet offered and feeding intensity: emissions were low in DPS with low feeding intensity and good (DPS-1) or average (DPS-2) diet quality. By opposition, emissions were high in DPS with high feeding intensity and an average quality diet (DPS-3 and DPS-4). Overall, methane contributed to 92% of the emissions and nitrous oxide 8 %. Milk offtake (kg ECM DPU^{-1} round-1) was low in DPS-2 (714), intermediate in DPS-1 (794) and high in DPS-3 (1260) and DPS-4 (1175). The CF (kg CO_2-eq per kg ECM), was lowest for DPS-1 (0.91) whereas significantly higher values ($P < 0.05$) were obtained for DPS-2 (1.21), DPS-3 (1.95) and DPS-4 (1.52). The results show that differences in local feeding strategies as shaped by the urbanisation level of a farming system's immediate neighbourhood impact emission intensities. Although they co-exist within the same geographical space of Greater Bengaluru, the individual dairy production practices of local DPU have distinctly different global environmental impacts.

Keywords: Dairy production, emission intensity, environmental impact, India, urbanisation

Contact Address: Marion Reichenbach, University of Kassel / Georg-August-Universität Göttingen, Animal Husbandry in the Tropics and Subtropics, 37213 Witzenhausen, Germany, e-mail: marion.reichenbach@uni-kassel.de

Genetic Characteristics of African Breeds: An Example of Multi-Breed GWAS for Morphometric Traits in Beninese Indigenous Cattle

Sèyi Fridaïus U. S. Vanvanhossou[1], Carsten Scheper[1], Tong Yin[1], Luc Hippolyte Dossa[2], Kerstin Brügemann[1], Sven König[1]

[1]*Justus Liebig University Giessen, Inst. of Animal Breeding and Genetics, Germany*
[2]*University of Abomey-Calavi, Fac. of Agricultural Sciences, School of Science and Technics of Animal Production, Benin*

In the African livestock breeding context, routine data recording systems for performance traits especially body weight rarely exist. In contrast, morphometric traits are more easily measurable and often correlated to performance traits. They are therefore potential indicator traits for novel breeding and conservation strategies in the large but poorly studied animal genetic resources in Africa. The aim of the present study was to estimate genetic parameters and to identify functional loci underlying the variability of six morphometric traits in four indigenous breeds (Borgou, Pabli, Lagune, Somba) kept in small herds in Benin. In total, 449 cattle with morphometric records for height at withers (HAW), sacrum height (SH), heart girth (HG), hip width (HW), body length (BL) and ear length (EL), were genotyped for 36,720 SNP. This dataset was used to estimate heritabilities, genetic and phenotypic correlations, and to perform a multi-breed GWAS.
Estimated SNP-based heritabilities for the six morphometric traits ranged from 0.46±0.14 (HG) to 0.74±0.13 (HW). The genetic and phenotypic correlations ranged from 0.14±0.10 (HW-BL) to 0.85±0.02 (HAW-SH) and 0.25±0.05 (HW-BL) to 0.89±0.01 (HAW-SH), respectively. The multi-breed GWAS detected two genome-wide and 25 chromosome-wide significant SNP associated with the morphometric traits. The identified SNP were located near (±25kb) or within a total of 15 genes among which 11 genes were related to morphological, growth and carcass traits like body weight and fat deposition in cattle as well as in other livestock species and humans. The genes were additionally involved in biological processes and pathways such as hemostasis and metabolism (PIK3R6, PIK3R1), immunity or inflammatory responses (PTAFR, LYPD8), chromatin organisation (PBRM1), DNA repair (EYA3), DNA binding (SSH2), and regulation of hepatocyte growth factor receptor signaling pathway (ADAMTS12). The association between all identified genes either with feed efficiency, stress or immune response suggests that adaptability to scarce food resources, disease and extensive managements plays an

Contact Address: Sèyi Fridaïus U. S. Vanvanhossou, Justus Liebig University Giessen, Inst. of Animal Breeding and Genetics, Ludwigstraße 21b, 35390 Gießen, Germany, e-mail: seyi.f.vanvanhossou-2@agrar.uni-giessen.de

important role in the phenotypic variability of the indigenous breeds. With regard to moderate to high heritabilities and the functional annotation for the associated SNP, the studied morphometric traits are suitable indicator trait candidates to develop breeding schemes for genetic improvements of the indigenous breeds.

Keywords: Genetic parameters, indigenous breeds, morphometric traits, multi-breed GWAS

Feasibility of Organic Certification of Sheep and Goats Produced in Pastoral Systems in Northern Kenya

FLORA VON STEIMKER[1], BRIGITTE KAUFMANN[1], HUSSEIN WARIO[2]

[1]*German Institute for Tropical and Subtropical Agriculture (DITSL), Germany*

[2]*Center for Research and Development in Drylands, Kenya*

In pastoral systems of Kenya, sale of small ruminants is the main regular income source of the majority of households. Although the meat of small ruminants produced in Marsabit county is renowned for its taste, up to now, no mechanisms are in place that allow for a respective price differentiation. Establishing value chains for branded sheep and goat meat would be an option to allow pastoralists to profit from their high process and product quality. The aim of the current study is to assess the requirements and feasibility to establish a value chain for branded products (e.g. origin labelled, organic certified) together with the primary and secondary actors of the Rendille pastoral small ruminant supply chain in Marsabit county.

Overall 28 group discussions with four Rendille women producer and marketing groups and 18 interviews with other relevant stakeholders, such as traders, veterinarians, government staffs, civil society organisations and organic certification bodies were conducted.

The results show that pastoralists' production is in large parts compatible with organic guidelines. For instance, livestock is fed on naturally growing fodder with no chemicals applied, and herding allows for animals' natural behaviour. However a number challenges hamper organic certification. These include; practice of ear notching and castrations without anesthesia, supplementary feeding of home-based animals with non-organic feeds during drought, unguided application and lack of records of veterinary drug use, challenges in record keeping and traceability system and financial difficulties to undertake internal control and monitoring due to the extensive nature of the production area. Improvement of veterinary infrastructure, innovative record keeping systems for illiterate producers, and further market studies to ascertain demand are recommended. While organic certification is desirable, labelling for designation of origin is a feasible starting point that would allow entry into higher priced markets as hurdles to organic certification are addressed.

Keywords: Organic cerification, pastoralism, social and policy issues

Contact Address: Flora Von Steimker, German Institute for Tropical and Subtropical Agriculture (DITSL), Hegelsbergstraße 18, 34127 Kassel, Germany, e-mail: fvon-steimker@gmx.de

Livelihood and Production Strategies of Smallholder Livestock Keepers in the Central Peruvian Andes

Maria Wurzinger[1], Marlene Radolf[1], Gustavo Gutiérrez[2]

[1]*University of Natural Resources and Life Sciences, Vienna (BOKU), Dept. of Sustainable Agricultural Systems, Austria*

[2]*National Agrarian University La Molina, Animal Science Faculty, Peru*

The living conditions of smallholders in the Peruvian Andes have always been marked by harsh climatic conditions. Nowadays farmers see their livelihoods more and more challenged by environmental and economic changes. This results in adaptations of their livelihood strategies that can be classified in on-farm production and income-generating strategies.

The objective of the study was to investigate income-generating and production strategies of livestock keepers in the Central Andes of Peru. In addition, farmers´ perception of their own livestock and perceived effects of climate change were investigated.

Therefore, semi-structured interviews were conducted in Spanish language with 46 livestock farmers from the provinces of Pasco and Daniel Carrión.

Most farmers diversify their livestock, as they keep llamas, alpacas, sheep and cattle in different combinations at once. Only a few farmers are specialised and keep alpacas in high numbers. A diversified production strategy can be seen as a means to decrease vulnerability regarding environmental and economic shocks and changes. The main reasons for a change in the herd compositions were due to economic and environmentally caused reasons as a lack of pasture and declining prices for especially sheep wool.

Climate change is strongly seen as a production constraint by all farmers and the ones that can afford it have already tried to cope via the adoption of diverse adaptation strategies. Farmers seem to plan a shift towards a higher number of llamas as they are seen to be more resistant to changing climate. But the market for llama products is small and prices are low, so farmers cannot rely on sufficient incomes by only keeping llamas.

More than half of the farmers work in non-farm activities. This shows that farmers experience a high economic pressure to look for work in other sectors than farm-related ones. For the young generation living in rural areas does not seem to be very attractive.

Therefore, investments in infrastructure, better extension services and capacity-building programs should be taken to support farmers to improve their livelihood. This can help to ensure that farmers are offered a perspective for their future in the High Andes.

Keywords: Andes, diversification, livestock, smallholders

Contact Address: Maria Wurzinger, University of Natural Resources and Life Sciences, Vienna (BOKU), Dept. of Sustainable Agricultural Systems, Gregor Mendel Straße 33, 1180 Vienna, Austria, e-mail: maria.wurzinger@boku.ac.at

Genomic Loci Affecting Somatic Cell Score in a Butana × Holstein Crossbreed Population in Sudan

Salma Elzaki[1,2], Danny Arends[2], Paula Korkuc[2], Gudrun A. Brockmann[2]

[1]*University of Khartoum, Dept. of Genetics and Animal Breeding, Sudan*
[2]*Humboldt-Universität zu Berlin, Albrecht Daniel Thaer Inst. for Agricultural and Horticultural Sciences, Germany*

Butana is an indigenous *Bos indicus* dairy cattle breed in Sudan, which is well adapted to extreme environmental conditions such as low feed intake, low water requirements, high temperature, and tropical diseases and parasites. Most dairy breeding programs aim at the improvement of milk production by selection, which consequently led to decreased udder health and increased genetic susceptibility to mastitis, a frequent and costly udder disease. Thus, genetic selection for mastitis resistance should be considered to maintain healthy and productive cows. As the somatic cell count in milk is increased when cows have mastitis, the somatic cell score (SCS) is frequently used as an indicator trait for mastitis.
Previously, 10 SNPs have been identified to be associated with SCS in German Holstein. In this study, we tested those SNPs in 109 purebred Butana and in 203 Butana × Holstein crossbreed cattle in Sudan. Allele-specific genotyping was performed using KASP assays. Association analysis was performed using linear mixed model implemented in R programming language.
We found that 1 out of the 10 previously reported SNPs was significantly associated with SCS in the Butana × Holstein crossbreed population ($p = 0.0054$). This SNP is located on chromosome X (at position 30341984 bp). Two additional SNPs located on chromosomes 6 and 19 show almost suggestive association with SCS in the same population ($p = 0.1557$ and 0.1423). No association was found between the 10 SNPs and SCS in purebred Butana cattle.
The marker associated with SCS on the X chromosome can be used for genetic improvement of mastitis resistance in Butana × Holstein crossbreed cattle, which are increasingly used in Sudan. The markers on chromosomes 6 and 19 have to be followed up in a bigger population. The genetic improvement of mastitis resistance and selection for lower SCS is consistent with the goal of maximising genetic improvement for total economic merit and should be included in breeding programs.

Keywords: Association analysis, *Bos indicus*, genotyping, mastitis, SNPs

---

Contact Address: Gudrun A. Brockmann, Humboldt-Universität zu Berlin, Breeding Biology and Molecular Genetics, Invalidenstraße 42, 10115 Berlin, Germany, e-mail: gudrun.brockmann@agrar.hu-berlin.de

In vitro Feed Digestibility Using Five Bacteria and their Crude Enzymes Obtained from Cow Rumen

Olusoji Adeyosoye[1], Kehinde O. Awojobi[1], Elizabeth Omokoshi Joel[2]
[1]Obafemi Awolowo University, Animal Sciences, Nigeria
[2]University of Ibadan, Dept. of Animal Science, Nigeria

Access to animal protein is inadequate in developing countries and this could be attributed to poor feed utilisation by animals. A large global supply of meat and milk comes from ruminants several breeds of cattle have been identified in Nigeria. The guts of a cow is a complex system that is responsible for animal's overall nutrition uptake and health because they are loaded with a consortium of microorganisms that allow utilisation of the simplest forms of lignocellulose component of the feed via fermentation for energy generation to drive several metabolic activities. Enzyme technology broadly involves production, isolation, purification and applications, therefore, this study focused on the application of culture of rumen bacterial isolates identified as *Klebsiella edwardsii, Photobacterium damselae, Pseudomonas aeruginosa, Stenotrophomonas matophilia, Burkhoderia cepacia* and their respective amylase, cellulase, pectinase, protease and lipase with specific activities on 24 h *in vitro* digestibility of cow feed concentrate via gas production technique at 3 h intervals. The test feed was analysed for its composition on dry matter (DM) basis. Rate of feed disappearance was also determined while digestibility characteristics such as metabolisable energy (ME), organic matter digestibility (OMD) and short chain fatty acids (SCFAs) were estimated using the standard methods. The results obtained revealed that of *K. edwardsii* had the highest gas volume (36.0 ml/0.5 g DM) and feed disappearance (98.0 %) while the least gas production was recorded for both *P. aeruginosa* and its crude pectinase (1.0 ml/0.5 g DM). The feed composed of 62.32 % carbohydrate (CHO) with estimated ME (8.02 MJ/kg, OMD (47.81 %) and SCFAs (0.80mmol). *K. edwardsii* and combined crude enzymes in this study showed exceptional ability in feed degradation but the actual digestibility was achieved by the synergistic effect of combined species, thus enhancing tissue accretion for increased supply of meat.

Keywords: Bacterial enzymes, feed digestibility, *in vitro* fermentation, rumen microorganisms, tissue accretion

Contact Address: Olusoji Adeyosoye, Obafemi Awolowo University, Animal Sciences, Ife-Ibadan Express Road, 220281 Ile-ife, Nigeria, e-mail: adeyosoye@oauife.edu.ng

Morphological and Genetic Diversity of Domestic Yak (*Bos grunniens*) at High Altitude Rangelands of Gilgit Baltistan, Pakistan

Asif Hameed[1], Eva Schlecht[1], Muhammad Tariq[2], Andreas Buerkert[3], Carsten Scheper[4], Sven König[4], Regina Rössler[1]

[1]*University of Kassel / University of Goettingen, Animal Husbandry in the Tropics and Subtropics, Germany*

[2]*University of Agriculture Faisalabad, Dept. of Livestock Management, Toba Tek Singh, Pakistan*

[3]*University of Kassel, Organic Plant Production and Agroecosystems Research in the Tropics and Subtropics, Germany*

[4]*Justus-Liebig University Giessen, Inst. of Animal Breeding and Genetics, Germany*

In the mountain range of Hindukush and Karakoram at northern Pakistan, yak is an important farm animal for providing milk, meat and other domestic and agricultural needs. However, the genetic studies on this farm animal are rare since it is a neglected animal species in the region. In this study, we examined the morphological and genetic diversity of yak populations involving 120 yaks from six different populations (20 from each; 14 females and 6 males). Body measurements and hair samples were taken in 2018. Thirteen microsatellite loci were used to determine the morphological difference and the level of genetic variation and relationship within and between populations. Heart girth, height at withers, body length, height at rump, horn length, muzzle circumference, tail and hair length, backline and head profile, as well as coat and horn colour differed significantly ($p < 0.01$) between sex and across different populations, while ear length and the distance between ends of horn were similar. The mean value of observed heterozygosity was 0.652 ± 0.02, whereas the mean expected heterozygosity (He) was 0.645 ± 0.01. Average genetic diversity was highest in Chipurson population (He = 0.667) and lowest in Khaplu population (He = 0.610). The results of genetic distance indicated the distance between Shimshal and Khaplu yak population was the most and the least between the populations of Phander and Hoper, the values were 0.2441 and 0.0722 respectively. Besides, three distinct genetic clusters were found. Overall, lower Fis value (-0.026 ± 0.02) indicated individuals in the populations were less related than expected under a model of random mating. Furthermore, highest genetic variation (94 %) was observed within individuals across populations, 2 % of the total genetic variation was among individuals within populations, and only 4 % was among populations, resulting in a relatively low Fst of 0.052. Our work underlines the importance

Contact Address: Regina Rößler, University of Kassel, Animal Husbandry in the Tropics and Subtropics, Witzenhausen, Germany, e-mail: regina.roessler@uni-kassel.de

of yak genetic diversity not only to obtain better understanding of the current domestic yak status but also to illustrate the importance for the development of biological genetic conservation strategies and specific breeding programs in yaks to enhance livelihood opportunities for subsistence level agro-pastorals in Gilgit Baltistan region.

Keywords: Genetic diversity, Gilgit Baltistan, microsatellite, yak

Alternative Feeding Options to Enhance Dairy Farm Sustainability in Bangladesh

MD SHAHIN ALAM[1,4], TIMOTHY J. KRUPNIK[2], SHANJIDA SHARMIN[3], JEROEN C.J. GROOT[4]

[1]*Hajee Mohammad Danesh Science and Technology University, Dairy and Poultry Science, Bangladesh*

[2]*International Maize and Wheat Improvement Center (CIMMYT), Sustainable Intensification Program, Bangladesh*

[3]*Bangladesh Agric.. Univ., Agric. Econ. and Rural Sociol., Bangladesh*

[4]*Wageningen University and Research, Farming Syst. Ecology - Plant Sci., The Netherlands*

Mixed-crop dairy farming systems are important contributors to income and nutrition security in Bangladesh. Dairy cattle production is however characterised by poor integration with primary farming systems, low productivity, mainly due to genotypic and feeding constraints. The latter is due largely to limited on-farm fodder production and dependency on external feed items. External feed costs are high, undermining farmers' potential profitability. When cattle and crop production are not tightly coupled, negative environmental externalities can result. Reconfiguration of non-integrated farming practices are needed to support increased appropriate feed production, soil organic matter (OM) balances, farm income, and to reduce the feed costs and off-farm externalities. Here, we investigated alternative feeding options for large ($\geq$ 10 cows), medium (4–10 cows) and small ($\leq$ 3 cows) integrated crop-livestock farm types to enhance economic and environmental performance in Jhenaidha and Meherpur districts in Bangladesh. Following focus group discussions from February to May 2018 with three farms in each size category, we considered baseline data from one representative average from each group for modelling in the software package FarmDESIGN. We optimised the operating profit, feed costs, OM balance, Soil N losses and self-reliance. In the results, we found the highest operating profit in Meherpur for medium-sized farms, as well as lowest feed costs per kg dry matter. Examining risks of soil N losses, OM balance and feed self-reliance (increasing the on-farm feed production), maximum improvements over baseline farm scenarios came from small farms in Meherpur, and large and medium-sized farms in Jhenaidha, respectively. Trade-offs and synergies between objectives were found. Increasing feed self-reliance reduces potential N losses, though this was also associated with lower operating profit and soil OM balances. Despite existing trade-offs, we found opportunities to improve economic and envi-

Contact Address: Md Shahin Alam, Hajee Mohammad Danesh Science and Technology University, Dairy and Poultry Science, Dinajpur, Bangladesh, e-mail: shahindps@hstu.ac.bd

ronmental performance simultaneously. It appears that feed self-reliance can be increased by intensifying cropping and substituting fallow periods with appropriate fodder crops. Integrated analysis of alternative feeding systems for sub-tropical dairy farms in Bangladesh can contribute and enhanced understanding of the combined economic and environmental benefits in mixed crop-dairy production systems, thereby overcoming apparent trade-offs and offering farmers a suite of management options to explore.

Keywords: Dairy farming, economic, environment, feeding strategies, optimisation, sustainability

Digestibility and Metabolisable Energy Intake Equations of Tropical Ruminant Forages Using Nutrient Concentration of Cattle Faeces

Alice Onyango, John Goopy

International Livestock Research Institute (ILRI), Kenya

Smallholder farmers of sub-Saharan Africa feed livestock almost exclusively on natural native pastures and forages whose nutritive value is often unknown or highly variable. This undermines optimisation of nutrition management. Accurate methods of determining the digestibility of feedstuffs is via *in vivo* experiments which are resource intensive. Substitute *in vitro* or *in sacco* methods have limitations like lack of surgically altered animals and instrumentation. Prediction equations based on chemical composition of feeds and faeces regressed against *in vivo* digestibility and metabolisable energy intake (MEI) values represent a viable alternative that is fast, cheap, routine and can be used under resource-constrained circumstances often found in developing countries. This study aimed at developing equations to predict apparent dry matter digestibility (DMD), apparent digestible organic matter in dry matter (DOMD) and MEI of tropical feedstuffs using proximate composition of forages and cattle faeces from three *in vivo* studies (n=42 steers) conducted at the International Livestock Research Institute, Kenya between 2014 and 2018. Faecal chemical composition from two studies were regressed against DMD, DOMD, and MEI. The third study provided a validation dataset.

Digestibility equations using combined dataset showed poor fit (r^2=0.05–0.10) and weak correlations (r=0.3–0.4) between predicted and actual values but had low prediction errors (PE=3–5 %) while equations of MEI showed moderate fit and correlation (r^2=0.4–0.5; r=0.7) but relatively high PE (i.e. 22 MJ/day). In contrast, equations developed using individual datasets separately had better fit, higher correlation, and lower PE. Using more datasets with varied diets and more test animals may improve these equations. Best predictors of digestibility and MEI were faecal dry matter, crude protein and fibre fractions in varied combinations. Analysis of these parameters is simple, cheap and routine.

Ideally, prediction equations are developed when feeding a balanced ration at maintenance level. However, prediction equations based on a large database of *in vivo* animal experiments with sub-optimal diets, feeding levels and quality may better reflect the prevailing conditions in smallholder farming. This study is a first step towards development of such a database in order to propose digestibility and MEI prediction equations for smallholder feeding situations.

Keywords: Digestibility, forages, metabolisable energy intake, prediction equations, ruminant, sub-Saharan Africa

---

Contact Address: Alice Onyango, International Livestock Research Institute (ILRI), Mazingira Centre, Old Naivasha Road, P.O Box 30709, 00100 Nairobi, Kenya, e-mail: onyangoalice@gmail.com

Arachis pintoi: Potential for Risk Reduction/Productivity Increase in Livestock Systems of the Colombian Orinoquía Region

Karen Enciso Valencia[1], Alvaro Rincón-Castillo[2], Alejandro Ruden[1], Stefan Burkart[1]

[1]*International Center for Tropical Agriculture (CIAT), Tropical Forages, Colombia*
[2]*The Colombian Agricultural Research Corporation (AGROSAVIA), Colombia*

In parts of the foothills of the Colombian Orinoquía region, bovine livestock production takes place on poorly drained soils. The region is dominated by extensive grazing system with *Brachiaira humidicola* cv. Humidicola. This grass has shown high adaptation potential under temporal waterlogging conditions. However, despite maintaining high yields in terms of biomass, its nutritional quality is low. In many cases, inadequate management practices and low soil fertility result in degradation. As a result of this, a lack of feed is a major constraint in the Orinoquía, particularly during dry season. According to climatic projections for the region, annual precipitation and maximum temperatures will increase and this will negatively affect quantity and quality of forages and increase waterlogging risks. Against this background, AGROSAVIA, in 2013, started agronomic evaluations of forage legumes. They selected *Arachis pintoi* CIAT 22160 (Arachis) as promising alternative for livestock production on soils with waterlogging problems. It showed good agronomic behaviour in terms of nutritional quality, persistence and compatibility with grasses such as Humidicola. Based on the agronomic evaluations, this study assesses milk profitability in the foothills of the Colombian Orinoquía from an economic perspective. We compared two production systems: T1, the association Arachis – Humidicola and T2, Humidicola as monoculture. We used a cashflow model and risk assessment under a Monte-Carlo simulation model to estimate economic indicators. The projections for economic returns consider changes in forage characteristics for both treatments, resulting from variations in the projected climatic variables under different climate change scenarios for the region RCP (2:6, 4.5, & 8.5). The LIFE-SIM model was used to simulate dairy production according to forage production, animal information and environmental characteristics. Results show that T1 incremented animal productivity by 11 %, among others. T1 also results in better grass persistence due to higher nitrogen (N) levels in soil resulting from the association with the legume. The legume also provides positive impacts on soil structure and composition. This helps improving the adaptation capacity of the system. Finally, producer incomes could increase as a result of lower vulnerability to (climate) risks and reduced production costs (due to lower N fertiliser use).

Keywords: Adoption, *Arachis pintoi*, climate change, legumes, Monte-Carlo simulation, risk analysis

Contact Address: Stefan Burkart, International Center for Tropical Agriculture (CIAT), Trop. Forages Program, Km 17 Recta Cali-Palmira, Cali, Colombia, e-mail: s.burkart@cgiar.org

Managing Feed Use Efficiency in Peri-Urban Dairy Herds in Pakistan

Muhammad Tariq[1], Andreas Buerkert[2], Eva Schlecht[3]

[1] *University of Agriculture Faisalabad, Dept. of Livestock Management, Toba Tek Singh, Pakistan*

[2] *University of Kassel, Organic Plant Production and Agroecosystems Research in the Tropics and Subtropics, Germany*

[3] *University of Kassel / University of Goettingen, Animal Husbandry in the Tropics and Subtropics, Germany*

Dairy farming is an important activity for many urban dwellers in Asia who try to serve the increasing urban demand for milk of the growing urban population. During 12 months, data on demographic events, types and amounts of feeds offered, milk offtake and body weight changes were collected in 15 mixed buffalo and cattle dairy herds in Faisalabad, third largest city of Pakistan. The farms were semi-commercial small-scale mixed (SSM), semi-commercial small-scale dairy (SSD) and commercial small-scale dairy (CSD); their animals were mainly stall fed. Regularly collected samples of feeds offered were analysed for their nutrient composition. The offer of feed dry matter and crude protein were significantly different ($p < 0.05$) between the three production systems across the four seasons of a year. The overall average body weight of female adult buffaloes and cattle was 579 kg (480–807) and 434 kg (339–630), respectively. Daily milk production (corrected to 4 % fat) per animal was 13.5 liters in buffaloes and 8.1 liters in cattle; while milk extracted for sale from buffaloes varied ($p < 0.05$) between seasons for the SSD and CSD production system, this was not the case in cattle. On a yearly basis, buffaloes received approximately 0.6 kg and 0.7 kg DM less roughage feed per day on SSM and CSD farms than on SSD farms. Gross margin of selling milk and occasionally young stock and culled females was higher on SSM and CSD farms than on the resource poor SSD farms, whereby the only variable costs accounted for were those of feed. It was concluded that more efficient feed utilisation in the studied dairy production systems is possible through separate feeding of groups of buffaloes and cattle, respectively, according to physiological and productive needs, and countering feed shortage periods with adopting silage making.

Keywords: Buffalo, cattle, energy balance, gross margin, milk production

---

Contact Address: Eva Schlecht, University of Kassel / University of Goettingen, Animal Husbandry in the Tropics and Subtropics, Steinstraße 19, 37213 Witzenhausen, Germany, e-mail: tropanimals@uni-kassel.de

Selfmedicative Behaviour in Gastrointestinal Parasite Infected Goats: Shift in Preferences for Tanniferous Fodder Plants

Marvin Heuduck[1], Christina Strube[2], Katharina Raue[2], Eva Schlecht[3], Martina Gerken[1]

[1]*Georg-August-University Goettingen, Dept. of Animal Science, Germany*

[2]*University of Veterinary Medicine Hannover, Inst. for Parasitology, Germany*

[3]*University of Kassel / University of Goettingen, Animal Husbandry in the Tropics and Subtropics, Germany*

Gastrointestinal nematode infections are a worldwide major cause of economic losses and a thread to pasture-based livestock due to their adverse effects on animal health and productivity. Especially regarding the trichostron-gyles, the most prevalent nematodes in ruminants, the excessive usage of conventional anthelmintic drugs over the last decades has led to an emergence of resistant nematode populations, and decreased the efficacy of available anthelmintics. This underlines the relevance of a paradigm shift towards a more sustainable and unconventional approach to control nematode infections. This study evaluated a possible change in feed preferences of trichostrongyle-infected Boer goats and hypothesised that infected goats increase their intake of tannin-containing fodder plants. Feed preferences of eighteen juvenile male Boer goats (3–4 months) were evaluated over a period of 12 weeks. The individually fed animals were assigned to three treatment groups: I) Non-infected + feed choice II) Infected + feed choice, and III) Infected without feed choice. At the beginning of the trial, all goats were free of trichostrongyles. After four weeks, group II) and III) were experimentally infected with third-stage trichostrongyle larvae representing the local species spectruM. Group I and II animals were offered a free choice cafeteria trial for 30 min prior to the usual daily feeding time. Four fodder plants (pelleted leaves of sainfoin, willow, walnut, blackberry) of varying tannin contents (low to high) and tannin-free hay pellets were simultaneously offered. The position of the pellets in the trough was randomised each day. During the cafeteria trial plant preferences were recorded by video surveillance and amounts of ingested pellets were measured. In addition, blood parameters, saliva composition and feces were analysed on a weekly basis in order to evaluate the course of infection and potential shifts in feed preferences due to changes in taste perception. The cafeteria trial revealed a shift from tannin-free (hay) and low tannin-containing feed (sainfoin) to plants with higher tannin-contents (walnut, blackberry) with proceeding trichostrongyle infection.

Keywords: Anthelmintic drug agents, condensed tannins, goats, nematodes, selfmedicative behaviour

Contact Address: Marvin Heuduck, Georg-August-University Goettingen, Dept. of Animal Science, Albrecht Thaer Weg 3, 37075 Goettingen, Germany, e-mail: marvin.heuduck@uni-goettingen.de

Why Small Ruminants Breeding and Meat Production Is a Challenge for Local Farmers in Attock Region of Pakistan

Muhammad Ali Mian, Daniela Lukešová

Czech University of Life Sciences Prague, Fac. Tropical Agrisciences, Dept. of Animal Science and Food Processing, Czech Republic

The Asian continent contributes 61 % of the total mutton and goat meat produced in the world when China produces 54 % and Pakistan ranks third in Asia for mutton and goat meat production after India. Sheep and goat production is one of the major economic activities under the arid and semi-arid condition of Pakistan, with census: 53.8 million of goats and 26.5 million of sheep. There are four main systems of production of livestock in Pakistan but still the local farmer is facing huge challenges in managing the benefits of breeding, rearing of goats and sheep. The research study deals with the assessment of factors affecting small ruminant's production in selected breeding farms of Pakistan in region Attock and evaluation of the socio-economic and marketing situation in this region. The data will be collected through self-administrated consumer questionnaire survey in the particular area and experimental observation of goat species rear under control condition importantly the cost analysis, husbandry practices, feeding and breeding of animals will be recorded and the data will be compare with survey conducted in selected areas. The technical data will be shared with local farmers to enhance production activities under present circumstances. The study will answer the challenges currently facing by farmers. The data will be published by comparing results. Due to multifactorial analysis, data obtained from farmers and markets, will be evaluated on basic food safety measures, which are a prerequisite for ensuring the maintenance of sustainable and safe meat production and adequate levels of Pakistani population well-being.

Keywords: breeding, feeding, husbandry practices, interview schedule

Contact Address: Muhammad Ali Mian, Czech University of Life Sciences Prague, Fac. Tropical Agrisciences, Dept. of Animal Science and Food Processing, F 101 Koleje Fgh. 16521 Prague 6- Suchdol., 160 00-169 99 Prague, Czech Republic, e-mail: mohammadalimian18@gmail.com

Effect of Feeding Graded Levels of *Moringa* Feed on Intake, Digestibility, Enteric CH_4 Emission, Rumen Fermentation, Milk Yield and its Quality of BLRI Cattle Breed-1 Dairy Cows

MUHAMMAD KHAIRUL BASHAR, NATHU RAM SARKER

Bangladesh Livestock Research Institute (BLRI), Animal Production Research Institute (APRD), Bangladesh

An experiment was carried out to evaluate the effect of feeding graded level of *moringa* feed on intake, digestibility, rumen fermentation, methane (CH_4) production, milk yield and its quality.

Fifteen BLRI Cattle Breed (BCB-1) dairy cows of third or fourth parity after wk 3 and 4 of calving were selected and divided into three dietary groups having five animals in each considering their live weight and ex-entry daily milk yield. A group of cows fed a control diet (T0) consisting 1:1 dry matter (DM) of napier silage and conventionally mixed concentrate. The other two groups were fed the control diet randomly replacing i) 50 % (T1) or ii) 100 % (T2) of its concentrate by *moringa* feed. All the three diets were iso-nitrogenous and formulated to supply daily energy and crude protein (CP) requirement of the cows according to BSTI standard.

The replacement of concentrate mixture by *moringa* feed to increase the feed efficiency and to reduce the DM or CP intake ($p < 0.05$) per 100 kg or metabolic body weight. The fresh milk and 4% fat corrected milk (FCM) yield were significantly higher ($p < 0.05$) in T2 group (4.39 kg d^{-1} and 4.59 kg d^{-1}) compared to T0 group (3.30 kg d^{-1} and 3.49 kg d^{-1}), respectively. It also revealed that the total volatile fatty acid (VFA's) concentration was increased ($p < 0.05$) and decreased the blood and milk cholesterol and ammonia-nitrogen (NH_3-N) when *moringa* feed was added in the control diet; without showing any significant($p > 0.05$) change in CH_4 production, fat, solid not fat (SNF), lactose or protein content of milk.

Therefore, *moringa* feed increased the productivity of dairy cow, replacing the whole concentrate diet.

Keywords: Digestibility, intake, milk production, milk quality, *moringa* feed, rumen environment

Contact Address: Muhammad Khairul Bashar, Bangladesh Livestock Research Institute (BLRI), Animal Production Research Institute (APRD), Dhaka, Bangladesh, e-mail: kbashar20@yahoo.com

Integrating Livestock into Agricultural Systems for Increased Livestock Productivity: Evidence from Smallholder Dairy Farmers in Babati District, Tanzania

Ben Lukuyu[1], Kevin Maina[2], Leonard Marwa[3], Alphonce Haule[4], Mateete Bekunda[5]

[1]*International Livestock Research Institute (ILRI), Feeds and Forages Program, Uganda*
[2]*International Livestock Research Institute (ILRI), Kenya*
[3]*Tanzania Livestock Research Institute (TALIRI), Tanzania*
[4]*Babati District Council, Livestock and Fisheries, Tanzania*
[5]*International Institute of Tropical Agriculture (IITA), Tanzania*

Smallholder farming system is characterised by production of forage and fodder in sub-optimal levels integrated with other aspects of agricultural production. By growing and utilising greater quantities of locally produced high quality forages, livestock production costs can be reduced without compromising productivity, thus increasing on-farm sustainability. Introducing improved forages into small-scale mixed farming systems would reduce the competition for land by utilising the same land for both crop and forage production. Smallholder farmers in Babati District keep an average of 3–4 heads of cattle per household. A feed assessment (FEAST) survey conducted in 2015 identified the availability of adequate feeds in terms of quantity and quality as one of the factors constraining smallholder dairy production. The current study tested the impact of feeding improved Napier grass and maize stover supplemented with bean haulms at different levels on milk yield under smallholder conditions. Using field trial experimentation, the study considered 2 genotypes (local and improved cattle) and 2 basal rations of Napier grass and maize stover supplemented with bean haulms at different levels (100, 80, 70 & 60 %). Using farms as experimental units and lactating cows as replicates, a total of 24 cows in early lactation were selected from two villages (Hysum & Bermi). Data was collected for 45 days with 7-day adjustment. Results from a regression analysis indicate that Napier grass supplemented at 70:30 & 60:40 levels yielded significantly higher milk output compared to other levels (100, & 80:20). Similarly, average increase in milk production was significant for Maize stover ration only at levels of 80:20 and 60:40. Other factors that significantly influenced increased milk production were frequency of feeding, water intake and the person feeding the animal (man). Feed supplementation with crop residues at different levels yields to increased milk productivity. However, this is not achievable in isolation. Therefore, there is an opportunity to promote improved forages and integration of crop residues in dairy production. There is need to upscale the technology and promote good dairy/livestock production practices for improved livestock productivity.

Keywords: Crop residues, feed supplementation, livestock productivity, sustainability

Contact Address: Kevin Maina, International Livestock Research Institute (ILRI), P. O. Box 30709-00100, Nairobi, Kenya, e-mail: mainakevin.km@gmail.com

Farming Crops to Increase Cattle Herd Sizes in a Sedentary System of Southwestern Uganda

Pius Nina[1,2], Patrick Van Damme[2]

[1]*Uganda Martyrs University, Fac. of Agriculture, Uganda*
[2]*Ghent University, Dept. of Plant Production - Lab. for Tropical Agronomy, Belgium*

This study investigates whether or not crop cultivation in semi-arid areas provides insurance for pastoral households to increase cattle herd sizes and hang on being pastoralists in a sedentary system. Several studies of pastoral livelihoods across Africa have reported diversification as a strategy sedentary pastoralists use to supplement households' income. Some of these studies have addressed the contribution of crops to pastoralist household food security. However, until now data remains scanty on how dry land cultivation and sales of crop surplus in particular might leverage pastoral cattle stocking. This study also provides the first evidence on diversification trends and contribution to incomes and livelihoods of sedentary pastoralists in southwestern Uganda, compared with diversification trends in other pastoralist societies and African rural settings in general. Data was collected from 366 pastoral households in southwestern Uganda, and analysed for variables such as crop types, household food situation and cattle herds owned in the past five years. Results show that crop farming makes pastoralist households more food secure and significantly reduce reliance on cattle stock off-take for petty cash. Besides cultivating crops, pastoralists also undertake other non-farm income activities to insure themselves against losses associated with livestock production. Diseases, low milk production and market failure for livestock products such as milk were perceived as major threats to obtaining income from livestock production. Furthermore, soaring food prices were also cited as a motivating factor for crop cultivation. Wealthy pastoralists (>100 cattle) cultivate crops largely for food to avoid stock-off-take in order to maintain stable large herd sizes. In contrast, poor pastoralists (<50 cattle) cultivate crops for both food and financial income in order to make a living, and as a pathway to rebuilding herd sizes. To achieve sustainable livelihoods of pastoralists in a sedentary system, our study recommends further research to investigate range resource efficiency and possible trade-offs where rangelands are used for both cattle grazing and crop cultivation.

Keywords: Adaptation, livelihoods diversification, rangelands, sedentarisation

Contact Address: Pius Nina, Uganda Martyrs University, Fac. of Agriculture, P.O. Box 5498, Kampala, Uganda, e-mail: piusm.nina@gmail.com

Effects of Different Production Systems on Growth Performance and Carcass Quality of Lohi Lambs in Pakistan

MUHAMMAD TARIQ

University of Agriculture Faisalabad, Dept. of Livestock Management, Toba Tek Singh, Pakistan

A study was conducted to evaluate the effects of different production system on growth performance and carcass composition of male Lohi lambs. Sixty-four Lohi breed lambs were divided in two groups (A and B) of thirty two animals in each group and kept in subgroups of eight in four pens. Half of the lambs were castrated. The group A was offered a concentrate diet (85g of DM/ kg of metabolic body weight/day) containing 15 % of crude protein and 3010 kcal kg^{-1} of ME along with hay (100 g per amb per day). Group B was offered fresh green forage *ad libitum* and supplemented with a concentrate (400 g per amb per day) containing 20 % of CP and 2940 kcal kg^{-1} ME. Feed offered and refused were sampled and their DM contents determined. Feed intake was measured/day and animals weighted/week. At the end the feeding period, all animals were slaughtered; a half carcass was sampled from each pen for carcass composition determination. The concentrate diet significantly affected daily DM intake ($p = 0.001$), FCR ($p < 0.05$), daily weight gain DWG ($p < 0.05$), final live weight ($p < 0.001$), carcass weight ($p < 0.001$), dressing percentage ($p < 0.001$), conformation scores ($p < 0.05$), internal viscera weight ($p = 0.001$), feet ($p < 0.05$), total bones ($p < 0.05$), buttock ($p < 0.05$) and fat score ($p < 0.001$) as compared to the fodder fed lambs. However fodder fed lambs had longer finishing period ($p < 0.001$), heavy total bones ($p < 0.05$), weight of *M. longissimus dorsi* ($p < 0.05$), Lean:bone ratio ($p < 0.05$) and more leaner carcasses ($p = 0.06$) than concentrate fed lambs. Entire lambs had significant daily DM intake ($p = 0.001$), FCR ($p < 0.05$), DWG ($p < 0.05$), liver weight ($p < 0.05$), lungs weight ($p < 0.05$) and Lean:bone ratio ($p = 0.05$) than castrated lambs however the castrated lambs had longer finishing periods ($p < 0.05$) and more total fat contents (intermuscular and subcutaneous fat) but these values were non significant. There was no interaction found between sexual status and feeding systems. In general, the concentrate feeding system showed good results for growth performance and carcass composition but economic decisions and intensive management must be regarded. The results of this trial show that the sex had an effect on growth and carcass composition and castration may only increase the fatness of carcass but lean remains lower.

Keywords: Carcass, growth, lamb, meat quality, production system, sex

Contact Address: Muhammad Tariq, University of Agriculture Faisalabad, Dept. of Livestock Management, Toba Tek Singh, 36050 Faisalabad, Pakistan, e-mail: tariqlm@uaf.edu.pk

High-Andean Oat (*Altoandina*) as Alternative for Colombia's High-Altitude Dairy Systems: An Economic Analysis

Karen Enciso Valencia[1], Javier Castillo Sierra[2], Luis Fernando Campuzano Duque[2], Luis Orlando Albarracín Arias[2], Stefan Burkart[1]

[1]*International Center for Tropical Agriculture (CIAT), Tropical Forages, Colombia*
[2]*The Colombian Agricultural Research Corporation (AGROSAVIA), Colombia*

In the Colombian high-altitude tropics (2200–3000 m.a.s.l.), Kikuyu grass (*Cenchrus clandestinus*) is the main feed source for the dairy system. This grass species has good characteristics regarding adaptability and productivity, but is affected by frost, grass bugs (*Collaria* sp.) and precipitation-related production seasonality. Forage deficits might thus be a problem at several times in a year. As a strategy to maintain production stable, dairy farmers use commercial feed concentrates increasing their production costs. AGROSAVIA, as a response to this, started in 2005 with the evaluation and selection of new forage species for the Colombian high-altitude tropics. The oat *Avena sativa* AV25T was identified as promising alternative to supply the requirements of dry matter in times of deficit and released as cultivar in 2018 under the name *Altoandina* (high-Andean oat). The objective of this study was to evaluate the economic viability of *Altoandina* in Colombia's high-altitude dairy systems. *Altoandina* (Aa) was provided as silage in two different diets: 35%Aa-65 % Kikuyu grass (yellow diet) and 65%Aa-35 % Kikuyu grass (red diet). The diet for comparison was traditional grazing with 100 % Kikuyu grass (blue diet). All diets were supplemented with 6 kg commercial feed concentrate, 0.5 kg cotton seeds and 0.5 kg Alfalfa flour per cow/day, respectively. To estimate economic indicators, we used a cashflow model and risk assessment under a Montecarlo simulation model. Including *Altoandina* incremented productivity per hectare by 82.3 % and 220 % in the yellow and red diets, respectively. According to the results of our economic model, the yellow diet is the best alternative. Its average NPV was superior in >80 % and showed a lower variability. The indicators Value at Risk (VaR) and probability (NPV<0) show the yellow diet to have the lowest risk for economic loss under different yield/market scenarios. The yellow diet also has the lowest unit production costs and uncertainty of productive parameters. According to our findings, supplementation with Altoandina at 35 %, i.e. during critical times, has high potential to improve efficiency and profitability. This information is key to the decision-making process of dairy farmers on whether or not to adopt this technology.

Keywords: Dairy system, forages, Monte-Carlo simulation, oat, silage, sustainability

Contact Address: Stefan Burkart, International Center for Tropical Agriculture (CIAT), Trop. Forages Program, Km 17 Recta Cali-Palmira, Cali, Colombia, e-mail: s.burkart@cgiar.org

Observations and Projections of Heat Stress for Livestock in Sub-Saharan Africa

Jaber Rahimi[1], John Yumbya Mutua[2], An Notenbaert[2], Karen Marshall[3], Klaus Butterbach-Bahl[1]

[1]*Karlsruhe Institute of Technology (KIT), Inst. of Meteorology and Climate Research, Atmospheric Environmental Research (IMK-IFU), Germany*
[2]*International Center for Tropical Agriculture (CIAT), Kenya*
[3]*International Livestock Research Institute (ILRI), Kenya*

Heat stress events for livestock species are expected to become more frequent due to climatic changes in the next decades. In this investigation, we assess the frequency of heat stress as well as the changes in consecutive days with heat stress events for different livestock species (dairy cattle, beef cattle, sheep, goat, swine, and poultry) in sub-Saharan Africa. Our analysis included observations as well as simulations from global and regional climate models from the Coupled Model Intercomparison Project Phase 5 (CMIP5) according to two representative concentration pathways (RCP4.5 and RCP8.5). We found that in recent decades the frequency of Severe/Danger heat events, i.e. events that result in significant decreases in productive and reproductive performances, has already significantly increased (at 95 % confidence level) in the region (e.g. for dairy cattle, it has increased in >1/5 of the study area). We also found that by the end of the 21st century, under both RCP scenarios, the frequency of Severe/Danger heat stress conditions and the mean average length of consecutive days with heat stress events are likely to significantly increase for almost all livestock species. As an example, such changes in frequency of dangerous heat stress condition may affect on average 15 % of our current livestock production (beef, milk, mutton, pork, poultry meat, and eggs) in East African countries by 2071–2100 climate period under RCP8.5. Our results highlight the hotspot regions where global climate change, in the absence of mitigation strategies, will significantly affect livestock productions in Sub-Saharan Africa in the future.

Keywords: Climate change, heat stress, livestock production, sub-Saharan Africa

Contact Address: Jaber Rahimi, Karlsruhe Institute of Technology (KIT), Inst. of Meteorology and Climate Research, Atmospheric Environmental Research (IMK-IFU), 82467 Garmisch-Partenkirchen, Germany, e-mail: jaber.rahimi@kit.edu

Effect of Willow Silage on Carcass Characteristics and Meat Quality of Fattening Awassi Lambs

Rawad Swidan, Sami Awabdeh

National Agricultural Research Center, Livestock, Jordan

The objective of this study was to evaluate the effect of willow silage (*Salix* spp.) on carcass characteristics and meat quality of Awassi lambs in Jordan. Twenty-six male Awassi lambs were randomly assigned to three groups with different levels of silages (0%; 50%; and 100 % of forage in the diet). Lambs in control (0 %) group were only fed wheat straw as forage; lambs in WS50 were fed 1:1 ratio of straw and willow silage, while WS100 group only fed willow silage as forage in their diet. Concentrate were formulated accordingly to each diet to fed iso-caloric, iso-nitrogenous rations. Lambs were fed high concentrate diet (20:80, F:C ratio) for 90 days after which 5 lambs from each group were sacrificed to study the effects of willow silage on carcass characteristics and meat quality in Awassi meat. Hot carcass weight, cold carcass weight, dressing percentages and weight of internal organs were all unaffected by feeding willow silage. Kidney fat for control lambs was significantly ($p < 0.05$) higher (141g) compared to the WS50 and WS100 (64 and 81g, respectively). As a percentage of cold carcass weights, control had significantly ($p < 0.05$) heavier loin cuts and lighter fat tail compared to other silage groups. Although loin width was significant ($p < 0.05$) higher in control lambs, total loin area was not significant. No different were observed in total muscle, total fat, subcutaneous fat and total bone of lambs' legs among all groups. Muscle to bone ratio and fat to bone ratio were also not effected by feeding willow silage. Lambs meat were fed WS100 were lighter than lambs fed WS50 and control groups but the different was not significant. On the other hand, shear force measurement on Longissimus muscle was significantly lower in WS100 compared to control (6.1 vs 8.8 newton), WS50 lambs were intermediate. In conclusion, feeding willow silage to weaned Awassi lambs increase fat deposition on the tail and increase meat tenderness (Lower shear force) compared to lambs fed common dry forage.

Keywords: Carcass characteristics, meat quality, willow silage (*Salix* spp.)

Contact Address: Rawad Swidan, National Agricultural Research Center, Livestock, Jarash Highway, 19381 Ain Basha, Jordan, e-mail: rawados@gmail.com

Comparative Performance of Sheep and Goats Raised in Water-Limited Areas

Helmy Metawi, Ahmed Ali
Animal Production Research Institute, Sheep and Goats Research, Egypt

This study was conducted in the northwestern coastal region of Egypt. The study area was intentionally chosen to include dry climatic conditions in order to assess the effect of water availability on the performance of both sheep and goats. The study area was subdivided into two agro ecological zones: Dabaa (DA) and Sidi Barani (SB). The annual rainfall in the study area averaged 108 mm for DA and 259 mm for SB during the period from 2005 to 2015. Data were collected from a total of 100 small ruminants keepers using a survey based on structured questionnaires. The results showed different contributions of the types of small ruminants to the livelihood of the family according to its capital assets. The share of small ruminants in the household economy ranged from 47.22 % in the SB to 73.9 % in the DA In DA district, sheep productivity was 30.3 % lower and total annual variable costs/ewe were 19 % higher in the DA district than in SB district. The corresponding figures in goats were 22.3 % and 14 %, respectively. The degree of dependence on rangelands is the main reason for the variability of feed costs in the two rain fall districts.The returns on capital invested in sheep production were 22.3 and 14.4 % for SB and DA, respectively. The corresponding figures for goat production were 20.4 and 16.7 %, respectively. Our findings support that goats raised under a less favourable environment perform better than sheep. In areas with limited water, household livelihoods can be improved by developing appropriate strategies to reduce the cost of animal feed.

Keywords: Water scarcity, goats, production, profitability, sheep, survey

Contact Address: Helmy Metawi, Animal Production Research Institute, Sheep and Goats Research, 5 Nadi Elsaid Str.Dokki, Giza, 16128 Dokki Cairo, Egypt, e-mail: hrmmetawi@hotmail.com

Fermentation Quality and Aerobic Stability of Maize Stover – Banana Pseudostem Mixed Silages in Ethiopia

Ashenafi Mitiku[1,3], Dries Vandeweyer[1,3], Bart Lievens[2,3], Sofie Bossaert[3,2], Sam Crauwels[2,3], Ben Aernouts[4], Yisehak Kechero[5], Leen Van Campenhout[1,3]

[1]*KU Leuven, Dept. of Microbial and Molecular Systems, Belgium*

[2]*KU Leuven, Laboratory for Process Microbial Ecology and Bioinspirational Management, Belgium*

[3]*KU Leuven, Leuven Food Science and Nutrition Research Centre, Belgium*

[4]*KU Leuven, Dept. of Biosystems, Belgium*

[5]*Arba Minch University, Animal Sciences, Ethiopia*

In Ethiopia, conservation of hay and crop residues are typically practised in an open space, where the sunlight, rainfall and deterioration might cause leaching of nutrients. This study was conducted to evaluate the fermentation and microbial quality of dry and fresh maize stovers mixed with banana pseudostem in different proportions. The dry maize stover was collected after full maturation of the maize plant for grain harvest and fresh maize stover at two-thirds of milking stage. The two agricultural byproducts were chopped (in 2–4 cm) separately using a chopper and mixed into six treatments based on their fresh weight basis. The banana pseudostems were ensiled with two levels (20, 30 %) of dry maize stover, two levels of (60, 80 %) fresh maize stovers. In addition, there was one level (100 %) of fresh maize stovers and one treatment of 95 % banana pseudostem with 5 % of molasses. The treatments are further referred to as T1, T2 T3, T4, T5 and T6 respectively. Samples were taken on 0, 7, 14, 30, 60 and 90 days for physicochemical and microbial analysis. The results showed there was significant ($p = 0.001$) reduction in pH for all treatments. However, in T1 and T2, the pH failed to drop below 4.5 throughout the entire fermentation period of 90 day, while there was no significant ($p = 1.000$) difference between these two treatments. The lactic acid bacteria counts reached a maximum while the Enterobacteriaceae counts decreased below the detection limit (2 log g/cfu) in the first 14 days. Although in all treatments the *Clostridium* spore counts were above the maximum level of 2 log g/cfu, this was not accompanied by a higher pH, except in T1 and T2. Surprisingly, the aerobic stability was higher in T1 and T2, while the well fermented (pH < 4.5) byproducts (T3, T4, T5 and T6) showed a lower aerobic stability after 60 and 90 days fermentation. This suggests that mixed silages improve the fermentation quality, but not always the aerobic stability of the byproducts.

Keywords: Aerobic stability, banana pseudostem, maize stover, microbial profile, mixed silages

Contact Address: Ashenafi Mitiku, KU Leuven, Dept. of Microbial and Molecular Systems, Kleinhoefstraat 4, 2440 Geel, Belgium, e-mail: ashenafiazage2010@gmail.com

The Potential of Willow Silage as Forage for Lactating Awassi Ewes and their Nursing Lambs

Sami Awabdeh, Rawad Swidan

National Agricultural Research Center, Livestock, Jordan

A study was conducted to evaluate the potential of willow silage (*Salix* spp) as a forage source for lactating Awassi ewes and their nursing lambs. Twenty-one Awassi ewes and their lambs were randomly assigned to one of the two dietary treatments; control group (Cont; n = 11) were ewes fed wheat straw, a common expensive forage source in Jordan and willow silage group (WS; n = 10) were ewes fed willow silage as a source of forage of the diets. Concentrate were formulated accordingly to each diet for iso-caloric, iso-nitrogenous rations. Lactating ewes were fed high concentrate diet ad libitum for 10 weeks of lactation. Milk yield and milk component were measured biweekly. Final body weight for ewes were measure at the end of experiment. Weaning weight for lambs were measured at 60 days of age. Blood samples were collected after 6 weeks to measure cell blood count. Neutral detergent fiber (NDF) and acid detergent fiber (ADF) was higher in control diet. There were no differences in final body weight (BW) and body condition score of ewes between two groups. Lambs nursing from ewes in WS diet had higher but not significant weaning BW and average BW gain than control. Ewes fed the willow silage diet had higher ($p < 0.05$) milk production than ewes fed control diets (1150 vs. 956 ml milk per day for WS and Control, respectively) with no differences among treatment groups in total solids, fat and protein content. No different in blood cell count except for platelets count in which WS group had higher platelets count compare with control as a result of higher salicylic acid in willow silage diets. Cost per kg milk production (US$) was higher ($p < 0.05$) in control group compared with WS. As a results, using willow silage in nursing ewe's diets will increase milk production, improve growth performance of nursing lambs and reduce cost of milk production, which demonstrate a potential to use as a forage source for ewes and their nursing lambs forage sources.

Keywords: willow silage, Awassi ewes, milk production, nursing lambs

---

Contact Address: Sami Awabdeh, National Agricultural Research Center, Livestock, Jarash Highway, 19381 Ain Basha, Jordan, e-mail: sami_awabdeh@yahoo.com

Renewable energy and food processing

Renewable energy supply in rural areas

Oral Presentations

Posters

Sociotechnical and Socioeconomic Constraints on Adoption of Small-Scale Biogas Plants in Indonesia: A Review

Ricardo Chandra Situmeang, Jan Banout, Jana Mazancová, Hynek Roubik
Czech University of Life Sciences Prague, Fac. of Tropical AgriSciences - Dept. of Sustainable Technologies, Czech Republic

The growing interest around the world for renewable energy to avoid climate change has increased demands for cleaner energy sources. Biogas in particular as an alternative energy source is not only sustainable but also offering economic, social, and environmental benefits. Currently, Indonesia has installed 43,896 biogas digesters in the past 10 years. This is quite low because biogas technology has been introduced since the 1970s and Indonesia has a great potential for adopting the technology. Challenges in developing this technology still depend on external financial support and its competition with non-renewable energy. Likewise, the technical installation and maintenance of biogas plants are poor due to the lack of knowledge and skills of biogas users. Early identification of barriers further helps to develop a rigorous approach in developing and implementing biogas technology in rural areas. These barriers identified individually for rural biogas systems and the results show that the type of barriers varies strongly between biogas systems due to the difference in technology maturity, feedstock availability and quality, supply chain, awareness level, and policy support. Therefore, it is very important to develop a mechanism that entails sociotechnical and socioeconomic perspectives. This paper explored the potential of biogas to help meet domestic energy needs and to comply with Indonesia's climate mitigation commitments and development planning. By analysing the previous literature on sociotechnical and socioeconomic constraints of the adoption of biogas in Indonesia, this paper provides an overview of the adoption of the technology that could lead to self-sufficiency in household energy provision for rural energy needs. In the future, their impact able to facilitate environmental management and economic development in Indonesia. This review highlights the need for support from central and local governments, NGOs and private sectors to better deliver the scheme without forgetting the proper training for technical maintenance. Cooperation between stakeholders needs to improve in order to increase the quality and quantity of biogas technology dissemination.

Keywords: Adoption, small-scale biogas technology

Contact Address: Ricardo Chandra Situmeang, Czech University of Life Sciences Prague, Fac. of Tropical AgriSciences - Dept. of Sustainable Technologies, Kamýcká 129, 16500 Prague, Czech Republic, e-mail: ricardositumeang@gmail.com

PESTEL Analysis of the Development of Small-Scale Biogas Technology in Sub-Saharan Africa: A Systematic Review

Chama Theodore Ketuama, Jana Mazancová, Jan Banout, Hynek Roubik
Czech University of Life Sciences Prague, Fac. of Tropical AgriSciences, Czech Republic

Fossil fuels and traditional energy sources still constitute a greater share of energy supply in rural households of sub-Saharan Africa (SSA). Transition to small-scale biogas technology, clean energy in SSA includes the understanding of the factors affecting the development of the technology in the region. This study was performed as a systematic review of peer-reviewed and grey literature to reveal the political, economic, social, technological, environmental and legal (PESTEL) constraints and their impact on the sustainable development of this technology in the period from 2000 to 2020. Data was collected through a survey of scientific and grey literature. A total of 64 publications; comprising of 49 (77 %) peer -reviewed articles and 15 (23 %) grey literature were selected. Selection was done by double screening of the titles and full texts based on criteria like the focus on the small-scale biogas plants, constraints, drivers and prospects of the development of the technology in SSA. After PESTEL analysis, the main results reveal that SSA countries are still mobilising for the development of renewable energy policies. While governments are expected to increase the institutional and financial support, the private sector remains a huge potential in increasing investments on small-scale biogas plants. Sustainable credits are absent but can greatly increase wider adoption. Renewable energy incentive mechanisms are still to be well applied in some countries of the region for small-scale biogas development. Poor quality, operation and maintenance have led to the low technological efficiency of the plants. The governments' regulatory and financial supports are needed by the users and the private sector to adopt and innovate small-scale biogas plants to reduce energy poverty and spur economic growth. Collective action involving all local and international stakeholders is still required to increase the energy generation capacity of the small-scale biogas plants and their capacity for climate change mitigation.

Keywords: Biogas, constraint, evolution, impact, prospect, sub-Saharan Africa

---

Contact Address: Hynek Roubik, Czech University of Life Sciences Prague, Fac. of Tropical AgriSciences, Dept. of Sustainable Technologies, Kamycka 129, 16500 Prague, Czech Republic, e-mail: hynek.roubik@seznam.cz

Using SWOT - AHP Approach in Determining the Dimensions of the Investment in Biogas Technology and its Location in Syria

Ghaith Hasan, Jana Mazancová, Jan Banout, Hynek Roubik

Czech University of Life Sciences Prague, Fac. of Tropical AgriSciences, Czech Republic

The need for reliable and renewable energy sources in Syria in the light of the civil war and the strong embargo is increasing. Therefore, current situation is opening interest in biogas technology as a solution for potential energy crises. The process of selecting the appropriate site for constructing biogas units is complex, and for its evaluation, it requires many different criteria. Therefore, the objective of this study is to determine the dimensions of the incubating environment for investment in biogas technology and to prioritise the criteria that affect investment in areas for establishing small scale biogas plants. Data collection (via questionnaire surveys) took place from June to August 2019 in the following provinces: Latakia, Tartus, Homs, Hama, Damascus, Sweida, and Daraa (provinces which are currently safely accessible). The research targeted 255 farms (227 livestock farmers and crop farmers and 28 owners of biogas units) to identify the strategic dimensions that must be exploited using the SWOT model and AHP (Analytic Hierarchy Process). AHP was divided into two levels, the first level is represented in the standards of the reality of biofuels in Syria. The second level is represented by three main alternatives that represent the study areas, which are the southern, central, and coastal regions. The study, using the SWOT model, found out the strategic direction of using biogas technology. The optimal strategy is to exploit the opportunities most likely to succeed based on the available strengths. These strengths represented by the Syrian society's acceptance of the technology and the desire to use it. The AHP model results show that the common approach to the use of biogas and the by-product organic fertiliser gained the highest weight among the criteria related to the community's adoption of biogas technology. The southern region was proved as the most suitable area for investment in biogas units followed by the central region and the coastal region. The study provides a structured general framework of the prospects for the application of biogas technology that helps decision-makers and planners make decisions based on a systematic method.

Keywords: Biogas technology, developing countries, organic waste, renewable energy

Contact Address: Ghaith Hasan, Czech University of Life Sciences Prague, Fac. Tropical AgriSciences, KamÃ½ckÃ¡ 129, 16521 Prague, Czech Republic, e-mail: hasan@ftz.czu.cz

Opportunities for Biogas Utilisation in East Africa: A Case Study of Uganda

Sören Richter[1], Noble Banadda[2], Kerstin Wydra[1]

[1]*University of Applied Science Erfurt, Fac. of Landscape Architecture, Horticulture and Forestry, Germany*

[2]*Makerere University, Dept. of Agricultural and Biosystems Engineering, Uganda*

Loss of nutrients and minerals along value chains of fruits and vegetables are a major constrain in East Africa and contributes to an increased food insecurity and social problems such as malnutrition, unemployment and poverty. Within the FruVaSe-Project ('Fruits and Vegetables for all season' supported by the German Federal Ministry of Food and Agriculture (BMEL)) improvements for fruit and vegetable processing are developed to address these problems. To improve cyclic processes, wastes from the processing should be utilised in a biogas process. This renewable energy technology is accepted as key technology for energy transition to reduce traditional wood fuel use and replace it by renewable energy sources especially for cooking purposes. Although the economy in East African countries (EAC) is strongly characterised by agriculture and high amounts of utilisable wastes, the biogas dissemination is still quite limited. In a mixed methodological approach, consisting of literature review, expert interviews and a SWOT analysis challenges in the implementation of biogas projects in the EAC and Uganda were identified.

The results show a high rate of failure and abandon of biogas installations on the household level due to a lack of sustainable availability of raw materials, a need for a high level of expertise, labour input, high initial costs and limited financial access. A further major obstacle is the increase in household water requirement of up to 88 % for installed systems. Additionally, the suitable pre-treatment of the substrate can be challenging for the user, e.g. due to changing mixing ratios within seasonal changes of available substrate. The project target to use jackfruit wastes as possible feedstock is considered to be limited due to nearly no cultivation on a commercial scale. Therefore, interviewed experts recommend the production of biogas from fruit and vegetable processing or from other organic wastes at institutional or commercial level, rather than in households. To improve the planning phase a stakeholder participation method should be used to create ownership through participation, and site selection tools should be used to evaluate biogas opportunities. Further investigations should focus on water-reduction and/or -recycling, improving pre-processing methods for varying feedstock and on detailed analysis of jackfruit for chemical compositions, amount of wastes and co-digestion opportunities, as well as on business models for commercial use of biogas systems.

Keywords: Abandon factors, biogas, domestic, failure, fruit processing, mixed methods, renewable energy, waste utilisation, water

Contact Address: Sören Richter, University of Applied Science Erfurt, Fac. of Landscape Architecture, Horticulture and Forestry, Leipziger Straße 77, 99085 Erfurt, Germany, e-mail: soeren.richter@dbfz.de

Chemical Pretreatments of Rice Straw for Anaerobic Digestion

YANS GUARDIA-PUEBLA[1], OLIVERA YASBEL[1], QUIRINO ARIAS[1], GERT MORSCHECK[2], BETTINA EICHLER-LÖBERMANN[2]

[1]*University of Granma, Chemistry, Cuba*
[2]*Rostock University, Agricultural and Environmental Faculty, Germany*

Rice (*Oryza sativa*) is an important staple food for approximately half of the world population. In addition to the grains also the straw can be used if not needed as organic fertiliser for the fields. Although rice straw has potential to be used for anaerobic digestion (AD) to produce biogas, its large scale application is still limited. The utilisation of rice straw for AD must particularly be developed regarding the pretreatment of the lignin biomass for the hydrolysis process and the biogas yield and quality. Among the pretreatment methods, the alkalisation method was found to be effective. In the present study the effects of rice straw pretreatment was investigated in a multifactorial approach. For this purpose, calcium hydroxide ($Ca(OH)_2$) and potassium hydroxide (KOH) were applied at different concentrations. Furthermore, the reaction time and inoculum-to-substrate ratio were included in the investigations. Results showed that the degradation of lignin, cellulose, and hemicellulose was higher after the chemical pretreatment. The main end-products identified were acetic and propionic acids. The highest hydrolysis yield was observed at a reaction time of 4 h, with an alkali concentration of 10 g per g rice dry matter, and an inoculum-to-substrate ratio of 50 %. Here also the maximum concentration of volatile fatty acids was observed with 187±ß0 mg L^{-1}. The composition of the acidified effluent was similar when both chemical compounds were used. Economically, the optimal amount of $Ca(OH)_2$ for the pretreatment of rice straw was 4 g per g rice dry matter. The results showed, that the additional effort of a pretreatment of rice straw can be worthwhile to improve the AD and to increase the biogas yield.

Keywords: Bioenergy, biogas, biomass, digestion

Contact Address: Bettina Eichler-Löbermann, Rostock University, Agricultural and Environmental Faculty, Justus-von-Liebig-Weg 6, 18059 Rostock, Germany, e-mail: bettina.eichler@uni-rostock.de

Regarding Environmental Issues, Are Inhabitants of Bandung Interested in Municipal Solid Waste Management?

Denisa Beňová[1], Kryštof Mareš[1], Tatiana Ivanova[1], Yayan Satyakti[2], Tereza Pilařová[3]

[1]*Czech University of Life Sciences Prague, Fac. of Tropical AgriSciences, Dept. of Sustainable Technologies, Czech Republic*

[2]*Padjadjaran University, Center for Economics and Development Studies, Dept. of Economics, Indonesia*

[3]*Czech University of Life Sciences Prague, Fac. of Tropical AgriSciences, Dept. of Economics and Development, Czech Republic*

The research conducted via questionnaire survey was aimed to evaluate an interest of inhabitants in the proper municipal solid waste management (MSWM) in Bandung, Indonesia. Nowadays, increasing waste pollution becomes a serious topic in developing countries and accumulation of municipal solid waste is turning out to be unsustainable in these areas. Aspects as population health, environment and agricultural production face huge threat from a side of an improper MSWM. The government of Indonesia has implemented laws and regulations dedicated to MSWM in the past years. Unfortunately, these are not well communicated to the public sector that has been determine as one of the key factors to successful, functional and sustainable MSWM in previous studies. Two main indicators were used to investigate a public approach and perception on the MSWM: public interest in MSWM in connection to the environmental issues, and willingness of inhabitants to pay for appropriate MSWM services. Binary probit model showed that age, educational level, locality and satisfaction with MSWM practices played significant role in respondents' interest in MSWM, which is a decisive factor in the public perception of MSWM. The results of Chi-square analysis confirmed that age, level of education and average monthly income were strongly associated with a willingness of inhabitants to pay a different sum as the tax for appropriate MSWM services. In addition, the results of descriptive statistics found other interesting specifics of MSWM in Bandung city. These results showed that 97 % of respondents are not satisfied with MSWM practices, also over 83 % respondents consider MSWM services in Bandung as insufficient. Awareness about important public waste operations like waste handling and recycling is at a low level. On the other hand, over 67 % of respondents are interested in MSWM in connection to environmental issues and are familiar with 3R concept. The complex of results brings a new perspective on the public perception and approach to the MSWM that should be considered in the planning and implementation of the sustainable MSWM, since this issue has not been deeply examined in Bandung yet and results show an imperceptible engagement of public sector in MSWM decision-making.

Keywords: Pollution, public awareness, waste generation, willingness to pay

Contact Address: Denisa Beňová, Czech University of Life Sciences Prague, Fac. of Tropical AgriSciences, Dept. of Sustainable Technologies, Kamýcká 129, 16521 Prague 6 - Suchdol, Czech Republic, e-mail: benova@ftz.czu.cz

Analysis of the Current Situation and Recommendation of Appropriate Waste Treatment Technologies in Bandung City, Indonesia

Kryštof Mareš[1], Denisa Beňová[1], Tatiana Ivanova[1], Yayan Satyakti[2], Tereza Pilařová[3]

[1]*Czech University of Life Sciences Prague, Fac. of Tropical AgriSciences, Dept. of Sustainable Technologies, Czech Republic*

[2]*Padjadjaran University, Center for Economics and Development Studies, Dept. of Economics, Indonesia*

[3]*Czech University of Life Sciences Prague, Fac. of Tropical AgriSciences, Dept. of Economics and Development, Czech Republic*

The present research was realised in Bandung, Indonesia and it was focused on waste services in the city and municipal solid waste management (MSWM) as a whole. Bandung's waste situation is not well documented and examined field as there is a small number of references. The study was based on questionnaire survey among Bandung's residents and interviews with both governmental and public sectors. The research was supported by an observation and photo documentation of the public MSWM services, transportation and collection services as well as temporary and final disposal sites. It was seen that the quality of MSWM is poor and improvement is in need. Despite the fact that Bandung city ranks high among Indonesian most economically developed cities, MSWM lacks investments in new and better waste treatment technologies. Moreover, there are no other plans besides constructing of further disposal sites such as unsanitary landfill. To avoid rejection or misunderstanding of alternative technologies' use, preferences and opinion of public are important to embrace in a planning of the new waste management. Questionnaire survey showed that the most preferred waste treatment method among respondents was recycling centre (40 % of respondents), which points the concern of people about MSWM situation in Bandung. Recommendation for MSWM is a complex system consisted of implemented variation of waste treatment technologies to fit the sustainable waste handling. Waste management can't be focused only on one specific technology and at least it should be inspired by the public attitude. Public knowledge and preferences of treatment methods are influenced by many factors such as cultural behaviour: accepted habits from past generations of waste burning or open dumping of waste (historically organic waste), and education: people are informed about different waste treatment methods, majority of respondents are aware of re-

Contact Address: Kryštof Mareš, Czech University of Life Sciences Prague, Fac. of Tropical AgriSciences, Dept. of Sustainable Technologies, Kamýcká 129, 16521 Prague 6 - Suchdol, Czech Republic, e-mail: mareskrystof@ftz.czu.cz

cycling as a suitable and sustainable method of waste handling. Multivariate probit model has shown that the acceptance of various technologies is interconnected, i.e. knowledge and acceptance of one technology lead to an acceptance of the other. The government could have the major impact on MSW handling considering the full support of the public in the target area.

Keywords: Incineration, landfill, open dumping, pollution, waste handling, waste management, waste treatment technology

Dairy Production and Energy Crisis in Goiás: Analysis of Rural Development and Solar Energy

Luciana Ramos Jordão[1], Thiago Henrique Costa Silva[2], Sybelle Barreira[3], Gutherrison Gonçalves das Chagas[4]

[1]*State University of Goias, Law College, Brazil*
[2]*Federal University of Goiás (UFG), Law College, Brazil*
[3]*Federal University of Goiás (UFG), Forest Engineering, Brazil*
[4]*University Center Alves Faria, Law College, Brazil*

Dairy producers in Goiás, a Brazilian midwestern state, have been facing troubles due to poor performance of it's energy company (ENEL), which was privatized in 2017. In 2019, family farmers made serious allegations about energy supply interruptions, instability and huge delays in responding to demands because of staff cuts. This scenario let to an investigation by Goiás state Legislative Assembly, which concluded that the company did not invest the contracted amount on maintenance and caused damage to rural activity, especially those dependent on refrigeration equipment, such as dairy producers. Goiás State is one of the greater producers of dairy in Brazil, which is the fifth greater dairy producer in the world according to FAO. In Goiás, almost 50 thousand families produced more than 1,4 billion liters of milk in year 2017. Even though these families live in a region with a great capacity photovoltaic energy, they cannot avoid using traditional and non-renewable energy matrices, such as oil and hydroelectric because access to solar panels are still too expensive. To understand the challenges for expanding solar energy use in Brazil, mainly in the countryside, specifically in Goiás, in order to promote social, environmental and economic development, this research analyses public policies for rural development and the obstacles that prevent solar energy from becoming a main source of energy in the countryside. After price analyzes, dairy farmers annual income and existing public policies to increase solar energy usage, it's deduces that the high costs for implementing photovoltaic systems and the absence of incentives by the government part are obstacles to the autonomy and better living conditions of family farmers in Goiás. Although farmers express the want to cease dependency on non-renewable energy matrices, the expansion of Goiás Solar Program is not enough to successfully resolve energy distribution crisis in the state, consequently generating a development that expands individual and collective freedoms.

Keywords: Clean energy, climate change, energy crisis, public policies

Contact Address: Luciana Ramos Jordão, State University of Goias, Law College, Av. Perimetral Norte N. 4356 C. 88b Cond. Alto da Boa Vista, 74445500 Goiais, Brazil, e-mail: lr.jordao@me.com

Sustainably Feeding Africa: Wood-Burning Technologies in the Food-Energy Nexus (FEN)

Mwemezi J. Rwiza[1], Matthias Kleinke[2]

[1]*The Nelson Mandela African Institution of Science and Technology (NM-AIST), School of Materials, Energy, Water and Environmental Sciences (MEWES), Tanzania*

[2]*Rhine-Waal University of Applied Sciences, Fac. of Life Sciences, Germany*

Can woodlands co-exist sustainably with the ever food- and energy-hungry human species? Are improved woodstoves a sustainable energy option for the kitchen and fireplace of rural and suburban Africa? There are many questions that surround ideas around woodfuel as an energy source. Many studies have been done to indicate how unsustainable reliance on woodfuel is to the forests, savannahs and woodlands. It should be borne in mind that > 90 % of rural/ suburban households in sub-Saharan Africa depend on biomass-based energy sources. Biomass-based technologies that focus on sustainability concepts may point to a better and healthier future for many families especially in sub-Saharan Africa. The objective of this study (this presentation) is to stimulate some discussion on the sustainable pathways in the bio-energy economy and how this relates to food security and environmental conservation. In this study a total of 26 stove users, non-users and promoters in Tanzania were interviewed using semi-structured, unstructured and focused interview methods. Results indicated that, at the farmer's level, proper application of improved woodstoves faces the following challenges: lack of techniques to evaluate stove's efficiency; farmers are not involved woodburning projects' evaluation; lack of technical capacity to make or repair woodstoves; and inherent woodstoves problem e.g. faults and non-versatility. In practice, women who are the primary users of stoves for cooking and heating are either passively or partially involved in decision-making related to the production and maintenance of woodstoves. At the policy-making level; the government is yet to make stove programs a priority and has settled for the NGO-led dissemination efforts. Results from this study will be a useful contribution for researchers, policy makers, NGOs and groups involved in promoting the sustainable adoption and use of improved woodstoves.

Keywords: Rural food-energy systems, sustainable futures, woodfuel economy

---

Contact Address: Matthias Kleinke, Rhine-Waal University of Applied Sciences, Fac. of Life Sciences, Marie-Curie-Str. 1, 47533 Kleve, Germany, e-mail: matthias.kleinke@hochschule-rhein-waal.de

Criteria for Assessment of Different Designs of Small-Scale Biogas Technology

Marek Jelinek, Jana Mazancová, Jan Banout, Hynek Roubik
Czech University of Life Sciences Prague, Fac. Tropical Agrisciences, Czech Republic

Biogas technology is capable of improving the living standards of rural households in developing countries and jointly reduce their environmental impacts. These potential benefits are often not fully achieved due to different technical or non-technical constraints related to specific local conditions and various aspects of different designs of biogas plants. Despite the addressing of the advantages and disadvantages of biogas in rural areas in many studies, the implementation of biogas technology has not been assessed in a holistic way. Therefore, the aim of this study is to develop a holistic assessment technique, which can be used for a comprehensive assessment of available designs of small-scale biogas systems. In addition, a decision-making tool for use by both researchers and technology implementers in the sustainable development of small-scale biogas plants will be designed. The research methodology is based on a multi-criteria sustainability assessment framework (SAF) involving four dimensional aspects - environmental, technical, economic and social - applied to evaluate different constraints of designs ensuring their suitability for use in selected target areas. The framework integrating multi criteria analysis principles use also the sub-criteria capturing the fundamental aspects of the sustainability dimensions, based on a critical examination of scientific literature (Technical: process efficiency, energy payback, operational reliability; social: health and sanitation, gender equality, cultural perception of biogas technology; economic: total investment costs, payback period; environmental: life-cycle analysis, global warming potential). Obtained secondary data will be further analysed in the evaluation matrix, using weights to indicate relative significance of each sub-criterion. The results presented in the study will provide suitable criteria for development of a decision-making tool, which will be available for researchers and practitioners in favour to implement the biogas technology according to the local conditions, while favouring the rural households and reducing their environmental impact.

Keywords: Anaerobic digestion, biogas, developing countries, small-scale biogas plants, sustainable development goals, waste management

---

Contact Address: Marek Jelinek, Czech University of Life Sciences Prague, Fac. Tropical Agrisciences, Pionyru 1337, 56401 Zamberk, Czech Republic, e-mail: marekjelinek@ftz.czu.cz

Food processing

Oral Presentations

Posters

Optimisation of Linamarin Extraction from Cassava Leaves by Ultra-Sonication and High-Performance Dispersing

Sajid Latif, Sonnal Lohia, Joachim Müller
University of Hohenheim, Inst. of Agric. Sci. in the Tropics (Hans-Ruthenberg-Institute), Germany

Cassava (*Manihot esculenta*) ranks as one of the most important crops, grown in 105 countries and a staple crop for nearly one billion people. It's mainly grown for starchy roots while other parts of the plant are considered as waste products, which lead to environmental pollution due to their disposal. Cassava leaves are rich in protein (17–38 %) but cyanogenic glucosides (linamarin) limit their use as food and feed. This research focuses on obtaining the optimal conditions for extraction of linamarin from cassava leaves by using response surface methodology (RSM). Conventional (Soxhlet) and non-conventional (ultra-sonic and high-performance dispersing) extraction methods were performed for two different organic solvents: ethanol and methanol. The process was carried out for solvent concentration (75, 85 and 99.9 %), solid-to-liquid ratio (01:10, 01:15 and 01:20 g ml^{-1}) and time (ultra-sonic – 10, 20 and 30 minutes; high-performance dispersing – 10, 15 and 20 minutes). The optimal conditions for an ultra-sonic extraction were found to be time 20.41 minutes, solvent concentration 86.58 % methanol at 35°C and 40 kHz, which yielded 81.2 % extraction efficiency. The parameter solid-to-liquid ratio had no significant effect on the ultra-sonic extraction and interestingly an inter correlation between time and solvent concentration was found. The optimal conditions for an high-performance dispersing extraction were found to be 19.95 minutes, 88.0 % methanol, 0.19 g ml^{-1} at 35°C and 11000 rpm, which yielded 88.8 % extraction efficiency. These two optimised methods are relatively simple, less time consuming, cost effective, feasible and environmentally sustainable for the extraction of linamarin from cassava leaves. This extracted linamarin can be utilised as a bio-pesticide and leading to a circular bio-economy.

Keywords: Bio-pesticide, Box-Behnken design, cassava leaves, high-performance dispersing, linamarin, ultra-sonic

---

Contact Address: Sajid Latif, University of Hohenheim, Inst. of Agric. Sci. in the Tropics (Hans-Ruthenberg-Institute), Garbenstr. 9, 70599 Stuttgart, Germany, e-mail: s.latif@uni-hohenheim.de

The Potential of a Sustainable Active Packaging Solution to Reduce Food Losses in Benin

Barbara Götz[1], Mathias Hounsou[2], Sylvain Dabadé[2], Thomas Havelt[3], Michaela Schmitz[3], Djidjoho Joseph Hounhouigan[2], Judith Kreyenschmidt[1]

[1]*University of Bonn, Inst. of Animal Sciences, Germany*

[2]*University of Abomey-Calavi, School of Nutrition, Food Science and Technology, Benin*

[3]*University of Applied Sciences Bonn-Rhein-Sieg, Germany*

Food security and food losses are of major concern in sub-Saharan Africa. The reduction of food losses can help to prevent malnutrition. Main reasons for food losses are damage and contamination during harvest, transport, processing, and storage, as well as an accelerated spoilage due to environmental and hygienic conditions. Active packaging solutions can prolong shelf life and increase food safety by inhibiting the growth of spoilage bacteria and pathogens. The implementation of locally produced, sustainable, biobased, and biodegradable packaging solutions is of great interest, especially in developing countries like Benin.

The aim of this study is the development of an active packaging solution made from local biobased materials with an active biogenic coating to reduce food losses in Benin. The active compounds are extracts of local plants with antimicrobial activity. Using the agar well diffusion method, the antimicrobial screening showed that all plant extracts (n=16, 70 % v/v Ethanol, 0.1 g ml^{-1}, 24 hrs., 25°C, 140rpm) are active against *Staphylococcus aureus* and other Gram-positive bacteria. Six plant extracts also showed activity against Gram-negative *Pseudomonas putida* and *Pseudomonas fluorescens*. The minimal bactericidal concentration (MBC) of these extracts identified *Gmelina arborea* (GA) with the lowest MBC of ≤ 10 μg ml^{-1} as most effective against Gram-positive bacteria. Additionally, GA shows lowest MBC of 20 and 50 μg/ml for *P. putida* and *P. fluorescens*. Besides, the total antioxidative capacity of GA was very high with 529.64 mg Teq ml^{-1} extract and a total phenolic content of 1,876.11 mg GAE l^{-1} extract. The determination of the antimicrobial activity (ISO 22196) showed a successful integration into the biogenic matrix of beeswax, shea butter, and coconut oil with an antimicrobial activity of 2,53 and a log reduction of > 2 cfu cm^{-2} for most bacteria. Matrices as base for this coating can be locally produced materials like paper, banana leaves and cloths.

The results of this research show that the developed biobased active packaging solution; with beeswax, plant oils and antimicrobial agents from Benin; has the potential to increase food safety and reduce food losses. Therefore, it can contribute to the resilience of food security in developing countries.

Keywords: Active packaging, antimicrobial activity, biobased, biodegradable, food loss, food security, sustainability, western Africa

Contact Address: Barbara Götz, University of Bonn, Inst. of Animal Sciences, Katzenburgweg 7-9, 53115 Bonn, Germany, e-mail: bgoetz@uni-bonn.de

Why Do Agronomists Measure Leaves and Not Food? A Neglected Food Security Issue

Sahrah Fischer[1], Thomas Hilger[1], Hans-Peter Piepho[2], Irmgard Jordan[3], Georg Cadisch[1]

[1]*University of Hohenheim, Inst. of Agric. Sci. in the Tropics (Hans-Ruthenberg-Institute), Germany*
[2]*University of Hohenheim, Inst. of Crop Science, Germany*
[3]*Justus-Liebig University Giessen, Center for International Development and Environmental Research, Germany*

Crop nutrient deficiencies are often determined based on leaf nutrient composition and their effect on yield, and less on edible part composition. Since leaves and edible parts of plants often have different plant functions and therefore different nutrient requirements, it remains unclear whether leaf nutrient composition can be used to draw conclusions on the nutrient composition of edible parts. Farm management and environmental factors such as water availability and soil fertility, may therefore have unknown consequences for the nutrient concentration of the edible part. Our main aim, therefore, was to investigate the relationship between trace element nutrient concentrations in leaves and edible parts (grain, tuber, and fruit) of three staple food crops.

Leaf and edible part samples of maize (*Zea mays* L.) and cassava (*Manihot esculenta* Crantz) from a poor fertility soil (Teso, Kenya), and maize and matooke (*Musa acuminata* Colla) from a higher fertility soil (Kapchorwa, Uganda) were collected. Concentrations of macronutrients Mg, P, S, K, and Ca and micronutrients Fe, Mn, Cu, and Zn were measured using a portable X-Ray Fluorescent Spectrometer (pXRF). Leaf and edible part nutrient concentrations were compared to yield and to each other using a bivariate linear mixed model fitted with residual maximum likelihood (REML) to calculate the marginal correlation.

Nutrient concentrations of leaf ($\rho\gamma(1{,}2) \leq 0.20$) and edible parts ($\rho\gamma(1{,}2) \leq -0.20$) showed the least similarity to yield in low fertility soils, meaning that leaf measurements can give less indication of edible part nutrient concentration on low fertility than high fertility soils. Nutrients with a low phloem mobility such as Fe, Mn, Ca, Cu, and Zn showed the largest difference in correlations to yield, compared to the other macronutrients. Correlations varied between nutrient concentrations in different plant parts depending on the type of nutrient, the type of crop, and the type of plant part in question. Leaves did not provide enough information to gauge both yields and quality of foods, particularly regarding micronutrients. Agronomists targeting the production of foods for human consumption should begin analysing the produced food quality to ensure adequate food nutrient concentrations.

Keywords: Food quality, human nutrition, plant nutrition, trace elements

Contact Address: Sahrah Fischer, University of Hohenheim, Inst. of Agric. Sci. in the Tropics (Hans-Ruthenberg-Institute), Garbenstr. 13, 70599 Stuttgart, Germany, e-mail: sahrah.fischer@uni-hohenheim.de

Effect of Cooking Time on Quality Traits of Tanzanian Vegetable Sauces

Amina Ahmed, Gudrun B. Keding, Elke Pawelzik

University of Goettingen, Dept. of Crop Science, Division of Quality of Plant Products, Germany

Vegetable sauces are an important part of the diet and are used as side dish or accompaniment with staple food in Morogoro, Tanzania. They can serve as a good source of micronutrients such as ascorbic acid and provitamin A. However, the composition and preparation of these sauces, which are offered on local markets, is not always known and seems to vary. It is therefore not always possible to assess whether they contribute to a healthy diet. In the present study, local recipes for making vegetable sauce from Morogoro region, Tanzania, were used as basis for preparation of tomato and carrot sauces. The basic ingredients of sauces consisted of either tomatoes or carrots with onion, ginger, garlic, salt and hot pepper (optional). Baobab pulp was added to increase viscosity and acidity of the sauces instead of starch and citric acid as used in local sauces. In addition, peanut paste was added to enhance bioavailability of carotenoids, which are important precursors of vitamin A. Ingredients for making sauce were blended using a kitchen mixer for 2 minutes to obtain the raw sauce. To study the effect of cooking time, they were cooked at 87 ± 3°C for 20, 25, and 30 minutes, respectively and physico-chemical properties, ascorbic acid and instrumental colour values ('L', 'a' and 'b' values) were determined. Moisture loss after cooking was higher in sauces from carrot than tomato. The tomato sauce had significantly ($p < 0.05$) higher ascorbic acid retention (69 %, 63 % and 49 %) than carrot sauce (16 %, 11.9 % and 11 %) when cooked for 20, 25 and 30 minutes, respectively. The instrumental colour values indicated that reduction of 'a' values (red pigments) was lower in tomato sauce than carrot sauce but all the formulations exhibited disparate 'b' values (yellow pigments). Generally, the pH of the sauces ranged from 3.3 to 3.7, which is within the recommended range. Cooking time for 25 minutes was promising for retention of ascorbic acid, and red and yellow pigments that reflects total carotenoids content. The studied recipes will be further evaluated for sensory qualities and optimal recipes will be identified based on nutrients retention and consumer acceptance.

Keywords: Cooking time, micronutrients, optimal recipe, sauce, sensory, vegetable

Contact Address: Amina Ahmed, University of Goettingen, Dept. of Crop Science, Division of Quality of Plant Products, 37075 Carl-Sprengel Weg 1, 37075 Goettingen, Germany, e-mail: aahmed1@gwdg.de

Impact of Processing Temperature on Drying Behaviour and Quality Changes in Stinging Nettle (*Urtica dioica*)

ZIBA BARATI, SEBASTIAN AWISZUS, SEBASTIAN REYER, LENA BUTZ, JOACHIM MÜLLER
University of Hohenheim, Inst. of Agric. Sci. in the Tropics (Hans-Ruthenberg-Institute), Germany

Stinging nettle (*Urtica dioica*), well-known as a medicinal plant, contains health-promoting compounds like flavonoids, phenolic acids, minerals, essential oils, vitamins and pigments. In order to prolong the shelf life, the stinging nettle must be dried after harvest, like most medicinal plants. The effect of different temperatures on the drying behaviour of stinging nettle and its properties was examined in this study. Drying experiments were conducted using the high precision laboratory dryer designed at the Institute of Agricultural Engineering, University of Hohenheim (Stuttgart, Germany). The stinging nettle was dried at temperatures of 30 °C, 50 °C and 70 °C. Air velocity and absolute humidity of the air were kept constant at $0.2\,m\,s^{-1}$ and $10\,g\,kg^{-1}$, respectively. The dried stinging nettle was analysed regarding the dry matter content, water activity, colour, caffeoyl malic acid, chlorogenic acid and pigments. The results indicate that the moisture content decreased slowly until desired moisture content of 10 % was achieved. The total drying time decreased noticeably by an increase of the temperature. It was observed that the colour values of stinging nettle were substantially influenced during the drying process. Drying at different temperatures had no significant influence on the total phenolic content, ranging from 900.75 ± 54.28 to 1192.19 ± 217.45 mg GAE $100\,g^{-1}$, and luteine content (54.32 ± 1.92 to 59.72 ± 2.65 mg $100\,g^{-1}$) ($p > 0.05$). Although, the stinging nettle dried at 70 °C contained significantly less caffeoyl malic acid (16.27 ± 1.27 $mg\,g^{-1}$), chlorogenic acid (1.27 ± 0.12 $mg\,g^{-1}$), chlorophyll a (237.92 ± 15.37 mg $100\,g^{-1}$) and chlorophyll b (116.67 ± 7.40 mg $100\,g^{-1}$) compared to the stinging nettle dried at 30 and 50 °C. Furthermore, a significant difference among the beta-carotene content of the stinging nettle dried at different temperatures ($p < 0.05$) could be measured. The results show that the choice of the drying temperature affects the colour and the content of valuable ingredients of the stinging nettle. For further studies, the optimum temperature for drying of stinging nettle needs to be determined according to the purpose of the drying process.

Keywords: Color, drying behaviour, medicinal and aromatic plants, phenolics, pigments, stinging nettle

Contact Address: Ziba Barati, University of Hohenheim, Inst. of Agric. Sci. in the Tropics (Hans-Ruthenberg-Institute), Garbenstr.9, 70599 Stuttgart, Germany, e-mail: barati@uni-hohenheim.de

Drying Kinetics and Quality Attributes of *Gardenia erubescens* Fruits as Affected by Slice Thickness, Pretreatment and Drying Air Temperature

Joseph Kudadam Korese[1], Matthew Atongbiik Achaglinkame[1], Oliver Hensel[2], Barbara Sturm[2]

[1]*University for Development Studies, Agricultural Mechanization and Irrigation Technology, Ghana*

[2]*University of Kassel, Agricultural and Biosystems Engineering, Germany*

The *Gardenia erubescens* (GE) fruit is a wild fruit common in the savannah zone including Ghana. It is rich in fiber, carbohydrates, minerals and health promoting bioactive compounds. However, due to its seasonality and perishability, the application of preservative methods such as drying is imperative to improve its shelf life. Unfortunately, there is currently limited information regarding the effect of drying conditions on quality retention of the dried fruit. As a response, systematic laboratory trials were undertaken to investigate the effect of slice thickness, pretreatment option and air temperature on the drying kinetics, colour, and bioactive compound (β-carotene, vitamin C, phenol, flavonoid and antioxidant activity) composition of GE fruits. Cleaned fresh GE fruits were sliced into 3 mm and 5 mm thicknesses, subjected to three different pretreatment options (steam blanching at 100°C for 3 min, dipping in 2 % w/v ascorbic acid solution for 3 min and control) and dried at temperatures of 40, 50, 60 and 70°C. Results indicated that the drying time increased with pretreatments in the order control□ascorbic acid□steam blanching and increase in slice thickness, but decreased considerably with increase in air temperature. Furthermore, the greatest colour change (ΔE =36.82) and chroma (C^*=51.85) were associated to 5 mm fruit slice thickness pretreated with ascorbic acid and steam blanching, respectively. However, the lowest β-carotene (42.70 μg/100 g), vitamin C (37.50 mg/100 g) and flavonoid (36.33 mg/100 g), phenol (97.33 mg/100 g) and antioxidant activity (21.04 mg/100 g) contents were obtained in fruits pretreated with ascorbic acid and steam blanching, respectively. GE fruits sliced to 3 mm thickness, without pretreatment and dried at 60°C was the best option considering most of the measured response variables. Findings of this study may therefore be useful to the food industry regarding selection of drying conditions for desired quality attributes of dried GE fruits.

Keywords: Drying conditions, drying kinetics, quality attributes

Contact Address: Joseph Kudadam Korese, University for Development Studies, Agricultural Mechanization and Irrigation Technology, Fac. of Agriculture, POB TL 1882, Nyankpala campus, Tamale, Ghana, e-mail: jkorese@uds.edu.gh

Effects of Processing Technologies on the Quality of Pasteurised Baobab Nectars Produced in Southern Benin

Ahotondji Mechak Gbaguidi[1], Flora Chadare[1], Bienvenu Gnanhoui[2], Achille Assogbadjo[3]

[1]*National University of Agriculture, Lab. of Food Science, Benin*
[2]*University of Abomey-Calavi, Lab. of Food Science, Benin*
[3]*University of Abomey-Calavi, Lab. of Applied Ecology, Benin*

Baobab is a multifunction tree used by many African rural communities. Its different parts (fruit, leaves, bark) are used in food industry and also for medicinal and cosmetic purposes. Baobab nectar, a product derived from baobab fruit pulp is widely consumed in Africa. The present research aimed to characterise the processing technologies of pasteurised baobab nectar, used by Beninese processors, and assess their effects on its quality. For identifying the technologies, a semi-structured interview was realised with the heads of the different processing units, based mainly in the cities of "Abomey-Calavi" and "Cotonou"; this survey was followed by a processing follow up in selected processing units. Based on identified processing technologies and to assess the processing's effect, pasteurised baobab nectars were produced at the laboratory scale and the derived nectars were analysed for dry matter, Brix value, pH, titratable acidity, colour (L*, a*, b*) and vitamin C. The survey revealed two different processing technologies based on the number of thermal treatments which are the bulk thermal unitary operation and the in-pack pasteurisation. The assessment of the technologies' effects revealed that the pasteurised nectars were characterised by Brix value, pH, titratable acidity, dry matter, lightness (L*), vitamin C of respectively, 12.2$\pm$0.0, 2.6$\pm$0.0, 735.0$\pm$ 22.9 g L^{-1} dw, 12.7$\pm$0.0 g/100 g, 74.3$\pm$0.1, 209.8$\pm$8.7 mg/100 g dw for the technology involving both thermal treatments, and 12.9$\pm$0.2, 2.7$\pm$0.1, 717.0$\pm$ 21.2 g L^{-1} dw, 13.2$\pm$0.3 g/100 g, 73.4$\pm$0.3, 180.1$\pm$7.0 mg/100 g dw for the technology involving only the in-pack pasteurisation. Both technologies lead to significantly identic ($p > 0.05$) nectars considering the colour, the titratable acidity, the dry matter and the vitamin C content; the two technologies are significantly different on the basis of their Brix value ($p \leq 0.05$). The vitamin C content decreased by about 40 % according to the technology with two thermal treatments, and by about 30 % when using the technology with a single pasteurisation. Further studies on this type of nectar produced by local processors, must be extended to sensorial properties for identifying the best pasteurisation techniques that meet scales adapted to processors and consumers' needs.

Keywords: *Adansonia digitata*, colour, food industry, survey, vitamin C

Contact Address: Flora Chadare, National University of Agriculture, School of Food Science Nutrition, BP114, Sakété, Benin, e-mail: fchadare@gmail.com

Quality of 'gari' as Affected by Age at Harvest, Cropping System and Variety of Cassava Roots

Michael Idowu[1], Abiodun Adeola[1], Rebecca Oyatogun[1], A. Adebowale[1], Olusola Adeola[2]

[1]*Federal University of Agriculture Abeokuta (FUNAAB), Dept. of Food Science & Technology, Nigeria*
[2]*University of Ibadan, Dept. of Chemistry, Nigeria*

Cassava is a foremost crop of food security in Nigeria since it is a source of many indigenous food products that are commonly eaten by the populace. Many resource-restricted farmers engage in its production, adopting diverse agronomic practices that could affect the quality of products derivable from cassava. 'Gari' is the most commonly traded and consumed cassava product. This study investigated the effect of age at harvest (AH), cropping system (CS) and variety on the quality attributes of 'gari'. Five cassava varieties (white- and yellow-fleshed) planted under two different CS (sole and intercropped) and harvested at different AH (12, 15, and 18 mo after planting) were converted into 'gari'. 'Eba', a popular traditional food derived from 'gari' was prepared. Quality of 'gari' such as proximate and mineral composition, total carotenoids, pH, total titratable acidity, bulk density, water absorption index, dispersibility, swelling power, solubility index, and pasting properties, as well as the sensory properties of both 'gari' and 'eba' were determined. Data obtained were analysed using a generalised linear model. The chemical composition, functional and sensory properties of 'gari' were significantly ($p < 0.05$) affected by AH, CP and variety. The moisture, ash, crude fibre, protein, fat and carbohydrate contents of 'gari' ranged from 3.70 to 11.60 %, 0.50 to 2.03 %, 1.44 to 2.40 %, 0.23 to 1.87 %, 0.20 to 1.32 %, and 83.12 to 90.65 %, respectively. The water absorption index, bulk density, dispersibility, swelling power and solubility index of 'gari' varied from 270.80 to 527.60 %, 0.41 to 0.77 g ml^{-1}, 1.50 to 45.50, 6.36 to 11.01, and 3.5 to 25.0 %, respectively. The moisture and carbohydrate contents and dispersibility of 'gari' decreased with increase in AH. The bulk density was highest at 15 mo AH. Preference for the colour and aroma of 'gari' increased with increase in AH. The overall acceptability of 'eba' increased with increase in AH. It is therefore desirable to harvest cassava for 'gari' and 'eba' preparation at 18 mo.

Keywords: Chemical composition, functional properties, gari, variety

Contact Address: Abiodun Adeola, Federal University of Agriculture Abeokuta (FUNAAB), Inst. of Food Security, Environmental Resources and Agricultural Research, Alabata Road, Abeokuta, Nigeria, e-mail: adeolaroni@yahoo.com

Drying Kinetics and Characterisation of Dried Osmotically Pretreated White-Flesh and Yellow-Flesh Cassava (*Manihot esculenta*)

Oluwatoyin Ayetigbo[1], Sajid Latif[1], Adebayo Abass[2], Joachim Müller[1]

[1]*University of Hohenheim, Inst. of Agric. Sci. in the Tropics (Hans-Ruthenberg-Institute), Germany*

[2]*International Institute of Tropical Agriculture (IITA), Tanzania*

Cassava is subject to rapid post-harvest physiological deterioration. Therefore, there is a need to process cassava into other shelf-stable forms. Dehydration by osmotic process followed by drying was conducted as a means to achieve this. In this study, the drying conditions for previously osmotically dehydrated (OSD) cassava was optimised. White-flesh and yellow-flesh cassava cubes were dehydrated at 45 °C in a 70 o brix salt-sugar osmotic solution and dried in a precision dryer at different temperatures (50, 65, 80 °C) and air velocities (0.5, 0.9, 1.3 m s^{-1}). Based on a 3-level factorial experimental design with optimisation criteria of minimum dryer energy consumption, shrinkage, drying energy, drying time (to 10 % moisture content) and maximum diffusivities, the optimum conditions for drying of pre-dehydrated (PDH) cassava of both varieties of was 80 °C at 0.5 m s^{-1}. The quadratic and linear models generated from the response surface were significant ($P1, = 1, < 1, 0.0001$ to 0.0142) and had insignificant lack of fit ($p = 0.1605$ to 0.2266). During drying, moisture ratio (MR) of PDH cassava cubes decreased with increase in drying temperature, while drying air velocity did not considerably affect MR. By fitting MR data to logarithmic, two-term, Henderson-Pabis, and Newton models, the Newton model was considered preferable to describe the MR. Diffusivities increased significantly with drying air temperature and air velocity during drying. Drying rate versus time, and drying rate versus moisture content relationships was fitted best by Peleg model. The colour, shrinkage, total cyanogenic potential and total carotenoids contents were significantly influenced by drying conditions. Electron micrographs of dried cassava cubes reveal cell wall collapse and loss of granular starch compared to the relatively intact cellular structure and starch granules of dried non-dehydrated (NDH) cassava.

Keywords: Air velocity, cassava, drying rate, moisture ratio, response surface method

Contact Address: Oluwatoyin Ayetigbo, University of Hohenheim, Inst. of Agricultural Engineering, Tropics and Subtropics, Garbenstr. 9, 70599 Stuttgart, Germany, e-mail: ayetigbo_oluwatoyin@uni-hohenheim.de

Discrete Element Modelling of Moisture-Dependent Grain Bulk Behaviour under Compressive Loading

IRIS RAMAJ, STEFFEN SCHOCK, SEBASTIAN ROMULI, JOACHIM MÜLLER
University of Hohenheim, Inst. of Agric. Sci. in the Tropics (Hans-Ruthenberg-Institute), Germany

Appropriate conception and construction of the in-storage bins are of great importance to the safety of stored grains. In this regard, a major problem to address is the substantial decrease in the bulk porosity imposed by the compressive stresses caused by the dead weight of bulk in the storage bins. The decreasing rate of porosity is mainly attributed to the mechanical deformation of grains and depends on their physical and viscoelastic properties, the aggregate stored mass as well as the bin properties. For practical purposes, the misestimation of the porosity contributes to ineffective aeration strategies and likely grain deterioration. Thus, this study aimed to investigate the bulk behavior of wheat (Pioneer A, DSV AG) during compression under controlled conditions. As the computational approach, the discrete element method (DEM) was used to analyse in detail the grain behaviour at alike conditions. The compression tests were carried out at two different moisture contents, explicitly 12.39 and 25.41 % w.b. The physical properties, such as dimensions, mass, solid density, Poisson's ratio, Shear modulus and Young's modulus were measured via laboratory experiments. The interaction properties of wheat, such as the coefficient of restitution, static and rolling friction, were empirically determined. A coherent set of yield normal stress from 0 to 300 kPa at a displacement rate of 1.25 mm min^{-1} was imposed under experimental and simulation frameworks. The increase of moisture content contributed to significant differences in physical properties and bulk compressibility at $p \leq 0.05$. A good agreement amongst the simulation and experimental results at different compressive loadings, in terms of stress, volumetric strain, and bulk density was observed. The spatial and ephemeral variations in particle-scale were modeled and compared to the real bulk formations, revealing a high potential of DEM for the design and optimisation of post-harvest handling processes.

Keywords: Calibration tests, mechanical properties, multi-sphere method, particle, physical properties, simulation, three-dimensional

Contact Address: Iris Ramaj, University of Hohenheim, Agricultural Engineering in the Tropics and Subtropics (440e), Garbenstr. 9, 70599 Stuttgart, Germany, e-mail: Ramaj@uni-hohenheim.de

Nutrient and Antinutrient Composition of Bouillon Cubes Developed from Fermented Condiments of *Ricinus communis* L. Seeds

Olaide R. Aderibigbe[1], Chizobam H. Igwe[1], Sunday F. Awe[1], Barbara Sturm[2]

[1]*National Horticultural Research Institute, Product Development Programme, Nigeria*

[2]*University of Kassel, Agricultural and Biosystems Engineering, Germany*

Fermented condiments are part of the diets of many African nations and, they serve as a good source of protein and other nutrients for rural populations. They are used as alternatives for expensive protein sources such as meat and egg in poor households. In Nigeria, these condiments are presented in slurry, paste or loose solid form and are usually wrapped in leaves. *Ricinus communis*, an oil seed plant that grows both in the tropical and temperate climate is one of the seeds processed into a fermented condiment and is locally known as "*ogiri*". The development of fermented bouillon cubes offers an opportunity to enhance the safety and attract more consumers to the product. This study therefore developed three samples of bouillon cubes (A, B and C) containing fermented *R. communis* seeds and binder (cassava starch) at ratio 20:5, 20:10 and 20:20, respectively. The bouillon cubes were subjected to proximate, total carotene, vitamin C, minerals and antinutrient analysis. The results showed that the protein and fat content decreased while the carbohydrate content increased significantly ($p < 0.05$) with increasing binder proportion. The protein content was 9.93 %, 7.04 % and 3.49 %, fat content was 29.33 %, 28.00 % and 22.77 % and carbohydrate content was 48.99 %, 52.46 % and 60.09 % for samples A, B and C, respectively. The total carotene increased with increasing binder concentration while the vitamin C did not vary significantly. As for the minerals, Na, K, Zn and Fe increased with increasing binder concentration while Ca did not vary significantly. Phytate was significantly higher in sample C (0.45 mg g^{-1}) than samples A and B (0.35 and 0.32 mg g^{-1}, respectively). In conclusion, the addition of cassava binder in the production of fermented bouillon cubes of *R. communis* produced products with decreased protein content, increased phytate content and, increased mineral and total carotene content. The use of high protein binders or blend of binders should be considered for the development of this novel product as they are known to be rich sources of protein.

Keywords: Bouillon cubes, fermented condiments, protein, *Ricinus communis*

Contact Address: Olaide R. Aderibigbe, National Horticultural Research Institute, Product Development Programme, Jericho G.R.A., 200031 Ibadan, Nigeria, e-mail: connecttolaide@yahoo.com

Comparison of Freeze-thaw and Enzymatic Pre-Treatments to Improve Peeling Process of Cassava Tubers

Ziba Barati, Sajid Latif, Sebastian Romuli, Joachim Müller

University of Hohenheim, Inst. of Agric. Sci. in the Tropics (Hans-Ruthenberg-Institute), Germany

Peeling is still considered as a main problem of cassava processing due to irregular shapes and sizes of cassava tubers. Different peeling methods comprising manual, mechanical, chemical and thermal methods to improve peeling process of cassava tubers have been reported. In this study, the effect of two different pre-treatments including freeze-thaw and enzymatic pre-treatment on the peeling process of cassava tubers was investigated and compared. A prototype cassava peeling machine, which had been constructed at University of Hohenheim, was used to peel the cassava tubers. Response surface methodology (RSM) was applied to optimise the freeze-thaw and enzymatic pre-treatment to increase the peeling performance. The variables were thaw temperature, thaw incubation time, peeling time and rotational speed in freeze-thaw pre-treatment and enzyme dose, incubation time and peeling time in enzymatic pre-treatment. Peeled surface area (%) and peel loss (%) were determined as the main responses, whereas multivariate correlations were well established through polynomial models. As a result, peeled surface area and peel loss were significantly influenced by freeze-thaw pre-treatment and enzymatic pre-treatment ($p < 0.05$). Under an optimal peeling condition, the peeled surface area and peel loss was 94.9 % and 21.7 % for freeze-thaw and 89.52 % and 24.61 % for enzymatic pre-treatment, respectively. Freeze-thaw pre-treatment could result in a higher peeled surface area than enzymatic pre-treatment, but also lead to a syneresis phenomenon that influenced the quality of the tubers in terms of texture and water content. On the other hand, enzymatic pre-treatment could increase the detoxification of the peels by hydrolysis. Application of both freeze-thaw and enzymatic pre-treatment can improve the peeling process of cassava tubers, but would add extra costs for cassava processing. The feasibility of these methods at industrial scale should be further investigated.

Keywords: Biochemical pre-treatment, cassava peeling, peeling efficiency, Response surface methods, thermal pre-treatment

Contact Address: Ziba Barati, University of Hohenheim, Inst. of Agric. Sci. in the Tropics (Hans-Ruthenberg-Institute), Garbenstr.9, 70599 Stuttgart, Germany, e-mail: barati@uni-hohenheim.de

Effect of Heating on Fatty Acid Composition of Edible Oils and Fat

Filip Vojacek

Czech University of Life Sciences Prague, Department of Sustainable Technologies, Czech Republic

Frying of food is one of the most popular methods for food preparation. Oil is during the frying process subjected to aggressive conditions such as high temperature, and the presence of oxygen and water. These conditions cause thermal-oxidative degradation changes in the oil according to the stability of the oil. These changes influence a fatty acid composition of the oil. In this study has been examined the influence of constant heating of oil on its fatty acid composition. Seven types of oil and one type of fat were heated up at 190°C in a commercial fryer. The heating process steadily continued for 24 hours and after every 4th-hour samples were collected. Fatty acid profiles were analysed after their transesterification according to the method of E.W. Hammon (2003) to methyl esters by gas chromatography coupled with mass spectrometry. The changes in the relative representation of fatty acids were monitored and analysed. The results were statistically analysed and graphically presented. Significant decreasing in the representation of unsaturated fatty acids due to the decomposition of double bonds was found. This decomposition occurs significantly after the eighth hour of heating. The most significant changes were in the content of linoleic acid and oleic acid. Their content after the heating period was decreased in all tested samples. The most significant increase occurred in stearic acid content. The results show that in all tested oils and the fat were the most affected the polyunsaturated fatty acids and that the oils become more saturated because of heating.

Keywords: Edible oil, fatty acids, transesterification

Contact Address: Filip Vojacek, Czech University of Life Sciences Prague, Department of Sustainable Technologies, StÅ□eleckÃ¡ 1063, 53701 Chrudim, Czech Republic, e-mail: vojacekf@ftz.czu.cz

Influence of Self-Compaction on the Airflow Resistance of Grain Bulks

Iris Ramaj, Steffen Schock, Shkelqim Karaj, Joachim Müller
University of Hohenheim, Inst. of Agric. Sci. in the Tropics (Hans-Ruthenberg-Institute), Germany

Aeration practices have been widely employed to force conditioned air through in-storage grain bulks to guarantee quality preservation and safe storage. Despite the attention given in the last decades as a principal grain management technique, many obstacles have been encountered in reducing the deteriorative effects of high moisture content and temperature throughout the in-storage bulk. This was attributed to the misestimation of the resistance of stored grains to the airflow, which led to deficiencies in bulk aeration. This resistance is complex and strongly dependent on airflow and grain properties. In this study, the airflow resistance of wheat grains (Pioneer A DSV AG, 12.37 % w.b moisture content) during storage was investigated. A cylindrical, stationary bed (0.5 m diameter and 3.6 m height) was used as an experimental basis. A coherent set of airflow velocities ranging from 0.01 to 0.15 $m\,s^{-1}$ and storage times ranging from 1 to 236 h at four grain depth levels were applied accordingly. The relationship between pressure drop and velocity was assessed experimentally and modeled theoretically with an overall goodness of fit of $R^2 = 0.99$, RMSE = 25.7, and MAPE = 10.4. Results demonstrated an increase of the airflow resistance throughout the depth of the grain bulk and storage time. This behavior was ascribed to the self-compaction of the bulk material arising from the burden pressures imposed by the dead weight of the bulk. The self-compaction decreased the porosity significantly, increased the bulk density, enlarged the airflow resistance and consequently, considerably increased the pressure drop. Hence, extra power supplies for aeration are prerequisites to overcome the resistance caused by self-compaction. The spatial and temporal effects of self-compaction in stored grain bulks should be accommodated in the design and analysis of aeration systems.

Keywords: CAD, compaction, grain aeration, mathematical modelling, physical properties, wheat

Contact Address: Iris Ramaj, University of Hohenheim, Agricultural Engineering in the Tropics and Subtropics (440e), Garbenstr. 9, 70599 Stuttgart, Germany, e-mail: Ramaj@uni-hohenheim.de

Management of Postharvest Fruit Softening and Quality of Peach by Calcium Chloride

Sami Ullah[1], Ahmad Sattar Khan[2], Kashif Razzaq[1], Ishtiaq Ahmad Rajwana[1], Muhammad Amin[1], Hafiz Nazar Faried[1], Gulzar Akhtar[1]
[1]*MNS-University of Agriculture, Dept. of Horticulture, Pakistan*
[2]*University of Agriculture, Inst. of Horticultural Sciences, Pakistan*

Application of calcium chloride ($CaCl^2$) has been reported to delay ripening of fruit. Hence, a study was conducted to examine the effects of postharvest CaCl2 application on peach fruit (*Prunus persica* L. Batsch. cv. 'Flordaking') softening and quality characteristics during shelf life and cold storage. The fruit after harvest were dipped for 5 min in aqueous solutions containing different concentration of $CaCl^2$ viz. 0, 2 %, 4 % or 6 % + Tween 20® (1 g L^{-1}). The treated fruit were kept at ambient conditions (25 $\pm$ 2 °C; RH 60–65 %), under cold storage (0 $\pm$ 1 °C; RH 80–85 %) and at ripening following by cold storage. The fruit were evaluated for physiological (ethylene production, respiration rate, weight loss, fruit firmness); biochemical fruit quality characteristics;, fruit firmness, activities of cell wall hydrolyzing enzymes and antioxidative during ripening, under cold storage and at ripening followed by cold storage as well. Postharvest application of 6 % $CaCl_2$ significantly reduced ethylene production and respiration rate during ripening, at ambient and following cold storage. Reduced fruit softening and activities of fruit softening enzymes including PE, EGase, endo-PG and exo-PG were exhibited by 6 % $CaCl_2$-treated fruit as compared with untreated fruit. Lowest fruit weight loss, SSC, and highest fruit firmness were observed in 6 % $CaCl_2$-treated peach fruit than untreated fruit during ripening, after storage and at ripening following cold storage. Application of 6 % $CaCl_2$ significantly enhanced the antioxidant scavenging activity (ASA), total phenolic contents (TPC) and activities of antioxidative enzymes including CAT, POD and SOD than untreated peach fruit during ripening, under cold storage and at ripening followed by cold storage. Although application of 6 % $CaCl_2$ significantly reduced ethylene production, fruit softening, fruit softening enzymes activities and increased anti-oxidative activity, peach fruit treated with higher concentration of $CaCl_2$ (6 %) exhibited skin discolouration and superficial pitting during shelf life ripening following by cold storage. Keeping the above deleterious effects of higher dose, 4 % $CaCl_2$ was more preferable to maintain fruit softening and quality of peach fruit during ripening and cold storage.

Keywords: Antioxidants, fruit ripening, low temperature storage

Contact Address: Sami Ullah, MNS-University of Agriculture, Dept. of Horticulture, 60000 Multan, Pakistan, e-mail: sami.ullah1@mnsuam.edu.pk

Response Surface Methodology Models for Optimisation of Traditional Fermentation of Cowpea Leaves

Joshua Owade[1], George Abong'[1], Michael Okoth[1], Agnes Mwang'ombe[2]
[1]*University of Nairobi, Dept. of Food Science, Nutrition and Technology, Kenya*
[2]*University of Nairobi, Dept. of Plant Science and Crop Protection, Kenya*

Cowpea leaves is one of the African leafy vegetables that has been promoted to mitigate food and nutrition insecurity in arid and semi-arid lands of western and eastern Africa. Spontaneous fermentation of this vegetable is one of the traditional processing techniques that have been heavily utilised to improve the keeping quality and sensory attributes while inadvertently imparting health benefits. However, the vegetable has been shown to have minimal fermentable sugars for optimal action of the bacteria thus the process tends to be slow and the product quality less optimal. The current study utilised the central composite design of the response surface methodology (RSM) to optimise the spontaneous fermentation of cowpea leaves. The RSM model generated 20 runs with the independent variables being concentrations of sugar and salt and period of fermentation in days whereas the pH and titratable acidity of the fermented vegetable were the response variables. The models showed significant ($p < 0.01$) changes in the pH and titratable acidity with R^2 of 0.89 and 0.60, respectively. Change in the concentration of sugar and period of fermentation significantly ($p < 0.05$) affected the pH and titratable acidity of the fermented vegetables. Salt concentration and interaction of the independent variables did not significantly ($p > 0.05$) influence the changes in the response variables. The RSM model generated the optimal fermentation conditions as 2 % salt and 5 % sugar concentration and 16 days fermentation period; optimal response variables were 3.8 and 1.23 % for pH and titratable acidity, respectively with a desirability of proportion of 85.9 %. Optimal concentrations of sugar and salt and period of fermentation improved the action of the natural culture in fermented cowpea leaves. To avert the challenges of poor product quality and slow fermentation process among the traditional communities, the addition of sugar and salt to the vegetables and optimisation of fermentation period should be observed.

Keywords: Desirability proportion, fermentation, response variables, titratable acidity

Contact Address: Joshua Owade, University of Nairobi, Dept. of Food Science, Nutrition and Technology, Nairobi, Kenya, e-mail: owadehjm@gmail.com

Rheological Properties of Orange Fleshed Sweet Potato, Pumpkin and Wheat Blended Flour Doughs and Quality Characteristics of Breads

Solomon Kofi Chikpah[1], Joseph Kudadam Korese[2], Oliver Hensel[1], Boris Kulig[1], Barbara Sturm[1], Elke Pawelzik[3]

[1]*University of Kassel, Agricultural and Biosystems Engineering, Germany*

[2]*University for Development Studies, Agricultural Mechanization and Irrigation Technology, Ghana*

[3]*University of Goettingen, Dept. of Crop Science, Division of Quality of Plant Products, Germany*

There is increased demand for new bread products with health promoting benefits and good eating qualities. Orange fleshed sweet potato (OFSP) and pumpkin are high in bioactive compounds and fibre that can enrich the nutritional values of bakery products. However, incorporating gluten free flours such as OFSP and pumpkin flours into wheat flour may affect the quality characteristics of dough and bread. The study assessed the rheological properties of OFSP-pumpkin-wheat flour blended doughs as well as physicochemical and textural properties of breads baked at different temperatures and times. The constraint mixture design was used for the experimental design. The flours limits in the mixtures were: 40%≤wheat≤ 90 %, 10%≤OFSP≤60 %, and 10%≤pumpkin≤40 % whereas breads were baked at temperature and time ranged between 150–200 °C and 15–25 min, respectively. Farinograph analysis of dough showed that optimum water absorption, dough development, stability and degree of softening of doughs varied significantly ($p < 0.05$) between 50.8–60.1 %, 2.2–33.8 min, 6.0–50.3 min and 9.0–138.0 BU, respectively. Moreover, increased incorporation of OFSP and pumpkin flours significantly ($p < 0.05$) decreased optimum water absorption and degree of softening whereas dough development and stability increased. The flour mixtures, baking temperature and time significantly ($p < 0.05$) influenced breads physicochemical and textural properties. Baking loss, loaf volume, loaf specific volume, crumb moisture and water activity of composite breads reduced while loaf weight increased as OFSP and pumpkin flours levels increased. Bread crusts and crumbs lightness (L*) reduced whereas redness (a*) and yellowness (b*) increased with increased addition of OFSP and pumpkin flours. Additionally, baking at higher temperature (>180 oC) for longer time (>19 min) caused a significant ($p < 0.05$) reduction in loaf weight, volume, specific volume, L* and b* of breads while a* increased. Furthermore, textural analy-

Contact Address: Solomon Kofi Chikpah, University of Kassel, Agricultural and Biosystems Engineering, Nordbahnhofstrasse 1a, 37213 Witzenhausen, Germany, e-mail: schikpah@uds.edu.gh

sis of bread crumbs showed that hardness and chewiness increased significantly ($p < 0.05$) whereas cohesiveness, springiness and resilience decreased as OFSP and pumpkin flours, baking temperature and time increased. Generally, OFSP and pumpkin flours have potentials to replace wheat flour for bread development at inclusions not exceeding 33 % and 12 % respectively and baking at 160–180 °C for 15–19 min.

Keywords: Baking temperature, composite bread, orange fleshed sweet potato, pumpkin, rheological properties, textural properties

Post-Harvest Handling Knowledge and Hygiene Practices of Cowpea Leaves in Kitui and Taita Taveta Counties

Ann Miano[1], George Abong'[1], Lucy Njue[1], Agnes Mwang'ombe[2]
[1]*University of Nairobi, Food Science, Nutrition and Technology, Kenya*
[2]*University of Nairobi, Plant Science and Crop Protection, Kenya*

Post-harvest losses contribute to about 50 % of total losses in cowpea value chain. Losses could be as high as 50–70 %. The study aimed at assessing the post-harvest handling and storage practices of cowpea leaves by farmers and traders in Kitui and Taita Taveta counties. About 405 respondents of equal distribution in both counties, were interviewed using a semi-structured questionnaire recorded on digital open data kit. Most of the respondents 65.9 % lacked basic handling and hygiene knowledge besides believing that harvesting of cowpea leaves during hot periods does not lead to any form of deterioration. In both regions low levels of education were observed with one out of every 20 persons having university and tertiary education while the highest percentage, having completed primary education (36.0 %). About 62.2 % of the respondents consumed cowpea leaves sourced from their farms while during scarcity, they obtain the vegetable from the local vendors. Cowpea leaves were packaged in sacks (54.6 %) as the primary and only packaging material then stored in granaries or transported to the market. Most farmers (95.8 %) incurred up to 10 % post-harvest losses while the rest incurred up to 30 % loss during transportation of the cowpea leaves to the markets. Post-harvest losses are majorly attributed by poor storage facilities, 58 % of the farmers lacking these facilities. About 45 % of the farmers do nothing to prevent these losses while some adopt different methods with the majority using sun drying (37.8 %) to extend the vegetables shelf-life. Farmers post-harvest handling and storage practices of cowpea leaves are poor leading to high losses. Improvement of good post-harvest practices to reduce losses is recommended. New and several other post-harvest handling and storage technologies should be developed and disseminated to the farmers in order to ensure quality and safety of cowpea leaves.

Keywords: Cowpea leaves, post-harvest, post-harvest handling, post-harvest losses

Contact Address: Ann Miano, University of Nairobi, Food Science, Nutrition and Technology, 29053, 00625 Kangemi-Nairobi, Kenya, e-mail: annmiano49@gmail.com

Effect of Flour Blends and Baking Conditions on the Quality of Orange Fleshed Sweet Potato-Pumpkin-Wheat Composite Breads During Storage

Solomon Kofi Chikpah[1], Joseph Kudadam Korese[2], Oliver Hensel[1], Barbara Sturm[1], Elke Pawelzik[3]

[1]*University of Kassel, Agricultural and Biosystems Engineering, Germany*

[2]*University for Development Studies, Agricultural Mechanization and Irrigation Technology, Ghana*

[3]*University of Goettingen, Dept. of Crop Science, Division of Quality of Plant Products, Germany*

Changes in moisture content and textural properties of bakery products during storage may be detrimental to the products' eating qualities and shelf-life. For example, excessive crumb moisture loss and firming can adversely affect eating quality. This study investigated the effects of flour blends, baking temperature and time on moisture, water activity (aw), textural properties and shelf life of orange fleshed sweet potato (OFSP)-pumpkin-wheat composite breads' crumbs during storage. The constraint design of the Design Expert software was used for the experimental design. OFSP-pumpkin-wheat flour blends were formulated using the flour limits: wheat (40–90 %), OFSP (10–60 %) and pumpkin (10–40 %). The breads were baked at different temperatures (150, 160, 170, 180,190 and 200 °C) and baking time of 15–25 min. Moisture, aw and textural profile analysis were conducted on the fresh bread crumbs and stored products at 24h interval for 5 days in a climatic chamber at 25 °C and 50 % RH. The rate of staling and mold growth were also assessed during storage. The study showed that flour blends, baking temperature and time significantly ($p < 0.05$) influenced crumb quality properties and shelf-life of OFSP-pumpkin-wheat composite bread during storage. It was observed that, bread crumb moisture content and aw decreased during storage by 5.5–34.7 % and 2.3–31.5 % respectively. Moreover, increased incorporation of OFSP and pumpkin flours slowed the rate of moisture loss from bread crumb and lowered aw during storage. Additionally, breads baked at higher temperatures (≥ 180 °C) for longer time (≥ 19 min) had a faster rate of moisture loss from crumbs. Furthermore, textural profile analysis of bread crumbs showed significant ($p < 0.05$) increased in crumb hardness and chewiness whereas cohesiveness, springiness and resilience reduced during storage. Generally, bread products that contained high levels of wheat flour (>70 %) and products baked at low temperatures for short time (≤ 170 °C and ≤ 19 min) had higher rate of staling and mold growth. The study showed that OFSP and pumpkin flours could be included in wheat flour at ≤33 % and ≤ 12 %, respectively as well as baking temperature of 170–180 °C for 17–19 min could be applied to enhance bread crumb quality properties and shelf-life during storage.

Keywords: Bread crumb, staling, textural profile

Contact Address: Solomon Kofi Chikpah, University of Kassel, Agricultural and Biosystems Engineering, Nordbahnhofstrasse 1a, 37213 Witzenhausen, Germany, e-mail: schikpah@uds.edu.gh

Valorisation of By-products from Coffee Processing to Improve the Sustainability of the Coffee Value Chain in Vietnam

HUYEN CHAU DANG[1], JUDY LIBRA[1], MARCUS FISCHER[1], CLAUDIA GLASER[2], CHRISTINA DORNACK[3]

[1]*Leibniz Institute for Agricultural Engineering and Bioeconomy (ATB), Post Harvest Technology, Germany*
[2]*Brandenburg University of Technology, Processing technology, Germany*
[3]*Dresden University of Technology, Inst. of Waste Management and Circular Economy, Germany*

Fulfilling the demand for high-quality agricultural products on the world market, while providing adequate income for farmers, is a challenge for many developing countries. In order to meet increasing worldwide quality standards, farming and processing practices often must be adapted. If the changes lead to intensification and centralisation in the agro-food value chains, traditional recycling pathways for agricultural residues between the field and processing can be disrupted, negatively impacting soil fertility, environmental and human health. Sustainable solutions for these by-products are required that exploit their valuable properties and create environmentally-sound products.

In the current project, CoffeeChar, a collaboration between Germany and Vietnam, coffee processing methods (wet and dry), their by-products, and disposal in Vietnam are being studied. The objective of the project is to develop innovative solutions for recycling wet solid by-products from 1) the wet-processing method for coffee berries, which is growing in importance to produce higher quality coffee beans and from 2) secondary processing of the coffee beans. The thermal conversion process, hydrothermal carbonisation (HTC), is being investigated for its potential to convert wet coffee by-products into carbon-rich material that can replace fossil fuel in coffee processing or households. In addition, the use of HTC can reduce the amount of solid waste treated in the (often overloaded) wastewater treatment plant of the factory. In this contribution, the results of the initial assessment of the status of coffee processing and by-products in Vietnam will be presented along with experimental results on the production of fuel pellets from HTC-char. Agglomeration and pelletizing experiments have been conducting to find out the optimal process for producing high-quality fuel pellets. The outcome of this project will support farmers and producers to improve the sustainability of the coffee value chain in Vietnam, and also provide a basis for its adaptation to other coffee production regions.

Keywords: Agglomeration, carbon-rich material, coffee by-products, coffee processing, coffee value chain, hydrothermal carbonisation (HTC), pelletizing, sustainability

Contact Address: Huyen Chau Dang, Leibniz Institute for Agricultural Engineering and Bioeconomy (ATB), Post Harvest Technology, Max-Eyth-Allee 100, 14469 Potsdam, Germany, e-mail: hdang@atb-potsdam.de

Quality of Baobab Products Across their Value Chain in Malawi

Kathrin Meinhold[1], Chimuleke R. Y. Munthali[2], Dietrich Darr[1]

[1]*Rhine-Waal University of Applied Sciences, Fac. of Life Sciences, Germany*
[2]*Mzuzu University, Department of Forestry, Malawi*

The baobab (*Adansonia digitata* L.) is an important indigenous fruit tree in sub-Saharan Africa. In Malawi the fruit is heavily commercialised, with a wide range of products available in informal and formal markets. Commercialisation has substantially intensified in the last decades, yet little information is available on the quality of baobab products on the market, as well as how quality might differentiate depending on the setup of the supply chain. In particular, it is unclear how handling of baobab may affect the high Vitamin C levels baobab is known for and whether microbiological contamination occurs. Therefore, this study aimed at characterising the baobab value chain in Malawi, determine important quality characteristics of baobab products at different stages of the supply chain as well as how supply chain configuration may affect product quality. To achieve this, a mixed methods approach was pursued, using semi-structured interviews with different members of the value chain (67 respondents in total), concurrent collection of baobab samples (fruit, pulp-on-seed, powder, 79 samples in total) and subsequent laboratory analysis (Vitamin C and mycotoxins). The main actors of baobab value chain in Malawi were identified as collectors, rural, semi-urban and urban traders, informal processors producing baobab ice lollies and juice, formal juice processors, as well as retailers. Preliminary results indicate that dryness and colour of fruit pulp are typically considered the most important quality indicators across different supply chain configurations; while other factors such as colour of or cracks in the baobab shell are more heavily disputed. Vitamin C levels in the samples varied greatly, with mean levels ranging from 127.7 $\pm$ 45.1 mg/100 g for fruit, 121.5 $\pm$ 44.9 mg/100 g for pulp-on-seed, to 68.7 $\pm$ 56.0 mg/100 g in powder samples. Mycotoxins were detected in one fruit and one pulp-on-seed sample (3.2 and 9.1 µg per kg, respectively). The results indicate that it is advisable to store baobab in the fruit and only process into powder as late as possible. Since mycotoxins can be present in baobab products and children are the dominant consumer group in Malawi, more awareness with regard to handling and processing baobab is required.

Keywords: Baobab, indigenous fruit tree, mycotoxins, quality, vitamin C

Contact Address: Kathrin Meinhold, Rhine-Waal University of Applied Sciences, Fac. of Life Sciences, Marie-Curie-Str. 1, 47533 Kleve, Germany, e-mail: kathrin.meinhold@hochschule-rhein-waal.de

Effect of Incorporating Peeled and Unpeeled Orange Fleshed Sweet Potato Flour on Rheological Properties of Dough and Quality Characteristics of Bread

Solomon Kofi Chikpah[1], Joseph Kudadam Korese[2], Oliver Hensel[1], Barbara Sturm[1], Elke Pawelzik[3]

[1]*University of Kassel, Agricultural and Biosystems Engineering, Germany*

[2]*University for Development Studies, Agricultural Mechanization and Irrigation Technology, Ghana*

[3]*University of Goettingen, Dept. of Crop Science, Division of Quality of Plant Products, Germany*

Utilisation of alternative flours from underutilised crops rich in essential nutrients in the bakery industry would help improve nutrition and reduce wheat flour importation in Sub-Saharan African countries. However, wheat flour substitution with gluten free flours could significantly affect quality properties of baked foods. Therefore, this study compared the effect of replacing wheat flour with peeled and unpeeled orange fleshed sweet potato (OFSP) flour on dough and bread quality properties. The experiments were designed using the constraint mixture design. Wheat flour was replaced with either peeled or unpeeled OFSP flour between 10–60 % and breads were baked between 150–200 °C for 15–25 min. Dough and bread quality properties were analysed using standard equipment and procedures. Farinograph results showed peeled OFSP composite dough had lower optimum water absorption for consistent dough development (54.5–60.1 %) and degree of softening (59–134 BU) but longer dough development time (2.7–11.2 min) and higher stability (6.5–13.6 min) than the unpeeled OFSP composite doughs whose values ranged between 59.8–63.0 %, 92–168 BU, 2.7–9.2 min and 4.7–7.3 min respectively. The optimum water absorption reduced greatly while dough development increased as peeled or unpeeled OFSP flour addition exceeded 45 %. Moreover, peeled OFSP composite breads had a significantly ($p < 0.05$) higher loaf volume (174–330 cm^3/100 g flour) and specific volume (1.573–2.613 cm^3 g^{-1}) than the unpeeled OFSP composite breads (152–280 cm^3 and 1.381–2.297cm^3 g^{-1}). The crumb moisture and water activity were slightly lower in the peeled OFSP composite breads (25.32–31.43 % and 0.866–0.917, respectively) than the unpeeled OFSP composite breads (25.73–36.50 % and 0.874–0.925, respectively). The crust and crumb colour L*, a* and b* values were similar ($p > 0.05$) among the corresponding peeled and unpeeled OFSP composite breads. Textural profile analysis of bread crumbs showed peeled

Contact Address: Solomon Kofi Chikpah, University of Kassel, Agricultural and Biosystems Engineering, Nordbahnhofstrasse 1a, 37213 Witzenhausen, Germany, e-mail: schikpah@uds.edu.gh

OFSP composite breads had significantly ($p < 0.05$) higher hardness, springiness, chewiness, resilience but similar cohesive values as compared with the unpeeled OFSP composite breads. Generally, loaf volume, specific volume, crumb cohesiveness, springiness and resilience decreased while hardness and chewiness increased as OFSP flour levels, baking temperature and time increased. Peeled OFSP flour could be used to substitute wheat flour for bread production at inclusion rate of ≤35 % and baking between 160–180 °C for 15–17 min.

Keywords: Crumb hardness, dough development, loaf volume, optimum water absorption, orange fleshed sweet potato, springiness

Food and nutrition

Food safety and security

Oral Presentations

Posters

Climate Variability and Rural Children Health Outcomes

Emily Injete Amondo, Alisher Mirzabaev
University of Bonn, Center for Development Research (ZEF), Economics and Technological Change, Germany

Children in rural farming households are often vulnerable to a multitude of risks, including health risks associated with climate change and variability. Cognizant of this, this study empirically traced the relationship between climate variability and nutritional health outcomes in rural children, while identifying the cause-and-effect transmission mechanisms. We combined four waves of the rich Uganda National Panel Survey (UNPS), part of the World Bank Living Standards Measurement Studies (LSMS) for the period 2009-2014, with long-term gridded and high frequency rainfall, and temperature datasets. In particular, rainfall and temperature datasets were from Climate Hazards group Infrared Precipitation Station (CHIRPS) and Moderate Resolution Imaging Spectroradiometer (MODIS) products respectively. Self-reported drought and flood shock variables were further used in separate regressions for triangulation purposes and robustness checks. Panel fixed effects regressions were applied in the empirical analysis, accounting for a variety of causal identification issues.

The results showed significant negative outcomes for children's anthropometric measurements due to the impacts of moderate and extreme droughts, extreme wet spells and heat waves. On the contrary, moderate wet spells were positively linked with nutritional measures. Agricultural production and child diarrhea were the main transmission channels, with heat waves, droughts and high rainfall variability negatively affecting crop output. Probability of diarrhea was positively related to increases in temperature and dry spells. Results further revealed that children in households who engaged in *ex-ante* or anticipatory risk reducing strategies such as savings had better health outcomes as opposed to those engaged in *ex-post* coping such as involuntary change of diet. These results highlight the importance of adaptation in smoothing the harmful effects of climate variability on health of rural households and children in Uganda.

Keywords: Agricultural production, children, diarrhea, drought, high temperatures, satellite weather data, undernutrition

Contact Address: Emily Injete Amondo, University of Bonn, Center for Development Research (ZEF), Economics and Technological Change, Genscherallee 3, 53113 Bonn, Germany, e-mail: emijete@gmail.com

Urban Food System of Kampala, Uganda: a Participatory Approach to Map Systemic Drivers for Healthier Diets

Andrea Fongar[1], Youri Dijkxhoorn[2], Beatrice Ekesa[1], Vincent Linderhof[2]
[1]*The Alliance of Bioversity International and CIAT, Healthy Diets from Sustainable Food Systems Initiative, Uganda*
[2]*Wageningen University and Research, Wageningen Economic Research, The Netherlands*

Africa is urbanizing, poverty and malnutrition hotspots are moving from rural to urban communities. The urban population is faced with malnutrition in all its forms. Personal and external food environments are changing. The questions "Who is" and "Why are they affected" remain unclear. The project NOURICITY seeks to answer these questions by engaging in a chain of multi-stakeholder participatory activities meant to deliver a blueprint for partnerships seeking to improve urban people towards healthier diets in Africa.
A series of participatory approaches are applied to understand the current food system of Kanyanya parish, Kawempe division in Kampala, Uganda. First, stakeholders at policy and parish levels were engaged in two separate workshops to collect information on urban food system challenges and to map with geo-references the common food outlets. Secondly, three Focus Group Discussions (FGD) were conducted with 30 parish representatives each (disaggregated into women, men and youth) in 2020.
Identified challenges within the food system included elements of the food supply chain (post-harvest losses, poor transport), food environment (availability, affordability, food safety) and consumer behaviour (waste, awareness). A key limitation to eat healthy, stakeholders mentioned the lack of cash, the price and availability of food, as well as the awareness of healthy and safe food. The main source of information regarding food safety and nutrition was the community radio and meetings. In total, 500 formal and informal food outlets locations were mapped, including kiosks, food markets, restaurants and hotels, butcheries, dairies, and supermarkets.
The participatory engagement of the community identified not only challenges and limitations but also gave opportunities and showed the willingness of the community members to act. An entry point to enhance awareness and dietary knowledge could be the local community radio, community meetings and leaders.

Keywords: Food environment, food mapping, Kampala, participatory approach, Uganda, urban areas

Contact Address: Andrea Fongar, The Alliance of Bioversity International and CIAT, Healthy Diets from Sustainable Food Systems Initiative, Plot 106 Katalima Road, Naguru, Kampala, Uganda, e-mail: a.fongar@cgiar.org

Analysis of Leafy Vegetables for Antioxidant Activity, Ascorbic Acid and Nitrate in Morogoro, Tanzania

AMINA AHMED, GUDRUN B. KEDING, ELKE PAWELZIK

University of Goettingen, Dept. of Crop Science, Division of Quality of Plant Products, Germany

Leafy vegetables are an important part of the diet in Morogoro region, Tanzania, and are used, among others, to make sauces. They are rich in several phytochemicals including antioxidants, ascorbic acid and dietary nitrate, which play important roles in human nutrition and health. However, a higher concentration of nitrate in leafy vegetables has been associated with health risks and bitter flavour that may affect acceptability. Knowing the nutritional value of these traditional vegetables would increase consumer awareness, intensify their inclusion in diets and contribute to food and nutrition security. The present study aimed to screen antioxidant activity, ascorbic acid and nitrate contents of African leafy vegetables from Morogoro, Tanzania. Fresh vegetable samples of African nightshade (*Solanum spp.*), cassava (*Manihot esculenta*) and cowpea (*Vigna unguiculata*) leaves were collected from 23 farms and 9 markets within 2 districts in Morogoro region. Antioxidant activity was analysed using 2, 2-diphenyl-1-picrylhydrazyl (DPPH) solution while the ascorbic acid and nitrate contents were analysed semi-quantitavely using strips. The antioxidant activity ranged from 17 % to 64 % which is comparable to citrus fruits. Cassava leaves showed the highest value (55.96 %) followed by cowpea leaves (43.02 %) and African nightshade leaves (29.95 %). The ascorbic acid content ranged from 491 to 7000 mg/L. Cassava leaves showed the highest value (55.96%) followed by cowpea leaves (43.02%) and African nightshade leaves (29.95%). The ascorbic acid content ranged from 491 to 7000 mg L-1. The highest ascorbic acid content was observed in cassava leaves (4419 mg L-1) followed by African nightshade (1464 mg L^{-1}) and cowpea leaves (1283 mg L^{-1}) which is much higher compared to citrus fruits (literature reports: 300 to 990 mg L^{-1}). Nitrate content ranged from 95 to 5000 mg L^{-1} being highest in African nightshade (2735 mg L^{-1}) followed by cowpea leaves (647 mg L^{-1}) and cassava leaves (295.98 mg L^{-1}). However, the observed nitrate contents were about 60 % to 90 % lower compared with spinach and about 15 to 90 % lower compared with collard greens as reported in the literature. The disparate nutrient values were related to fertiliser application, leaf age and harvest time as reported by farmers and market vendors where the leaves were collected. The high contents of these essential nutrients and comparably low values of nitrate show the nutritional importance of these African leafy vegetables.

Keywords: Antioxidants, ascorbic acid, health, leafy vegetables, nitrate, nutrition

Contact Address: Amina Ahmed, University of Goettingen, Dept. of Crop Science, Division of Quality of Plant Products, 37075 Carl-Sprengel Weg 1, 37075 Goettingen, Germany, e-mail: aahmed1@gwdg.de

From Small-Scale Food Production to Consumption in Peru: First Effects of COVID-19

Liza Melina Meza Flores, Karla Gabaldoni Arias-Schereiber

Slow Food Peru, Slow Food Community Lima, Peru

Peru is a country of food diversity, which is directly linked to its 36 ecosystems, varying from Andean, Amazonian , coastal, dry forests, moors and aquatic ecosystems. This richness derives in a diversity of cultures and food consumption habits. In this context, family farmers feed 80 % of the Peruvian population and micro and small enterprises represent 95 % of all enterprises nationwide. But despite having legislation for promoting and developing these small organisations, they face many limitations regarding the maintenance of production, distribution and marketing. The situation has been aggravated due to COVID-19. However, since the pandemic started, the agro-exporting companies are the ones that have increased their sales by 6 %. Additionally, Peru is importing potato and potato flour from European countries, while family farmers of native potatoes are being paid cents per kilo of their products. In this paper we argue that the food system in Peru should be developed at the small scale level with short chains, guaranteeing diversity all year around. To achieve it, first, the governmental agencies must purchase healthy and nutritious foods from local family farmers, livestock owners, harvesters, fishermen and processors within the framework of Food Complementary Programs. Second, stakeholders at all levels must support solidarity baskets development and consumers' coops formations. With these two proposals, local economies are revitalised, while guaranteeing a nutritious, regional and even native food, directly from small producers and in accordance with the traditions of the localities . We found that it is also key to improve the cold chain, working in cooperation with restaurants and processing plants, minimising food waste. Governmental Agencies, in cooperation with NGOs and civil society organisations, such as Slow Food, must combine and generate an accessible and updated database of producers and formal carriers. Finally, as an operational support, simple manuals on the documents, sanitary and formalisation requirements and procedures for the transport and commercialisation of food, in Spanish and local languages should be created and shared all around the country. Finally, to protect the highly nutritious native biodiversity, the moratorium on GMOs must be maintained to guarantee food sovereignty and adaptation to climate change.

Keywords: COVID-19, direct purchases, food distribution, food sovereignty, food systems, short chains

Contact Address: Liza Melina Meza Flores, Slow Food Peru, Slow Food Community Lima, Av. Panamericana Sur 251, Lima 4 Lima, Peru, e-mail: lizameflo@yahoo.com

Potential Sources and Types of Food Safety Hazards in Selected Tomato Supply Chain of Ethiopia

Eshetu Abrham Gebre[1], Yetenayet Bekele Tola[2], Sirawdink Fikreyesus Forsido[2]

[1]*Jimma University, Horticulture and Plant Sciences, Ethiopia*
[2]*Jimma University, Post-harvest Management, Ethiopia*

Increased consumption of fresh (raw) fruit and vegetables is recommended for a healthy life. However, in countries where hygiene standards may be low, the chances of consuming contaminated fresh food, particularly vegetables, are high. In line with this, a survey was conducted to assess potential sources and types of food safety hazards of tomato fruits in its supply chain from small scale growers in central rift valley areas to central markets of Ethiopia.
Data were collected from both primary and secondary sources. The primary data were generated by a survey using a pre-tested structured questionnaire, key informant interviews using checklists and focus group discussions. Data on potential sources and types of food safety hazards were collected from a total of 146 respondents (70 farmers, 27 transporters and 59 traders). The data were analysed using SPSS software. According to the respondents, 11.5 % of the land used for tomato production was previously used for animal production, 29 % of the adjacent farmland is used for a residential area, and most of the groundwater holes were not properly constructed and protected. Regarding surplus pesticide disposal system. 41.4 % indicated that they respray, 5.7 % dispose of it in the soil, and 45.6 % pour into irrigation pond. Regarding disposal of empty pesticide container, 98.6 % hang it on a wooden stalk or throw it in the far*m.* Similarly, 47.1 % and 91.4 % of the farms did not have a toilet and hand-washing facilities, respectively. All of the tomato growers did not receive any training related to food safety. Additionally, 46.7 % of the respondents use neither clean nor appropriate containers. A majority (75 %) of the product is loaded with any domestic good, and the produce is not covered during transportation; similarly, 75 % of product flow zones are not protected from contamination at the market place.
The food safety systems in the tomato supply chain were found to be inadequate. Therefore, the establishment of specific food safety steps and procedures and awareness creation to the actors is needed.

Keywords: Ethiopia, food safety, supply chain, tomato

Contact Address: Sirawdink Fikreyesus Forsido, Jimma University, Post-harvest Management, Jimma University College of Agriculture, 307 Jimma, Ethiopia, e-mail: sirawdink@gmail.com

Nutritional Quality and Food Safety of Processed Small Indigenous Fish Species from Ghana

Laura Wessels[1], Astrid Elise Hasselberg[2], Inger Aakre[1], Felix Reich[2], Amy Atter[3], Matilda Steiner-Asiedu[4], Samuel Amponsah[5,3], Marian Kjellevold[1], Johannes Pucher[2]

[1]*German Federal Institute for Risk Assessment, Germany*
[2]*Institute of Marine Research, Norway*
[3]*Council for Scientific and Industrial Research, Food Research Institute, Ghana*
[4]*University of Ghana, Ghana*
[5]*University of Energy and Natural Resources, Ghana*

Malnutrition is a severe issue in low-and middle-income countries, although fish has the potential to mitigate this burden. In Ghana, fish is a central part of the diet, but data on nutritional value and food safety parameters in processed indigenous fish species are scarce. On markets in five different regions of Ghana, samples of smoked, dried or salted small fish species were collected including *Engraulis encrasicolus* (European anchovy), *Brachydeuterus auritus* (Bigeye grunt), *Sardinella aurita* (Round sardinella), *Selene dorsalis* (African moonfish), *Sierrathrissa leonensis* (West African (WA) pygmy herring) and *Tilapia* spp. (tilapia). The samples were analysed for the content of nutrients (crude protein, fat, fatty acids, several vitamins, minerals, trace elements), microbiological contamination (total colony counts, *E. coli*, coliforms, and *Salmonella*), and levels of contaminants (PAH4, heavy metals). All sampled processed fish species (except for tilapia) showed the potential to significantly contribute to the recommended nutrient intakes of vitamins, minerals, and essential fatty acids. However, high levels of iron, mercury and lead were found in certain fish samples, which necessitates further identification of the sources along the food value chains. In all samples, the total colony counts were in an acceptable range while *E. coli* was found only in one sample. In addition, high numbers of coliform bacteria were found in the samples which calls for improving the hygiene conditions along the value chains. *Salmonella* was not detected in any of the samples. PAH4 in smoked samples reached high concentrations and necessitate the improvement of smoking practices and smoking equipment. This study provides valuable data with respect to nutritional value and potential food safety hazards of small processed fish as food in Ghana. Future research should seek to identify potential sources of contamination and critical points along the value chains, and develop applicable mitigation strategies to improve the quality and safety of small processed fish in Ghana.

Keywords: Fishery products, food quality, food safety, processed fish, nutrition

Contact Address: Johannes Pucher, German Federal Institute for Risk Assessment, Max-Dohrn-Straße 8–10, 10589 Berlin, Germany, e-mail: Johannes.Pucher@bfr.bund.de

Postharvest Handling, Hygiene Knowledge and Practices of Guava Fruit Farmers: A Comparative Study of two Counties of Kenya

Judith Ndeme, Jasper Imungi, George Abong', Charles Gachuiri

University of Nairobi, Dept. of Food Science, Nutrition and Technology, Kenya

In many parts of Kenya, guava (*Psidium guajava*) grows unattended to in the farms and forests. The fruit is a climacteric fruit characterised by high perishability and huge postharvest losses attributed to neglected postharvest management. Little information therefore exists on postharvest handling properties. This study aimed at establishing the postharvest handling, hygiene knowledge and practices of guava farmers. A cross-sectional study of farmers (n=417) was done in arid and semi-arid lands of Kitui and Taita Taveta counties, Kenya. Data on postharvest handling, knowledge on hygiene and practices was collected through semi-structured questionnaires using Open Data Kit (ODK). The results showed that the main indicative maturity indices of guavas to farmers in Kitui and Taita Taveta were skin colour (98.5 %, 92.1 %) and full ripe level (38.7 %, 18.7 %) respectively. Packaging was not a common practice in Kitui and Taita Taveta and did not differ significantly ($\chi 2=8.717$, $p = 0.003$). Storage of the fruit was low (Kitui=41.6 %, Taita=55.2 %) with no significant difference among the two ($\chi 2=7.699$, $p = 0.006$). Shelf life of guavas differed between kitui and Taita Taveta with 3.4 ± 1.8 and 4.2 ± 1.9 days respectively ($t=44.264$, $p < 0.001$). Pest and diseases were more rampant in Kitui (95.33 %) than Taita Taveta (77.8 %), however, both did not apply pesticides for control. A cluster analysis indicated that Kitui farmers (71.9 %) had more knowledge on hygiene and postharvest than those in Taita Taveta (49.7 %). Additionally, female farmers (65.4 %) were more knowledgeable than males (55.4 %). Poor postharvest handling knowledge of guava farmers can be attributed to the huge postharvest losses reported in Kitui and Taita Taveta.

Keywords: Guavas, hygiene, post-harvest handling, post-harvest losses, preservation

Contact Address: Judith Ndeme, University of Nairobi, Dept. of Food Science, Nutrition and Technology, Nairobi, Kenya, e-mail: judithkatumbi@gmail.com

Human nutrition and diets

Oral Presentations

Posters

Dietary Patterns and their Association with Overweight and Obesity among Rural Populations in East Africa

Jacob Sarfo, Elke Pawelzik, Gudrun B. Keding

University of Goettingen, Dept. of Crop Science, Division of Quality of Plant Products, Germany

Nutritional concerns had been focused mainly on hunger and in addressing underweight in sub-Saharan Africa. However, socio-economic, cultural, societal and behavioural change patterns in developing countries have brought about the "nutrition transition" – shift towards more modern diets and its associated health conditions. Studies done so far reveal that dietary patterns contain more and more processed foods in urban areas with less emphasis on rural settings. Also, patterns associated with processed foods do not examine what level of processed foods they are. The "Fruits and Vegetables for all seasons" (FruVaSe) project, as one of its work components, seeks to assess dietary patterns – exploring the level of processed foods within the patterns, if any – that exist among rural populations in East Africa and their association with overweight and obesity.

Household surveys focusing on women of reproductive age (15–49years) were conducted in six study sites in Kenya, Tanzania, and Uganda. Four repeated 24-hour dietary recalls were collected across two seasons (plenty and lean), demographic and socio-economic data, and anthropometric measurements were conducted in only one season. In all, 1800 households were surveyed, however, final analysis consisted of 1153 households. Standard food recipes were prepared to determine intake quantities for the 24-hour recalls.

The percentage of overweight and obese women in Kenya and Tanzania was estimated at 47 % and 42 %, respectively. Overweight and obese data for Uganda was almost twice as little as that of either Kenya or Tanzania (26 %). Dietary patterns across the three countries will be analysed using principal component analysis. Dietary patterns examined will be related to the overweight and obesity data. Possible confounding factors such as wealth, age, nutrition education received, and educational status are to be adjusted for during analysis. It is expected that women with higher BMI follow dietary patterns which contain a greater amount of processed foods.

Keywords: Dietary patterns, East Africa, obesity, overweight, rural communities, women

Contact Address: Jacob Sarfo, University of Goettingen, Division Quality of Plant Products, Carl-Sprengel-Weg 1, 37075 Goettingen, Germany, e-mail: jsarfo@gwdg.de

Nutritional Opportunities of Edible Insects in Madagascar, Case of *Borocera cajani* and *Bombyx mori* Chrysalis

Christian Tolojanahary Ratompoarison, Felamboahangy Rasorahona, Jean Rasoarahona

University of Antananarivo, Department of Food Science and Technology, Madagascar

In Madagascar, food consumption is marked by the predominance of cereals (rice and corn) and roots and tubers (cassava) and the low diversity of the diet. Food population intake is nutrient-poor. Several recent literatures show that insect consumption is common in rural areas. The lack of data on the nutritional composition of insects minimises the attention that nutrition stakeholders may pay to insects and constitutes a brake on the development of its consumption. In the present study, the nutritional potential of the chrysalis of two edible insects, *Borocera cajani* and *Bombyx mori*: the most locally used as food was determined and evaluated for the first time. *B. cajani* and *B. mori* contain 63.98 and 54.37 % crude protein, 29.84 and 35.78 % fat, 4.10 and 5.67 % ash, and 2.06 and 4.36 % carbohydrate, respectively. The essential amino acid (EAA) content shows that the amino acid scores of both insects exceeded the recommendation given by FAO/WHO/UNU (score >100). Amino acid profile shows that polyunsaturated fatty acid (PUFA) was the most predominant fatty acid found in both insects, followed by saturated fatty acid (SFA) and monounsaturated fatty acid (MUFA). Mineral content was generally higher than that of conventional meat types. Both species fulfiled the recommended daily allowance (> 100 %) for Zn and Cu. The energy content based on 100 g of insect flour of *B. cajani* and *B. mori* were 532.78 and 557.05 kcal. Knowledge of the nutritional quality of those edible insects allows to be better integrate them into the diet. We suggest that the two insects could be considered as good alternative sources of protein and fat in addition to traditional meats. The use of edible insects products as food diversification and food fortification of starchy based diet is recommended to enhance nutrition of vulnerable groups in Madagascar.

Keywords: Entomophagy, fatty acids, Madagascar, nutrition, protein quality

Contact Address: Christian Tolojanahary Ratompoarison, University of Antananarivo, Department of Food Science and Technology, Campus Universitaire Ankatso, 101 Antananarivo, Madagascar, e-mail: christianratompoarison@gmail.com

Preferences and Consumption of Pigeon Peas among Rural Farming Households as Determinants for Development of Diversified Products

Zahra Majili[1], Cornelio Nyaruhucha[1], Kissa Kulwa[1], Khamaldin Daud Mutabazi[2], Stefan Sieber[3,4], Constance Rybak[4]

[1]*Sokoine University of Agriculture, College of Agriculture, Tanzania*

[2]*Sokoine University of Agriculture, School of Agric. Economics & Business Studies, Tanzania*

[3]*Humboldt-Universität zu Berlin, Thaer-Institute, Dept. of Agricultural Economics, Germany*

[4]*Leibniz Centre for Agric. Landscape Res. (ZALF), Inst. of Socio-Economics, Germany*

Pigeon peas are legumes with a high nutritional value. Existing studies of pigeon peas in Tanzania mainly examine production and marketing, but little has been documented with respect to consumer preferences and the consumption of pigeon peas. This study assesses the preferences surrounding pigeon peas and their consumption as bases for the development of diversified and shelf-stable products for nutrition and income improvement. Consumption study was conducted among 303 randomly selected farming households. Furthermore, 60 farmers participated in six focus group discussions in the Lindi region. A structured questionnaire and checklist with guided questions were used for data collection. Information about household characteristics preparation, preference and consumption of pigeon peas were collected. The analysis uses SPSS (V. 21) software. Data were tested for normality using the Shapiro-Wilk test. Differences between groups were established using Kruskal-Wallis and Mann-Whitney tests. The association were tested using Spearman's ρ at $p < 0.05$. The mean age of farmers was 35.8 ± 8.5 (SD) with 95 % aged between 15-49 years. About 61 % of farmers were male, 71 % were married and 85 % had primary education. The mean pigeon peas consumption during harvesting and lean seasons was 80 g per person per day and 18 g per person per day respectively. The frequency of consumption was higher during harvesting (92 %) than the lean (29 %) season. During harvesting, 44 % of farmers consume pigeon peas more than 5 days compared to the lean season (4 %). Five pigeon pea recipes exist in the area. The majority of farmers (91 %) preferred to consume the local variety, with 84 % of them consuming pigeon peas as a stew. Consumption difference was observed in terms of availability, taste, source of income and familiarity as the factors determining pigeon pea consumption and preferences. Consumption preference was significantly affected by educational level, availability, and taste ($p < 0.05$). With limited recipes and other barriers limiting consumption, the creation of innovative ideas for the development of diversified and shelf stable products fitting their consumption preferences is needed.

Keywords: Consumption, pigeon peas, preference

Contact Address: Zahra Majili, Sokoine University of Agriculture, Food Technology, Nutrition and Consumer Sciences, Sua, Morogoro, Tanzania, e-mail: majilizahra11@gmail.com

Determinants of Children's Fruit Intake in Teso-South Sub-County, Kenya

Eleonore C. Kretz[1,2], Annet Itaru[3], M. Gracia Glas[2,1], Lydiah Maruti Waswa[3], Sahrah Fischer[4], Samuel M. Mwonga[5], Thomas Hilger[4], Irmgard Jordan[2]

[1]*Justus-Liebig University Giessen, Inst. of Nutritional Sciences, Germany*
[2]*Justus-Liebig University Giessen, Center for International Development and Environmental Research, Germany*
[3]*Egerton University, Dept. of Human Nutrition, Kenya*
[4]*University of Hohenheim, Inst. of Agric. Sci. in the Tropics (Hans-Ruthenberg-Institute), Germany*
[5]*Egerton University, Dept. of Crops, Horticulture and Soils, Kenya*

Inadequate child feeding practices remain a challenge in some sub-Saharan African countries despite various attempts to introduce age-appropriate diets. This study aims at identifying facilitating and hindering factors to improve the consumption of fruits among children in Western Kenya.
Trials of Improved practices (TIPs) were carried out from August-October 2019 in eight villages in Teso-South Sub-County, Kenya, targeting 53 households with children under eight years of age. The trials included three household visits with counselling to improve children's diets and to negotiate to test household specific recommendations. Interview guides were used to capture experiences and perceptions of the improved practices. The responses were analysed by performing a structuring qualitative content analysis using QDA-Software.
Factors identified to facilitate fruit consumption among children included the availability of fruits or fruit seeds for planting, the influence and positive reactions of family members towards the practices, lifestyle and routines that enabled the mothers to be in charge herself as well as getting assistance. Further drivers were the perception of an easy implementation of the recommendation, the tastiness of the foods and the mother's experience with positive outcomes like a sustained satiety of the child or an increased food intake. The factors mentioned to hinder fruit consumption were the unavailability of fruits, partly due to seasonality or financial constraints, child's preferences, lifestyle and routines like time constraints or periods of sickness and perceptions towards the practices (unnecessary, not tasty). Unfavourable weather conditions, pests, and children uprooting seedlings hindered the planting of fruit trees within the homegardens.

Contact Address: Irmgard Jordan, Justus-Liebig University Giessen, Center for International Development and Environmental Research, Senckenbergstr. 3, 35390 Gießen, Germany, e-mail: Irmgard.Jordan@ernaehrung.uni-giessen.de

The unavailability of fruits and the inability to plant (more) fruit trees were the main challenges to enhance fruit consumption among children. Field trials are needed with less focus on yield of staple crops but testing on how best fruit trees within homegardens and on farms can be included to enhance fruit availability throughout the year.

The study was conducted within the EaTSANE-project financially supported by BMEL/ptble (Germany) and MOEST (Kenya) within the LEAP-Agri initiative.

Keywords: Child feeding, fruit consumption, trials of improved practices

Pathways Leading to Diversified Diets: A Retrospective Analysis of a Participatory Nutrition-Sensitive Agricultural Project

Julia Boedecker[1], Carl Lachat[2], Dana Hawwash[2], Patrick Van Damme[3], Marisa Nowicki[1], Céline Termote[1]

[1] *Bioversity International, Healthy Diets from Sustainable Food Systems, Kenya*

[2] *Ghent University, Dep. of Food Technology, Safety and Health, Belgium*

[3] *Ghent University, Dept. of Plant Production - Lab. for Tropical Agronomy, Belgium*

Our previous studies in Western Kenya have revealed that participatory farm diversification and nutrition education leads to increased dietary diversity of women and young children. In this present study, we have determined the pathways that led to an increased dietary diversity. Moreover, we have identified possible aftereffects of the intervention, such as factors related to sustainable behaviour change and factors that increased the participants' overall wellbeing beyond their nutrition (ex: empowerment). The overall objective of the study is to assess the extent to which widely used theoretical frameworks on agriculture-nutrition linkages applied to a real world, community-based, participatory project.

The study design consists of a qualitative, cross-sectional study of 10 focus group discussions and 5 key informant interviews. We conducted focus group discussions with community members who engaged in the participatory farm diversification program. The key informant interviews were conducted with local authorities that worked with these communities during the project. The data collection and analysis techniques used in this study were inspired by a widely used framework related to nutrition-sensitive agriculture. We then used social models to better identify and quantify potential aftereffects – which can be beneficial outcomes in and of themselves. Gender disaggregation during data collection and analysis enabled more nuanced interpretation regarding gender equality and women's empowerment.

Based on the data analysis, we developed our own framework showing the pathways from our participatory nutrition-sensitive agricultural intervention to improved diets. All three conventional pathways that are shown in widely used theoretical frameworks were well utilised in this project, including food production, agricultural income, and women's empowerment. Instead of a linear relationship, as shown in existing frameworks, our framework is a complex network of branched pathways with back-loops. In particular, the framework reveals details regarding empowerment as a pathway. Key components

Contact Address: Julia Boedecker, Bioversity International, Healthy Diets from Sustainable Food Systems, c/o ICRAF, P.O. Box 30677, 00100 Nairobi, Kenya, e-mail: j.boedecker@cgiar.org

of this pathway include women's income, women's time, and men's support in agriculture. Identified aftereffects include empowerment, community cohesion, and increased harmony within households.
In the future, we hope to see more nutrition-sensitive agricultural projects that develop frameworks beforehand and to measure aftereffects that can be outcomes in and of themselves.

Keywords: Agricultural-nutrition pathways, community-based participatory approach, dietary diversity, nutrition-sensitive agriculture

Analysis of Consumer Behaviour on Cowpea Quality Attributes in Gombe State, Nigeria

Isah Bakoji[1], Isiaka Mohammed[2], Murtala Nasiru[3], Salawu A. Jibril[3]

[1]*Nigerian Correctional Service, Agricultural Economics, Nigeria*

[2]*Federal University Kashere, Agricultural Economics and Extension, Nigeria*

[3]*Abubakar Tafawa Balewa University, Agricultural Economics and Extension, Nigeria*

The study analysed consumer behaviour on cowpea quality attributes in Gombe State, Nigeria. Two hundred and sixty seven (267) respondents were randomly selected for the study while data were collected using structured questionnaire. The results were analysed using descriptive statistics and hedonic regression models. Finding from the result shows that majority of cowpea consumers fell within the age bracket of 36–60 years and 1–5 persons in their household. Based on cowpea preference farmers (53.85 %) and household consumers (50.42 %) preferred medium size cowpea, processors (42.11 %) preferred small seeded size while retailers (100 %) claimed that they traded all the three sizes of cowpea. With regard to consumers' reasons for consumption farmers (55.45 %), household consumers (50.45 %) and processors (55.70 %) showed that they consumed cowpea because of its nutritive. Based on skin texture more than half of processors (52.63 %) preferred smooth texture over rough texture presumably due to good swelling capacity while household consumers prefer rough texture over smooth texture because of taste and ease of cooking respectively. The findings further revealed that seed size, colour, skin texture, grain colour mixes and grain size mixes are the significant variables affecting cowpea consumers in the area. It is concluded that consumers in the study area have different levels and forms of preference for cowpeas. The characteristics and socio-economic features of consumers also have influence on the level of preference. It is recommended that efforts to improve upon grain size, grain colour, skin texture by crop breeders and good grain standardisation by cowpea traders will be worthwhile in the area as consumers seems to be interested in large seeded cowpea, white skin colour, rough skin textured cowpea as well as cowpea without grain size and grain colour mixes. There should also be more research into the disliking intrinsic characteristics of cowpea by crop breeder

Keywords: Consumer, cowpea, preference, quality, texture

Contact Address: Isah Bakoji, Nigerian Correctional Service, Agricultural Economics, Tsohon Kamfanibauchi, +234 Bauchi, Nigeria, e-mail: bakojiisah@gmail.com

How Did Diets in Urban and Rural Uganda Develop Over Time?

Vincent Linderhof[1], Nina Motovska[1], Valerie Janssen[1], Andrea Fongar[2], Beatrice Ekesa[2]

[1]*Wageningen University and Research, Wageningen Economic Research, The Netherlands*

[2]*The Alliance of Bioversity International and CIAT, Healthy Diets from Sustainable Food Systems Initiative, Uganda*

In 2010, FAO concluded that the main components of the Ugandan diet were root and tubers (cassava, sweet potatoes and cooking banana) and cereals (maize, millet, sorghum). Pulses, nuts and green leafy vegetables complement the diet. Overall, diets were poor in micronutrient-rich foods, such as fruits and fish. In urban areas, FAO found that food consumption patterns are changing, and rice is gaining importance. In this study, we analysed food security in urban and rural Uganda in the period after 2010. We used household food consumption information from 4 household surveys conducted by the Uganda Bureau of Statistics (UBOS) in collaboration with the World Bank Living Standard Measurement Study – Integrated Surveys on Agriculture (LSMS-ISA) namely 2009–2010, 2011–2012, 2013–2014, and 2015–2016. Food security indicators were the share of households consuming food product from food groups, and the household dietary diversity score (HDDS), where we distinguished 12 food groups: i) cereals; ii) root and tubers iii) vegetables; iv) fruit; v) beans and ground nuts; vi) meat; vii) fish; viii) eggs; ix) dairy products; x) sweets; xi) beverages; and xii) condiments cereals, starches and roots ,amongst others. The preliminary results show that high share of households consume starches and roots as well as cereals. This holds for rural and urban households. The food groups eggs, dairy products, fruits, meat and fish were consumed by less than 50 percent of the households in rural and urban settings in all surveys. The share of households eating fruits (nutrient-rich food) and meat increased over the years with a peak consumption in 2013/14. The share of households eating fish and dairy products remained constant over time. Dietary diversity increased between 2010 and 2014, where the HDDS was slightly higher for urban population. Based on our results, we conclude that diets of urban and rural people in Uganda have not changed much since 2010. However, more research is necessary to analyse the nutritional aspects of the dietary transition in more detail.

Keywords: Food security, Household Dietary Diversity Score (HDDS), LSMS-ISA, Uganda

---

Contact Address: Vincent Linderhof, Wageningen University and Research, Wageningen Economic Research, Prinses Beatrixlaan 582, 2595 BM The Hague, The Netherlands, e-mail: vincent.linderhof@wur.nl

What Hinders Men to Participate in Project Activities Linking Nutrition and Agriculture?

Paulina M. Kossmann[1], Daisy Alum[2], M. Gracia Glas[3,1], Cedric Masai Cheptoek[2], Margaret Kabahenda[2], Sahrah Fischer[4], Johnny Mugisha[5], Thomas Pircher[6], Thomas Hilger[4], Irmgard Jordan[1]

[1]*Justus Liebig University Giessen, Center for International Development and Environmental Research, Germany*
[2]*Makerere University, Dept. of Food Technology and Nutrition, Uganda*
[3]*Justus-Liebig University Giessen, Center for International Development and Environmental Research, Germany*
[4]*University of Hohenheim, Inst. of Agric. Sci. in the Tropics (Hans-Ruthenberg-Institute), Germany*
[5]*Makerere University, School of Agricultural Sciences, Uganda*
[6]*University of Hohenheim, Research Center for Global Food Security and Ecosystems, Germany*

Women and children form the most vulnerable group when it comes to dietary intake and thus are the focus of most nutrition programmes. However, at household level men are most often considered to be the decision maker. Therefore, to improve dietary intake mobilisation of both men and women is considered essential.

Women from 50 farming households in Kapchorwa District, Uganda, took part in trials of improved practices (TIPs) to improve local nutrition education messaging. Qualitative data collection was carried out through private counselling with the women at their homes. Part of the programme was a solar-dryer construction session together with the women and their husbands participating in the TIPs. Finally, group discussions were held with women and men separately. No incentives were provided for joining the solar-dryer construction sessions, while a travel allowance was paid to all participants for joining the group discussions.

While 37 husbands participated in the workshops, only 8 participated in the solar-dryer session. Those who joined the solar-dryer session joined out of their own interest or because they felt obliged to come. Some of the main reported reasons for not participating were being busy with farm activities, the weather conditions, and late invitations. In addition, some husbands thought the invitation was just for women because the project deals with vegetables, which are considered crops for females. Lastly, women reported their men stating the project was a waste of time and thus not participating because

Contact Address: Paulina M. Kossmann, Justus Liebig University Giessen, Center for International Development and Environmental Research, Giessen, Germany, e-mail: paulina.m.kossmann@nu.uni-giessen.de

there was no financial allowance. However, seeing the solar-dryer at the workshops, men became very interested to learn how to construct it.
The results imply that men may not feel the necessity to join the research activities because they believe the topics like vegetable production address women only. Further, allowance is seen as a bigger motivation than knowledge gain which is questionable because allowance impacts the sustainability of the activity. Hence, nutrition-agriculture projects need to specifically identify alternative incentives to improve participation of both, men and women, to enhance impact of its activities.
The study was conducted within the EaTSANE-project funded by BMEL/ptble within the joint AU/EU-LEAP-Agri initiative.

Keywords: Agriculture, gender, incentives, nutrition, nutrition-agriculture linkages, trials of improved practices, Uganda

Association of Body Mass Index with Physical Fitness among Small-holder Famers in Malawi and Kenya

Johanna L. Piotrowski[1], M. Gracia Glas[1,5], Lydiah Maruti Waswa[2], Paul Falakeza Fatch[3], Gabriella Chiutsi Phiri[3], Thomas Hilger[4], Sahrah Fischer[4], Elizabeth Kamau[2], Michael B. Krawinkel[5], Ernst-August Nuppenau[6], Irmgard Jordan[1]

[1]*Justus-Liebig University Giessen, Center for International Development and Environmental Research, Germany*

[2]*Egerton University, Dept. of Human Nutrition, Kenya*

[3]*Lilongwe University of Agriculture and Natural Resources, Extension Department, Malawi*

[4]*University of Hohenheim, Inst. of Agric. Sci. in the Tropics (Hans-Ruthenberg-Institute), Germany*

[5]*Justus-Liebig University Giessen, Institute of Nutritional Sciences, Germany*

[6]*Justus-Liebig University Giessen, Inst. of Agric. Policy and Market Res., Germany*

The double burden of malnutrition which is characterised by the co-existence of various forms of under- and overnutrition is a common health problem predominantly affecting households in low- and middle- income countries. The study objective was to investigate whether double burden of malnutrition affects the field work capacity of affected households.

Cross-sectional surveys were conducted in 2017 in Kenya and Malawi targeting households with children below six years of age (n=432 and n=355, respectively). Anthropometric measurements were assessed to determine nutritional status of the mother-child pairs. Hand-grip-strengths (HGS) for the mother was measured and used as a proxy indicator for physical fitness. Anthropometric data was used to identify households with double-burden of malnutrition. Mean HGS of the mothers was determined and comparisons made between mothers living in households with (DB-group) and without double-burden (comparison-group) of malnutrition. The comparison group were differentiated in households with only one and without any form of malnutrition.

The body-mass-index (BMI) and HGS of the women in Malawi were significantly correlated ($p < 0.05$). There were significant differences between women's HGS of case-households with women's HGS living in a household with an overweight/obese family member. In Kenya, a significant but weak non-linear correlation ($p < 0.05$) was identified between women's HGS and BMI. HGS was lower among underweight and obese women compared to

Contact Address: Johanna L. Piotrowski, Justus-Liebig University Giessen, Center for International Development and Environmental Research, Senckenbergstr. 3, 35390 Giessen, Germany, e-mail: johanna.l.piotrowski@nu.uni-giessen.de

normal weight, but no differences were found between DB and non-double burden households.
The results of the countries vary in form and significance. However, programmes are needed in both countries addressing double-burden of malnutrition also in rural areas although a link between HGS and double burden was not proven. However, underweight and overweight women showed lower physical capacity. Thus, there is a link between the HGS and BMI that may impact on food production systems due to lower physical fitness of the farming women herself.
This study was conducted within the HealthyLAND project which was funded by BMEL/ptble.

Keywords: BMI, children, double burden of malnutrition, handgrip strength, Kenya, Malawi, smallholder farmers, women

Developing and Testing Guava- and Jackfruit-Nut-Bars to Bridge Seasonal Food Gaps in East Africa

Sirui Xing, Gudrun B. Keding, Elke Pawelzik
University of Goettingen, Dept. of Crop Science, Division of Quality of Plant Products, Germany

Guava (*Psidium guajava*) and jackfruit (*Artocarpus heterophyllus*) are two naturalized fruits in East Africa. Both have high contents of nutrients that are essential for human bodies. However, through improper post-harvest treatments many nutrients or even whole surplus fruits are lost while at the same time fruit consumption in East Africa is far below the recommended amounts. Besides, especially guava is not available all year round. In the framework of a joint project on "Fruits and vegetables for all seasons" with partners in East Africa and Germany, this study aimed to explore the development of nutritional fruit-nut-bars with a long shelf-life. The effects of common processing methods (cooking and oven drying) and varying ingredients (mango, lemon juice, desiccated coconut, peanut and cashew nut) on nutrient composition in fruit-nut-bars were measured. Results showed that bars with guava and lemon juice contained the highest content of ascorbic acid (81.19 $\pm$ 0.37 mg/100 g FM). For phenolic content, most samples of jackfruit (range from 437.95 $\pm$ 37.9 mg GAE/100 g FM to 931.90 $\pm$ 106.63 mg GAE/100 g FM) contained more than guava products (range from 508$\pm$ 2.8 mg GAE/100 g FM to 563 $\pm$ 2.0 mg GAE/100 g FM), while lemon juice made not much difference. Fruit bars with lemon juice had higher acidity, yet, samples of guava and jackfruit showed quite different tendency for before- and after cooking, as well as after drying. For mineral content, potassium showed the highest value (11.5 mg g^{-1} DM and 14.1 mg g^{-1} DM for guava- and jackfruit- based samples, respectively) among macro minerals, followed by phosphorous, sulphur and magnesium. Copper contents were relative higher (0.08 mg g^{-1} DM) among micro minerals, iron and zinc showed similar values (0.06 mg g^{-1} DM) while the manganese content was the lowest one (0.02 mg g^{-1} DM). Water content of final products was <10 %, which indicated a longer shelf-life. In conclusion, cooking and drying procedures decrease, as expected, nutrient contents, yet, to varying extent. While the consumption of fresh fruits is always the better choice, the fruit-nut-bars still provide a good option to process surplus fruits and bridge seasonal gaps.

Keywords: East Africa, fruit-nut-bars, guava, jackfruit, nutrient content

Contact Address: Elke Pawelzik, University of Goettingen, Dept. of Crop Science, Division of Quality of Plant Products, Carl-Sprengel-Weg 1, 37075 Göttingen, Germany, e-mail: epawelz@gwdg.de

Nutritional Characterisation of Traditional Preserved Cowpea Leaves Consumed in Coastal Drylands of Kenya

Joshua Owade[1], George Abong'[1], Michael Okoth[1], Agnes Mwang'ombe[2]

[1]*University of Nairobi, Dept. of Food Science, Nutrition and Technology, Kenya*

[2]*University of Nairobi, Dept. of Plant Science and Crop Protection, Kenya*

Consumption of African leafy vegetables such as cowpea leaves is high among communities in the arid and semi-arid lands in sub-Saharan Africa. However, seasonal availability of these vegetables has often limited their utilisation. Consequently, some households often resort to the traditional preserved forms of the vegetables to enhance their availability. It is not sufficient to be dismissive of these techniques as less efficient ways of enhancing nutrient intake among communities, thus the current study sought to characterise the nutritional quality of the traditional preserved forms of cowpea leaves consumed and utilised in the arid and semi-arid lands of the coastal region of Kenya. Twenty samples of the preserved forms of cowpea leaves were obtained from farmer groups in Taita Taveta County. Similar forms of preserved leaves from the same group were mixed and homogenized; twelve samples were then subjected to nutritional analysis. The blanched cowpea leaves had significantly ($p < 0.05$) lower crude protein and ash contents than the fresh and unblanched shadow-dried and sundried leaves. The unblanched shadow-dried leaves had significantly ($p < 0.05$) the lowest crude fat content, 0.2±0.0 %. Traditional preservation techniques induced a change of between between 88 % and 98 % in beta-carotene content. The unblanched shadow-dried leaves had higher beta-carotene content (2.60±0.21 mg/100 g) than both the blanched and unblanched sundried leaves, 0.54±0.03 and 0.40±0.10 mg/100 g, respectively. Zinc, iron and calcium contents of preserved leaves had no significant ($p > 0.05$) difference compared to that in fresh forms. Maximum variability (100 %) in the data was explained by nine principal components. Extracting the first two components, accounting for 56.5 % variability in the data; revealed higher correlation among the three variables, ash, protein and fibre on one hand, while on the other hand calcium, moisture and beta carotene content. As much as the preservation techniques induced nutrient deterioration in the vegetables, the techniques still helped avail the key nutrients in cowpea leaves: protein, zinc, iron and beta carotene. Promotion of traditional dehydration technique for alleviation of seasonal availability of the vegetables can therefore not be ruled out as cost-effective strategies of addressing food and nutrition security in the arid and semi-arid lands.

Keywords: Blanched, cowpea leaves, dried, principle components, proximate composition

Contact Address: Joshua Owade, University of Nairobi, Dept. of Food Science, Nutrition and Technology, Nairobi, Kenya, e-mail: owadehjm@gmail.com

(Socio)economics

Rural transformation, gender and livelihoods dynamics

Oral Presentations

Posters

Women's Role in Small-Scale Aquaculture Sector and Implications for Technical Efficiency: Empirical Evidence from Myanmar

Yee Mon Aung[1], Manfred Zeller[1], Ling Yee Khor[1], Nhuong Tran[2]

[1]*University of Hohenheim, Inst. of Agric. Sci. in the Tropics (Hans-Ruthenberg-Institute), Germany*
[2]*WorldFish, Myanmar*

Myanmar's aquaculture sector has grown rapidly over the last two decades and plays an essential role in national fish supply. However, the technical and economic characteristics of this sector, mainly small-scale aquaculture, have been poorly studied. This study addresses this knowledge gap by measuring the overall technical efficiency from the input-oriented approach, slacks associated with each input, and its determinants for 440 small-scale fish farms in the Delta region, Myanmar. In order to account for both bias and serial correlation of conventional DEA (CDEA) scores, the two-stage double bootstrap DEA (BDEA) is used to correct for the bias of CDEA scores in the first stage and then produce the valid statistical inference in the second stage.
Results reveal that most small-scale fish farmers in the study areas perform their fish production below the production frontier. Since there is evidence that efficiency scores through CDEA are biased and highly correlated, the efficiency score derived from this model (0.55) is higher than that from BDEA model (0.44). Therefore, farmers can reduce their input use by over 50 percent while still producing the same output levels. All the inputs used contained slacks such that all of them are over-utilized in inappropriate ratios. Regarding its determinants, women's participation in decision-making process within the household and female labor contribution in fish production activities have a significant and positive effect on technical efficiency. Also, the use of fish farming practices, particularly polyculture and integrated agriculture and aquaculture (IAA) and climate-change adaptation practices are the most significant efficiency shifters. Together, the findings highlight the important need to promote interventions targeted at improving technical efficiency of small-scale aquaculture producers. Programs and policies aimed at increasing aquaculture productivity would benefit by including interventions to reduce gender inequality and promoting equity.

Keywords: Bootstrap truncated regression, data envelopment analysis, Myanmar, slacks, small-scale aquaculture, technical efficiency, women

---

Contact Address: Yee Mon Aung, University of Hohenheim, Inst. of Agric. Sci. in the Tropics (Hans-Ruthenberg-Institute), Karlshofstr.67, 70599 Stuttgart, Germany, e-mail: yeemonyau@gmail.com

The Changing Role of Agriculture in Rural Villages of an Emerging Market Economy: The Case of Thailand

Chompunuch Nantajit, Hermann Waibel

Leibniz University Hannover, Development and Agricultural Economics, Germany

This paper analyses the long-term changes in the livelihoods and the role of agriculture in rural households in Thailand, based on a panel dataset collected in 2009 and 2018 from a census of a village in Phetchaboon province, Thailand. Results are shown by a descriptive analysis of the panel data and a Positive Mathematical Programming (PMP) model. The descriptive analysis shows that per capita household income has increased by 8.5 % on average per year while income inequality increased, with the Gini coefficient rising from 0.47 to 0.59. Nevertheless, poverty decreased from 54.3 % to 28.9 % head count ratio. The share of agriculture in household income decreased from almost 60 % to just over 40 % within the ten-year period. Wage employment instead of own agriculture has become the main occupation due to largely temporary migration. Pronounced inequality in land holding did holding not change much and was already high in 2009 with a Gini coefficient of 0.65. In the analysis, we split households into two groups, namely those where agriculture exceeds 50 % of household income and those below. Clearly, we see that the farmer have declined. The demographic structure of the village shows an ageing population with strong implications for the organisation of agriculture. The PMP village model helps to explore future village development trends, in particular the development of agriculture. Key variables are the price of the maize, as the main crop, and the costs of labour. Results show that even high price incentives would not bring more labour back into agriculture. Higher wages will shift more labour out of agriculture but there is no effect on the distribution of land holding. Villagers hold on to their farm and engage in part-time farming as a safety measure. To a considerable extent, this impedes structural change in agriculture. However, as shown by the recent Corona crisis, it is perhaps a good livelihood strategy of rural households in emerging market economies like Thailand.

Keywords: Panel data, positive mathematical programming, role of agriculture, rural livelihood, Thailand

Contact Address: Chompunuch Nantajit, Leibniz University Hannover, Development and Agricultural Economics, Stöckener Str. 119, 30419 Hannover, Germany, e-mail: chompunuch@ifgb.uni-hannover.de

Forest Dependence and Household Food and Nutrition Security in Kenya

Josiah Ateka, Robert Mbeche, Esther Waruingi
Jomo Kenyatta University of Agriculture and Technology, Dept. of Agricultural and Resource Economics, Kenya

About 1 billion extremely poor people globally depend on forests for part of their livelihoods, with about 350 million depending heavily on forests. While there is growing recognition that forests and tree-based systems complement farmland agriculture in providing food security and nutrition (FSN), few studies have focused on this dimension of livelihood. This paper assesses the effects of forest dependence on household food security in rural western Kenya. Data for this paper was collected using a cross sectional survey of 924 households in Mt Elgon, Western Kenya. Forest dependency was measured in terms of both a count of total number and aggregate value of forest products extracted while FSN was measured using three indicators – Food Insecurity Experience Scale (FIES), Household Dietary Diversity (HDD) and the share of food expenditure. The influence of forest dependency on FSN was analysed using a set of Poisson and two-stage residual inclusion (2SRI) regressions. The results show that about half (48.9 %) of the households were engaged in forest extraction. The mean annual value of forest products extracted was approximately KES 31930 (about US$ 320) which accounts for about 23 % of overall household annual expenditure. Of the 73.5 % of the respondents who were food insecure, 13.9 % and 21.7 had slight, moderate levels of insecurity while 40.9 were severely food insecure. The results further show that although dietary patterns on a weekly basis were rather diversified, (6.3), consumption of Vitamin A (2.7) and iron rich foods (0.92) was limited suggesting possibility of high micronutrient deficiencies. The share of expenditure spent on food was generally on average accounting nearly half (47.5 %) of the household expenditure. The econometric results show that while the value of forest products extracted had a positive influence on household food security, forest dependency measured as a count of products extracted was associated with higher levels of food insecurity. The findings suggest that while forests can play an important role in the food system, household forest use decisions can have a mixed influence on food security, depending on the nature, type and extent of extraction.

Keywords: Food expenditure, food security and nutrition, forests, household dietary diversity, Poisson regression, two-stage residual inclusion (2SRI)

Contact Address: Josiah Ateka, Jomo Kenyatta University of Agriculture and Technology, Department of Agricultural Resource Economics, Nairobi, Kenya, e-mail: atekajm56@gmail.com

Fisheries in Burkina Faso: Implications for Livelihoods and Food Security

Vincent-Paul Sanon[1], Patrice Toé[2], Hamid El Bilali[3], Erwin Lautsch[1], Stefan Vogel[4], Andreas Melcher[1]

[1]*BOKU - University of Natural Resources and Life Sciences, Inst. for Development Research, Dept. of Sustainable Agricultural Systems, Austria*

[2]*University Nazi Boni, Inst. for Rural Development, Dept. of Sociology and Rural Economy, Burkina Faso*

[3]*International Centre for Advanced Mediterranean Agronomic Studies (CIHEAM-Bari), Italy*

[4]*BOKU - University of Natural Resources and Life Sciences, Inst. for Sustainable Economic Development, Dept. of Economics and Social Sciences, Austria*

Burkina Faso (BF) has attempted to establish an economically efficient fisheries sector to cope with poverty and food insecurity. However, the fishing sector is still unable to meet the country's fish demand. The interest in research about fisheries in population's livelihoods and the actors' strategies for developing the sector along the value chain is big and still increasing. However, data and evidence about this topic are scarce and often outdated. This paper analyses fisheries sector contribution to sustainable livelihoods and, consequently, food security in BF. Quantitative data were gathered through a representative survey performed from August to September 2018 with 204 fishermen's households in three study areas in BF, and complementary, qualitative data collected by interviews with different stakeholders (e.g. fishermen, traditional authorities, decision makers, politicians and officers) across the country.

The results show that fishermen's households rely on diversified livelihoods as a way to cope with uncertainties, risk and vulnerability. Indeed, fishing, agriculture and animal breeding are the main activities in fishing communities. Surprisingly, most fishermen allocate more time to agriculture, fish selling then animal breeding. Likewise, most of the household income of the interviewees is obtained from agriculture followed by fish selling and animal breeding; the share of trade, fish processing and craft in income seems marginal. The importance of fisheries to fishermen's livelihoods and food security lies in the fact that income generated from fishing is primarily allocated to investment in other income-generating activities (e.g. crop production, animal breeding) and for covering food, education and health expenditures. It is surprising that although the majority of fishermen have fairly low income,

Contact Address: Vincent-Paul Sanon, BOKU - University of Natural Resources and Life Sciences, Inst. for Development Research, Dept. of Sustainable Agricultural Systems, Peter-Jordan-Straße 76/i, 1190 Vienna, Austria, e-mail: vincent.sanon@boku.ac.at

they dedicate a large share of income from fishing to investments, rather than consumption, which shows fishing importance as a source of capital in the economic development of entire communities in BF. Considering the central role of fisheries in the livelihoods of an important share of Burkinabe combined with the decreasing fisheries capture, one of the most important recommendation is to make the case for developing fish farming to address fish resources overexploitation, meet the increasing fish demand, and to provide more opportunities to the stakeholders.

Keywords: Burkina Faso, fish farming, fisheries, food security, innovation, livelihoods, sustainable development strategies, sustainable livelihoods approach

Labour Markets and Income Stability of Smallholder Rubber Farmers: Evidence from Southwest China

Haowen Zhuang, Hermann Waibel
Leibniz University Hannover, Institute of Development and Agricultural Economics, Germany

The expansion of rubber cultivation has been a driving force for poverty reduction in Xishuangbanna Dai Autonomous Prefecture (XSBN), Southwest China. However, monoculture causes income variance due to rubber price fluctuations. In addition, since 2011, rubber price has decreased constantly. Hence, smallholder rubber farmers are challenged to diversify their income portfolio. With the rapid growth of labour markets in industry and service sector in China, complementary off-farm employment, which includes non-farm self-employment and off-farm wage employment, becomes an option for farmers. In this paper, we investigate the role of off-farm employment in stabilising household income of smallholder rubber farmers in XSBN. The specific objectives are: (1) to examine the factors influencing the participation in off-farm labour markets; (2) to explore the impacts of off-farm employment on rubber farmers' income stability.

Our analysis is based on a unique panel data set of three waves of comprehensive socioeconomic surveys of 612 rubber farmers in XSBN in 2013, 2015 and 2019. We apply two models, namely a random effects logit model for the first objective and a propensity score matching with difference-in-differences approach for the second question. To facilitate the analysis, we grouped the sample households in those with and without off-farm employment. The variance in net farm income from 2013 to 2019 is used to indicate income stability.

Our results show that just over one third of household income stems from off-farm employment. The factors that determine participation in off-farm labour market are education, age of household head, ethnicity, elevation, number of mobile phones and the distance to village. We also find a significant difference in income stability between households with and without off-farm employment. As long as the economy is growing, off-farm employment offer stable employment opportunities and contribute to income stability.

Keywords: Income stability, off-farm employment, smallholder rubber farmers, Southwest China

Contact Address: Haowen Zhuang, Leibniz University Hannover, Institute of Development and Agricultural Economics, Koenigsworther Platz 1, 30617 Hannover, Germany, e-mail: zhuang@ifgb.uni-hannover.de

Female Farmers' Participation in Off-farm Activities and their Determinants in Rural Bauchi State, Nigeria

Mustapha Yakubu Madaki[1], Abbas Shehu[2], Miroslava Bavorova[1], Jadeera Mahmood[2]

[1]*Czech University of Life Sciences Prague, Fac. of Tropical AgriSciences, Dept. of Economics and Development, Czech Republic*

[2]*Abubakar Tafawa Balewa University Bauchi, Dept. of Agricultural Economics and Extension, Nigeria*

Women plays a vital role in different economic sectors however, they are constraint to the developmental contribution due to the lack of access to productive resources, as in agricultural sector in most of the developing countries. This forces them to take part in off-farm activities to augment their little farm income. This study investigated the different type of off-farm activities engaged by women farmers and their determinants in rural Bauchi state, Nigeria. Purposive sampling method was used in selecting three local government areas out of seven in western agricultural zone of the state and random sampling method was used in selecting ten wards and 5 % of registered women famers in each selected ward making 134 respondents. Semi-structured questionnaires were administered to women farmers in 2017. The data collected included socio-economic characteristics of the respondents, institutional factors and types of off-farm activities. The data were analysed using descriptive statistics and logistic regression models. The result revealed that food processing (25.6 %), farm products sales (24.6 %), trading (21.6 %) and tailoring (14.6 %) are the major off-farm activities engaged by the rural female farmers. The regression result shows that land ownership, remittances, access to extension service, access to credit, access to electricity and market increase the likelihood of rural women farmers to engage in off-farm livelihood activities significantly while increase in age and farm size affecting their participation in off-farm livelihood activities negatively. As the study found out the significant influence of access to extension services, credit, electricity and market, we hereby emphasise the provision of extension services, facilitating rural credit programmes, provision of rural electricity and markets.

Keywords: Bauchi state, Nigeria, off-farm activities, women farmers

Contact Address: Miroslava Bavorova, Czech University of Life Sciences Prague, Fac. of Tropical AgriSciences - Dept. of Economics and Development, Kamýcká 129, 16521 Prague - Suchdol, Czech Republic, e-mail: bavorova@ftz.czu.cz

Livelihood Assessment and Characterisation of Charcoal Producers in Bauchi and Toro LGA Bauchi State, Nigeria

Isah Bakoji[1], Uchendu Kelvin Iroegbute[2], Yidi Mohammed Suleiman[2], Joseph Nanloh Jatbong[3]

[1]*Nigerian Correctional Service, Agricultural Economics, Nigeria*
[2]*Federal University Kashere, Agricultural Economics, Nigeria*
[3]*Taraba State University, Agricultural Economics, Nigeria*

In recent years the rate of use of charcoal for alternative fuel had grossly increase due to problem associated with alternative fuels (wood, kerosene and gas).The study therefore analyses livelihood and characterisation of charcoal producers in Bauchi and Toro Local Government Area of Bauchi State. Purposive and multi stage sampling was used to select eight villages. Simple random sampling was then employed to select 160 charcoal producers. The analytical technique used was descriptive statistics. The finding revealed that male (100.00 %) involved in charcoal production. Charcoal production process was male-dominated with no record of any women participating in the production in the study area. Result further shows that 61.25 % of the respondents combined charcoal production with farming.How an appreciable number venture in charcoal business to make money for specific expenditure while 18.36 % due to lack of alternative job opportunities. Major constraints affecting charcoal producers are as follows; charcoal productions are much harder and dangerous to their health, insufficient fund, charcoal breakages, and confiscation and collecting of bribe by forestry official amongst others. It was concluded that that charcoal production in the study area largely depend on natural forest where by natural regeneration is the source of forest recovery. Deforestation is source of concern in the area. It was recommended that the study draw data from only two Local Government Areas in the State, there is need for additional empirical research on charcoal base livelihood and characterisation of charcoal producers for the entire State. There is need to organise the existing informal structure of charcoal producers and find ways for supporting them financially. Creation of job opportunities to our teeming youth will reduce the rate of energetic youth going in to charcoal production industry.

Keywords: Deforestation and charcoal production, livelihood

Contact Address: Isah Bakoji, Nigerian Correctional Service, Agricultural Economics, Tsohon Kamfanibauchi, +234 Bauchi, Nigeria, e-mail: bakojiisah@gmail.com

Effects of Livelihood Assets on Food Security of Rural Households in the COVID-19 Condition

Masoud Yazdanpanah[1], Savari Moslem[1], Katharina Löhr[2], Stefan Sieber[2]
[1]*Agricultural Sciences and Natural Resources University of Khuzestan, Iran*
[2]*Leibniz Centre for Agric. Landscape Res. (ZALF), Sustainable Land Use in Developing Countries (SusLAND), Germany*

The outbreak of coronavirus in December 2019 resulted so far in about 3 million virus-infected cases and a significant number of human deaths worldwide and a global stage of emergency got declared by the World Health Organisation (WHO). On a global level, the pandemic shows negative impacts on all sectors, including: economic, social, cultural and educational, with a growing body of studies underway. So far, the studies tend to focus on health policies and technical initiatives (i.e., hardware aspects), such as the causative agent of the disease, pathogenesis epidemiology, diagnosis, treatment and preventive interventions. These interventions are provided for the urban citizens, encompassing a substantial portion of the human population. However, people in rural areas are so far neglected while they usually have weaker infrastructure, and fewer health and economic facilities to cope with such crisis. To mitigate the lack of knowledge on effects of coronavirus particularly in rural areas in the developing world, this study presents findings on the effect of livelihood assets on food security during the Corona period in rural areas of southern Iran. The results show that households are not in a favourable position in terms of food security because more than 70 % of them are in a state of food insecurity and their biggest problem during pandemic has been their inability to prepare a balanced diet and worry about running out of food. Also, results regarding the effects of corona virus on livelihood assets (financial, psychological, physical, human, social and natural) show that coronavirus has the highest effect on financial and psychological resources. Furthermore, results show a significant difference between the effects of coronavirus and food security levels. Moreover, the results of rural households' livelihoods assist on the levels of food security in corona virus conditions showed that four assets including psychological, physical, human and financial have positive and significant effects on rural household food security.

Keywords: COVID-19, food security, sustainable livelihoods, vulnerability

Contact Address: Masoud Yazdanpanah, Agricultural Sciences and Natural Resources University of Khuzestan, Mollasani, 744581 Ahvaz, Iran, e-mail: masoudyazdan@gmail.com

How Is the Issue of Overageing of Cocoa Farming Households Influenced by their Endowment with Livelihood Capitals?

Aline Roth[1], Sonja Trachsel[1], Sabine De Castelberg[2], Monika Schneider[3]

[1]*Zhaw Life Sciences und Facility Management, Inst. of Natural Resource Sciences, Switzerland*

[2]*Zhaw Life Sciences und Facility Management, Inst. of Chemistry and Biotechnology, Switzerland*

[3]*Research Institute of Organic Agriculture (FiBL), Dept. of International Cooperation, Switzerland*

The ageing of rural populations will negatively affect agricultural productivity and impact the socio-economic structure of rural communities. In Alto Beni, Bolivia, 70 % of the cocoa producers are over 50 years old and this ageing of cocoa farmers could lead to less productive cocoa farming and accelerate migration to urban areas. This study aims to describe the problem of ageing cocoa farmers in Alto Beni, how coping with old age is linked to households' endowment with livelihood capitals and proposes measures to address the problem of ageing. 11 qualitative interviews were used to identify different perspectives on the problem of ageing in cocoa production in Alto Beni. Based on a survey with 120 farming households, it was assessed how households are equipped with different livelihood capitals and how a household's livelihood endowment influences the way they deal with old age.

The analysis of the qualitative interviews showed that older cocoa farmers work as long as their health permits. The main reasons for continuing working in old age are insufficient financial resources, a lack of formal pension and social security systems. If elderly farmers lived with their children, they were taken care of. Such farmers living alone belonged to the most impoverished group in Alto Beni with the lowest livelihood endowment. In addition, it is difficult for the older cocoa farmers to secure the farm take-over. The younger and partly also the older generation do not consider cocoa farming to be a livelihood-securing activity. Reasons for that perception are mainly low and fluctuating cocoa market prices and the use of inefficient cocoa production practices. The analysis of the livelihood survey showed that how cocoa farming households cope with ageing does not depend on the family's endowment with livelihood capitals. The main influences on coping with old age are socio-cultural perceptions of cocoa farming and the insufficient pension system. Training and education in cocoa farming practices, more collaboration between cocoa farmers, their cooperatives and actors from international organisations, science and government could offset the ageing of cocoa farmers in Alto Beni.

Keywords: Ageing of rural population, Alto Beni, Bolivia, cocoa production, livelihood assessment, rural development

---

Contact Address: Sonja Trachsel, Zhaw Life Sciences und Facility Management, Inst. of Natural Resource Sciences, Grüental, 8820 Wädenswil, Switzerland, e-mail: trso@zhaw.ch

The Perception of Deficiencies of Smallholder Coffee Farmers, a Panel Analysis of a Rural Community in Eastern Uganda

Anna Lina Bartl

Georg-August-Universität Göttingen, Agricultural Economics and Rural Development, Germany

For most smallholder farmers in the Mount Elgon region of Uganda, Arabica coffee cultivation is the major income-generating activity. It is well-known that smallholder coffee farmers often live under conditions that do not surpass existential needs. Even if development activities are partly implemented in the region, the perception of deficiencies of coffee farmers is not well examined.

Quantitative and qualitative answers of 431 smallholder coffee farmers, living in three sub-counties in the Mount Elgon region, interviewed in 2018 and in 2019, are used to identify categories and intensity of deficiencies. In addition, changes in perceptions deriving from a comparison of the two surveys are analysed and discussed, as well as the potential causes for improvements or deteriorations. Results from the explorative Factor-analysis identified five factors which have been grouped into two main topics (1) constitution for farm management activities and (2) general life quality. Whereas the perception of deficiencies in the constitution for farm management activities differs significantly between the three investigated sub-counties, the perception of variables linked to general life quality was more equally distributed. Interestingly, most of the farmers and their wives do not perceive the lack of health insurance as one of their major deficiencies. Rather, water quality and mosquitos are mentioned as the main favouring appearance factors for sickness, especially during the first interview round in 2018. Improvements of water quality in the second survey round 2019 could be explained by the implementation of tap water and water filters. However, water availability and nutrition supply are not perceived to have high impacts on health conditions. Furthermore, a two-way ANOVA showed improvements in access to health care centres nearby for HHs based in the Simu district, whereas coffee farmers living in the two other sub-counties recognised deteriorations in comparison to the previous year of the survey. For all factors, improvements but also deteriorations were perceived to result from external forces such as NGOs, the government and climate change.

Keywords: Deficiencies, life quality, smallholder coffee farmers, Uganda

Contact Address: Anna Lina Bartl, Georg-August-Universität Göttingen, Agricultural Economics and Rural Development, Göttingen, Germany, e-mail: anna.bartl@uni-goettingen.de

Effect of Land Degradation on Farmers' Food Security and Poverty Status Nexus Livelihood Diversification in Nigeria

Yusuf Oladimeji, Adunni Sanni, Abdullahi Hassan, Abubakar Sani
Ahmadu Bello University, Dept. of Agricultural Economics, Nigeria

The objective of this study was to assess the effect of land degradation on smallholder farmers' food security and poverty status nexus livelihood diversification in Kwara State, north central, Nigeria. Primary data with the aid of structured questionnaire was employed to collect the relevant data. A multistage random farming household survey resulted in four Local Government Areas, eight villages, 240 farmers which were filtered to 92 and 148 land graded farmers (LDF) and non-degraded farmers (NDF) respectively. Descriptive statistics, land degradation perception index, food security and poverty index, dichotomous regression models were used to achieve the aims of the study. The result indicates that 81.3% of the sampled farmers identified erosion as the main cause of land degradation with perception index of 4.3. Only 14.1% of LDF more than two-fifth, (42.6%) of NDF were food secured. The poverty status revealed that only 9.8% of LDF fall under the threshold of 0.00-20.00 category implying non-poor while 41.2% of NDF fall under this threshold. The results showed that the parameters that affected food security and poverty status of LDF varied from those that affected NDF. If the parameter was the same, the magnitude and direction differed. The average livelihood security composite index of LDF and NDF were 0.33 and 0.72 respectively implying that NDF had a low diversification value. The factors influencing livelihood diversification to off- and non-farm activities by LDF and NDF also differed in magnitudes, coefficients and directions. The results revealed that intercropping and mixed cropping are the most common strategies adopted by farmers in mitigating land degradation.

Keywords: Food security, land degraded farmers (LDF), mitigating strategies, poverty depth

Contact Address: Yusuf Oladimeji, Ahmadu Bello University, Dept. of Agricultural Economics, Department of Agricultural Economics / Institute for Agricultural Research, Zaria, Nigeria, e-mail: yusuf.dimeji@yahoo.com

Close the Gaps!-Losses in Local Food Systems: Evidence from Smallholder Vegetable Farmers in Southwest Nigeria

Albright Anozie, Oluwafunmiso Adeola Olajide
University of Ibadan, Dept. of Agricultural Economics, Nigeria

The demand for richer and better quality diets in Africa is increasing with the emergence of a larger middle class. This puts pressure on current production practices to deliver outputs sustainably; but can the current local food systems and supply chains support the increase in production that could occur? This study provides some evidence by examining the vegetable supply chain in Oyo state with a focus on the production node; output was assessed against Post-Harvest Loss (PHL); the PHL value was computed and its effect on the poverty status of farmers was examined. A two-stage sampling technique was used to select 120 vegetable farmers in peri-urban communities of Oyo state from whom data were collected through oral interviews with the aid of a structured questionnaire. The data were analysed using descriptive statistics, Foster-Greer-Thorbecke Index, gross margin and regression analyses. The results indicate that the node was dominated by men (94 %). The mean age was 44 years and average farm size was 3 hectares. Almost all (96 %) depend on vegetable production as the only source of income. At least 50 percent of the farmers experienced High-Post Harvest Loss (over 40 percent of output). The tobit analysis shows that PHL was likely to worsen the poverty status of farmers. The study recommends intensive training of farmers on harvesting and postharvest handling techniques; and the need for technology in processing and preservation techniques. Additional research to examine other sources of leakages across the supply chain and the interventions required to close the gaps is recommended.

Keywords: Farmers, livelihood, post-harvest loss, poverty, supply chain

Contact Address: Oluwafunmiso Adeola Olajide, University of Ibadan, Dept. of Agricultural Economics, Ibadan, Nigeria, e-mail: preciousfunso@yahoo.com

Rural Entrepreneurship: Motives and Barriers to Small Business in Ghana: A Gender Analysis

Bernard Kwamena Cobbina Essel, Miroslava Bavorova

Czech University of Life Sciences Prague, Fac. of Tropical AgriSciences - Dept. of Economics and Development, Czech Republic

The high incidence of youth unemployment and poverty in recent years threatens both the socio-economic growth and security of countries in Sub-Saharan Africa. In response, several policy interventions have been implemented to enhance job creation, value addition and to reduce poverty. Hence, this study explored one of these initiatives, that is, Promotion of Rural Based Small Enterprises in the Sunyani Municipality of Ghana. The study examined the gender perspective of Ghanaian rural entrepreneurs with respect to motivation, enterprise characteristics and the challenges faced. More specifically, it aims to identify whether men and women have different motives for becoming entrepreneurs. A cross-sectional survey was conducted among 200 small scale business owners selected using multi-stage sampling in Sunyani municipality of Ghana in 2015. Data were analysed using descriptive technique such as mean, frequency and cross-tabulations and chi-square test to test hypothesis. The descriptive analyses show three forms of small-scale enterprises vis-à-vis processing (46 %) (dominated by females), and artisans (37 %) and craft makers (17 %) (dominated by males). Female entrepreneurs (51 %) marginally dominate entrepreneurship. Further, business operation favours sole proprietorship (93 %) with insignificant numbers of family ownerships (5 %) and partnerships (2 %). Findings indicated men (54 %) are more likely to engage into business for economic benefits compared to women (46 %) but the difference is not significant (X^2 (1)6.613, $p < 0.010$). The need for independence to ensure balance between family and work relations are the motivations for women (67 %) into business start-ups. Further, it seemed more challenging for women (75 %) to acquire capital for business compared to men (25 %) the difference is significant at (X^2 (1)7.776, $p < 0.005$). Furthermore, women (61 %) found it challenging to access market for their products. An important consequence of the findings is the need for improved support of access to financial capital and to market for women starting entrepreneurship.

Keywords: Entrepreneur, entrepreneurship, gender, rural enterprises

Contact Address: Bernard Kwamena Cobbina Essel, Czech University of Life Sciences Prague, Fac. of Tropical AgriSciences - Dept. of Economics and Development, 16500 Kamýcká 129 , Czech Republic, e-mail: esselb@ftz.czu.cz

The Prestige of Farmer Occupations Perceived by Russian Youth: The Case of Altai Krai

Lenka Hofierková

Czech University of Life Sciences Prague, Faculty of Tropical AgriSciences, Department of Economics and Development, Czech Republic

All agricultural regions in Russia are facing an out-migration of youth, possibly also due to generally negative attitudes towards work in agriculture and the opinion that the agricultural sector is not prestigious enough. Migration represents a significant outflow of human capital from the agriculturally important areas causing a chronic shortage of workers at most of the farms. Government, as well as private institutions, try to re-engage young people in agricultural activities, but they are often targeting mainly economic incentives in their programs. The role of the overall low prestige of agricultural occupations, a multidimensional indicator involving economic as well as non-economic dimensions, is often neglected in policies as well as in research. The diploma thesis was focused on the prestige of farmer occupations as perceived by 350 young students in Altai Krai, the Russian largest agricultural region, and factors influencing their prestige perception. The following farmer professions were involved in the study: Farm manager, Private farmer, Small-holder farmer. The mean prestige of the involved farmer professions was perceived rather low, compared to other occupations. By running four multiple linear regression models (M1 – Small-holder farmer, M2 – Private farmer, M3 – Farm manager, M4 – Mean prestige of farmer occupations), factors influencing the prestige level of the agriculture-related occupations were identified. Attitudes towards work in agriculture revealed the strongest effect on the prestige level of farmer professions, of which the variable agriculture is an 'Exciting work' was the best predictor for all models. The opinion that agriculture is exciting influenced prestige level even stronger than the opinion that agriculture is 'Low-income work', although this predictor was found important too. Besides, the prestige level of all farmer professions was negatively influenced by the opinion that agriculture is a 'Men work', which was a surprising finding. Based on these results, in order to strengthen the prestige level of agriculture, the government could use information campaigns to show the exciting aspects of the work in agriculture and support the women in the agricultural jobs.

Keywords: Labour shortage, occupational prestige, youths' career

Contact Address: Lenka Hofierková, Czech University of Life Sciences Prague, Faculty of Tropical AgriSciences, Department of Economics and Development, Štefánikova 654, 27801 Kralupy Nad Vltavou, Czech Republic, e-mail: hofierkova.lenka@gmail.com

Effect of Community-Based Natural Resources Management Programme on Poverty Status of Fishing Households in the Riverine Areas of Ondo State, Nigeria

Oluwakemi Oduntan

The Federal University of Technology, Akure, Dept. of Agricultural and Resource Economics, Nigeria

With 2030 just ten years away, it seems unlikely that Nigeria meets the Sustainable Development Goal (SDG) targets for eradication of extreme poverty and hunger. Despite the huge fisheries resources Nigeria is endowed with, low yield, deplorable state of the fisheries resources in the country due to human activities, poor data statistics for proper planning and management especially in the face of climate change effects pervade the fisheries sector of Nigeria. The article therefore examines the effect of community-based natural resources management programme on poverty status of fishing households in the riverine areas of Ondo State, Nigeria. One hundred and twenty respondents were selected using a multi-stage sampling procedure. Data were collected from the respondents and analysed using a combination of descriptive statistics, Foster-Greer-Thorbecke (FGT) poverty measure, and probit regression model. Results of the descriptive statistics revealed that majority of the fish farmers are females and are in their economically active age, with an average household size of 6 and 7 for participants and non-participants respectively. FGT results showed that 35 % of the participants and 47 % of the non-participants respectively fell below the poverty line. Participants needed 10 % of the poverty line to get out of poverty while the non-participants needed 15 % of the poverty line to get out of poverty. Poverty was a bit more severe among non-participants (7 %) than the fish farmers who participated in the programme (4 %) in the study area. The results of probit regression model revealed that household size, years of education, share of farm income, access to credit and participation in the community-based natural resources programme are the determinants of poverty among the respondents. The study recommended that Government should encourage the non-participants to participate in the community-based natural resource management programme and also introduce policies that facilitate increased level of education, increased access to credit facilities are essential to help reducing poverty among fishing households in the study area.

Keywords: Fishing households, Nigeria, poverty status, riverine areas

---

Contact Address: Oluwakemi Oduntan, The Federal University of Technology, Akure, Dept. of Agricultural and Resource Economics, PMB 704, Akure, Nigeria, e-mail: ooduntan@futa.edu.ng

Crop Diversification for Improved Livelihoods of Organic Cotton Growers in India

Eva Goldmann, Amritbir Riar, Tanay Joshi
Research Institute of Organic Agriculture (FIBL), Dept. of International Cooperation, Switzerland

India is the biggest producer of organic cotton, most of it produced by smallholder farmers who sell it as a cash crop for in-season income generation. For these smallholder farmers, organic cotton-based farming systems need to be economically viable. Diversification is one option that is often proposed as a means to improve the livelihoods of smallholder farmers. It holds great potential to increase and diversify their income and mitigate economic risks. Diversifying the farming system can be achieved through optimising crop rotation, including intercrops and accompanying crops and using green manure, among others. Compared to monocultures, these practices bear agronomical and environmental benefits - such as increased productivity, additional lever in pest management, and fostering biodiversity.
We initiated a study in early 2020 to investigate the impact of crop diversification practices on the livelihood of organic cotton growers in India. A mixed-methods approach (Analysis of existing data sets, stakeholder interviews, and participatory horizontal farmer-stakeholder workshops) applied at landscape level enables us to identify existing diversification measures, economic, agronomic, and environmental strengths and weaknesses of existing and potential diversification measures to monoculture production of cotton. Particular attention is paid to market access and the local agro-climatic conditions as well as other contextual factors that influence the well-being of farmers.
The holistic appraisal of the different options to diversify cotton-based farming systems enables the recommendation of solutions suitable for the local conditions. Therefore, the study will contribute to a better understanding of the importance of diversification to improve the livelihood of organic cotton farmers.

Keywords: Cotton, crop diversification, India, livelihood, organic agriculture, smallholder farmers

Contact Address: Eva Goldmann, Research Institute of Organic Agriculture (FIBL), Dept. of International Cooperation, Ackerstrasse 113 Box 219, 5070 Frick, Switzerland, e-mail: eva.goldmann@fibl.org

Understanding Farmers' Shock-Coping Capacity in Dealing with Corona Virus Pandemic: An Iranian Case Study

Masoud Yazdanpanah[1], Tahereh Zobeidi[2], Nadejda Komendantova[3], Katharina Löhr[4], Stefan Sieber[4]

[1]*Agricultural Sciences and Natural Resources University of Khuzestan, Iran*

[2]*Zanjan University, Dept. of Agricultural Extension, Communication and Rural Development, Iran*

[3]*International Institute for Applied Systems Analysis (IIASA), ASA, Austria*

[4]*Leibniz Centre for Agric. Landscape Res. (ZALF), Sustainable Land Use in Developing Countries (SusLAND), Germany*

Rural households are frequently hit by severe shocks which can be classified as either idiosyncratic or covariate shocks. While the former is specific, affecting individuals or households, for example, illness, injury or unemployment of household members, the latter is correlated across households within a community (e.g. village), such as floods, droughts or epidemics like the Coronavirus. Both these shock categories can have devastating consequences on rural household welfare. The emergence of the novel Coronavirus pandemic presents an unprecedented health communications challenge. Most of the affected populations live in rural areas (or are farmers) with very minimal resources but at high risk of COVID-19 due to their surrounding and livelihoods. A detailed understanding of households' shock-coping capacity is needed to design appropriate social safety net programs and interventions. Therefore, the aim of this study is to investigate the factors affecting shock-coping capacity of farmers using Cognitive Theory of Stress Model. To investigate farmers' shock-coping capacities, a self-reported questionnaire survey was carried out in March 2020 in southern Iran. Participants were farmers who filled in a self-report online questionnaire. In total 310 questionnaires were completed and 294 were found useful for this empirical analysis after excluding incomplete responses, with a successful response rate of 17 %. Structural equation modelling (SEM) was applied to examine relationships between variables of the stress theory. The structural model provided a good fit to the data (Chi square/df = 2.683, IFI = 0.917; CFI = 0.916, RMSEA = 0.076). Results of structural equation modelling showed the adequacy of the stress theory in predicting the shock coping (R^2=0.90). Moreover, demand appraisal ($\beta = 0.83$, $p < 0.0001$) and collective efficacy ($\beta = 0.35$, $p < 0.0001$) emerged as the most powerful predictors of farmers' shock-coping.

Keywords: Iran, shock-coping capacity, theory of stress model

Contact Address: Masoud Yazdanpanah, Agricultural Sciences and Natural Resources University of Khuzestan, Mollasani, 744581 Ahvaz, Iran, e-mail: masoudyazdan@gmail.com

Food policy and governance: Examining linkages and food security outcomes

Power, Legitimacy, and Participation in the Co-Creation of Water Policy

Helga Gruberg[1], Joost Dessein[1], Jean Paul Benavides[2], Marijke D'Haese[1]

[1]*Ghent University, Dep. of Agricultural Economics, Belgium*

[2]*Instituto de Investigaciones Socio Económicas (IISEC); Bolivian Catholic University San Pablo, Dept. of Economics, Bolivia*

Integrated water resources management is considered a complex sustainability problem since it requires the active and effective participation of diverse communities of knowledge for the collaborative identification and implementation of transformative solutions. Therefore, the construction of a water related policy demands a highly participative and integrative approach, with the involvement of different actors. Recently, transdisciplinarity and collaborative governance have been promoted as ideal approaches to empower participants through their meaningful engagement in the process of co-creation of policy. However, empowerment does not necessarily occur in hierarchical structures where power relations may obstruct a more inclusive and equitable experience. Therefore, it is crucial to recognize power imbalances in early phases of the process to try to level up the floor for all participants. Otherwise, it is possible that some stakeholders may end up being excluded, overruled and, in worst cases, abused by dominating actors. Consequently, it is important to study power relations in the co-creation of natural resources policy. We take on a case study of the construction process of a water policy in Tiraque (Bolivia). Data was collected between 2017 and 2020 and analysed using stakeholder and power analysis tools. Findings show that although the construction of this policy was a common need prioritised by multiple local actors, the agenda was set by a local NGO. Instead of a process of co-creation, it was mainly a socialization process of a policy draft, developed by the NGO, at meetings with only strategically chosen people with decision-making power. Attendees had to transmit the information to their grassroots. However, due to language and other barriers, it was not possible for them and for their communities to critically analyse it. To conclude the process, the policy was presented for its approval. This failed, as irrigation organizations managed to challenge and block the process, which led to a new agenda, starting all over again. This case shows that effective participation depends largely on access to timely and adequate technical and legal information. Moreover, it reflects the importance of collaborative frameworks that can transcend grounded local, top-down, decision-making power structures.

Keywords: Empowerment, legitimacy, policy, power relations, sustainability

Contact Address: Helga Gruberg, Ghent University, Dep. of Agricultural Economics, Ghent, Belgium, e-mail: gruberghelga@gmail.com

Linking Agency to Food and Nutrition Security (FNS): The Role of Agroecology

Chukwuma Ume, Stephanie Domptail, Ernst-August Nuppenau

Justus-Liebig University Giessen, Inst. of Agric. Policy and Market Res., Germany

Food agency concerns are sparsely considered in Food and Nutrition Security (FNS) policies and analyses to date. To achieve FNS especially in regions with deeply entrenched power inequalities in agri-food systems requires agency responsiveness, empowering vulnerable and marginalised groups, and strengthen stakeholder engagement. This paper aims to develop a framework for investigating the role of agency of households in building their food security. Importantly, the study considers farming households as an element of the production and food system they are embedded in. Thus, we look at food insecurity as one outcome of the food system, under the current dominance of the corporate food regime, which aim is not the achievement of rural food security, but the production of global welfare. Thereby, the novelty of the framework is its focus on framing how especially agroecological farming practices and market systems shape the agency of farming households. Our claim is that rural households gain food system agency and independence from the dominant corporate food regime expressed in gaining control over productive resources and production decisions by producing agroecologically and using agroecological market systems. The paper reviews and builds upon the newly published thorough description of agency by the HLPE and the large body of literature on agroecology and agroecological food systems. It brings the two bodies of literature together by suggesting a conceptual framework and identifies pathways in which agroecology increases agency and leads to food security among smallholder farmers. The framework highlights the centrality of and interplay between agroecology and agency in achieving food security, especially among rural farming households, in a political ecology context of the dominance of the corporate food regime. Understanding how FSN is linked to governance, equity, and power dynamics within the agri-food system at multiple scales is important for the operationalisation of agency in empirical research.

Keywords: Agroecology, food agency, food security, food system, framework

Contact Address: Chukwuma Ume, Justus-Liebig-University Giessen, Inst. of Agricultural Policy and Market Research, Giessen, Germany, e-mail: chukwuma.ume@agrar.uni-giessen.de

Urban Food Policies and Governance in Kenya: Examining the Linkages

Robert Mbeche, Josiah Ateka
Jomo Kenyatta University of Agriculture and Technology, Dept. of Agricultural and Resource Economics, Kenya

Urban food systems have increasingly been recognised as a topic that needs to be better understood, in order to address issues of urban food security and urban poverty. This is particularly so in Africa, which has the fastest rates of urban population growth and high levels of urban food insecurity. There has, however, been surprisingly little work on examining the existing processes and policy framework through which urban food systems are governed. Using a case of Nairobi city, in Kenya, this paper assesses the policy, legislative and administrative environment to identify if it provides an enabling environment for sustainable urban food system. Data were collected through an extensive review of existing urban food policies in Kenya complimented with 27 key informant interviews involving actors in the urban food system. In addition, five Focus Group Discussions (FGDS) with government, NGO officials, traders and other actors in the food system were conducted. The paper finds that food policy in Kenya continues to treat rural and urban areas the same way despite the recognition that food insecurity, malnutrition and income opportunities are highly context specific and that diverse geographic areas (rural or urban) react differently to specific policies. The governance of Nairobi food system is highly complex with a multiplicity of actors with competing agendas which limit the implementation of existing policies. While there have been some attempts to enhance coordination among the key actors, actions to follow enactment have largely been lacking. The implication is that greater coordination is needed among the multiple actors and institutions involved in the formal and informal governance of the urban food system. The paper reflects on how to improve the policy environment to address the challenges of food insecurity for the most vulnerable in cities such as Nairobi. We propose a territorial approach to food security and nutrition as an alternative framework for capturing the variety of conditions across urban and rural regions as well as across different territories.

Keywords: Food and nutrition security, governance, Kenya, urban food policies

Contact Address: Robert Mbeche, Jomo Kenyatta University of Agriculture and Technology, Dept. of Agricultural and Resource Economics, P.O Box 62000-00200 , Nairobi, Kenya, e-mail: rmbeche@jkuat.ac.ke

Southern Agricultural Growth Corridor in Tanzania Programme and Food and Nutrition Security in COVID-19 Crisis

Benedict Mongula, Hezron Makundi, Rose Shayo
University of Dar Es Salaam, Institute of Development Studies, Tanzania

Southern Agricultural Growth Corridor in Tanzania is a programme to promote agriculture in the southern part of Tanzania from the eastern coast through the southern highlands. The programme was evolved by private investors, the government of Tanzania and development partners, in line with partnership framework proposed by G20, and along Responsible Agricultural Investments (RAI), endorsed by UN General Assembly in 2012. SAGCOT was expected to bring over 350,000 hectares under productive cultivation by large scale and smallholder farmers, with benefits in employment creation, increased outgrower activities and Corporate Social Responsibility. It was assumed investors guided by RAI principles with wide scale support of Government, Development Partners, international finance and civil society and farmers organisations will contribute to a huge success of the programme. But this was not the case, as the principles became flouted by a number of investors, unforeseen bottlenecks arose and Government support turned lukewarm. Undermined also by low commitment to the Partnership, poor communication and politics, the programme requires concerted efforts to live to its expectations and the expectations of the New Alliance for Food and Nutrition (NAFSN). In the ensuing COVID-19 crises of economic recession, capital and labour, this is all the more important in order for SAGCOT to contribute to global food and nutrition security for which NAFSN was conceived. This presentation is an evidence based paper examining the inclusive character of six SAGCOT large scale investments and their implications on food security and nutrition in the context of COVID-19 crisis. It proposes urgent national and international interventions to enable the programme to effectively contribute to global food security and nutrition.

Keywords: Agriculture, COVID-19, food and nutrition security, inclusive large scale agriculture, outgrower farming, principles of responsible agricultural investments

---

Contact Address: Benedict Mongula, University of Dar Es Salaam, Institute of Development Studies, University of Dar Es Salaam, Dar Es Salaam, Tanzania, e-mail: bmongula@gmail.com

Staple Food Consumption Influencing Economic, Social, Cultural, Nutritional, and Public Health Aspects in Brazil

Carlos Magri Ferreira[1], Alcido Elenor Wander[1], Reginaldo Santana Figueiredo[2]

[1]*Brazilian Agricultural Research Corporation (EMBRAPA), Brazil*

[2]*Federal University of Goiás (UFG), School of Agronomy, Brazil*

In 2016 the Brazilian Agricultural Research Corporation (Embrapa) launched an initiative (movement) entitled "Rice and Beans: the Brazilian Food" aiming to demonstrate the nutritional importance of diets based on these products via the dissemination of customized information about their nutritional and functional aspects. This initiative was launched because in the last decades the food industry has offered several products and services reshaping the eating patterns. A substitution of culinary preparations based on fresh or raw grains/cereals for almost ready or semi-ready foods was the consequence. In Brazil, the change affected the consumption of rice and beans, the staple foods of the population, with negative consequences in economic, social, cultural, nutritional, and public health aspects. This situation is not restricted to Brazil. The World Health Organisation recommends to governments to guide populations towards more healthy food choices. The movement was planned in three stages: 1) consumption diagnostics, 2) evaluation of the data from the first stage by a multidisciplinary team, 3) disclosure of clarifications to the population of critical points identified in the surveys. The data was obtained by diagnosis was carried out with consumers in the states of Goias and Mato Grosso. The diagnosis identified aspects related to changes in eating habits that affect the consumption of rice and beans. However, they remain essential food, regardless of age and family income. On the other hand, it was detected the need to better inform the population, with scientific arguments, about the benefits of consumption, with special attention to people with diseases, such as obesity, arterial hypertension, diabetes, high levels of triglycerides and cholesterol. The results corroborate and subsidise the main proposal of movement that institutions and agents in the production chains promote the dissemination of information on nutritional and functional characteristics of rice and beans to different profiles of Brazilian consumers, encouraging them to prefer more healthy diets. Summarising, steps one and two were performed as planned. To perform step three, it still is necessary to gain the support of more food chain members to overcome some obstacles related to funding and dissemination of information to consumers.

Keywords: Consumption patterns, food culture, food security, nutrition security, sustainable diets

Contact Address: Alcido Elenor Wander, Brazilian Agricultural Research Corporation (EMBRAPA), Rodovia GO-462, km 12, 75375-000 Santo Antonio de Goias, Brazil, e-mail: alcido.wander@embrapa.br

Linkage of Irrigation and Nutrition in Sub-Saharan Africa: A Review of Kenya's Irrigation Arrangements

Nixon Murathi Kiratu, Martin Petrick, Eefje Aarnoudse

Justus Liebig University Giessen, Center for International Development and Environmental Research, Germany

For the last five years, the number of undernourished people in the world has been increasing and is almost at levels last seen a decade ago. While prevalence in the percentage of undernourished people varies by region, in sub-Saharan Africa, the number is almost double the global number. In spite of the investments done in the agricultural sector in sub-Saharan Africa from the commencement of the Millennium Development Goals and the Comprehensive Africa Agriculture Development Programme about two decades ago, evidence shows a minimal impact on nutritional outcomes.
With irrigated agriculture having come back in Africa's policy agenda in the last few years the means of how this agricultural development is implemented to achieve the intended outcomes in food and nutrition security is a critical aspect of research and development arena. As governments and the development community cite irrigated agriculture's ability to utilise marginal areas, mitigate effects of climate change and possible solution to increasing food production in the region, it is imperative for researchers to provide clear means in which it can be linked to nutritional outcomes. This research seeks to give an understanding pathways through which irrigation can affect nutrition. This paper reviews recently published frameworks on irrigation and nutrition outcomes and innovatively looks at the socioeconomic aspects of farmers and the different institutional aspects of irrigation arrangements. In addition, the paper focuses on the aspect of neglected area of farmer initiated irrigation using secondary data from Kenya. Apart from documenting knowledge of the linkage between of irrigation and nutrition and building upon recent irrigation-nutrition frameworks, the paper also provides insights into the possible irrigation-nutrition pathways. In addition, through the use of secondary data, the paper provides insights on the aspects of irrigation linkages to nutrition in a sub-Saharan Africa country.

Keywords: Irrigation, Kenya, nutrition security, pathway, smallholder

Contact Address: Nixon Murathi Kiratu, Justus Liebig University Giessen, Center for International Development and Environmental Research, Unterhof, 35392 Giessen, Germany, e-mail: nixon.kiratu@agrar.uni-giessen.de

Link of Ecuadorian Universities with the Rural Sector through Pre-Professional Practices

Bélgica Bermeo Cordova[1,3], José Luis Yagüe Blanco[1], Maritza Satama Bermeo[2,1], Angel Satama Tene[3]

[1]*Universidad Politécnica de Madrid, Spain*

[2]*CEIGRAM, Spain*

[3]*Universidad Técnica del Norte, Ecuador*

During academic training, agricultural professionals acquire skills that are a fundamental pillar for revitalizing the agricultural sector, generating social change, meeting the needs of the productive sector and sustainable growth in rural areas. To promote this, Ecuadorian legislation requires the establishment of pre-professional practices that effectively bring students closer to the rural sector. The Organic Law on Higher Education (LOES) in Ecuador establishes the obligation to link producer organisations, public and private companies through a pre-professional practice program, which must be implemented and managed from the University. The present investigation tries to find out if this linkage is being carried out successfully in universities. For the purpose of the study an on-line survey was carried out in five of the main universities in Ecuador that are offering agricultural studies. In order to explore perceptions of different stakeholders four study groups were identified: managers of agri-food companies, students of agri-food studies, professionals already working after training, and finally, public and private institutions that have included pre-professional practices in their curricula. Data analysis was carried out by using multivariate hierarchical cluster technique. As expected, results show that Ecuadorian university complies with regulations regarding roles and functions of teaching and research. However, a direct relationship with the agricultural-rural environment is not well established yet, as the student's training is being undervalued in the framework of pre-professional practices. Several interesting strong points and deficiencies have been identified, and together they can very well guide high education decision makers in enhancing their practices programs.

Keywords: Ecuador, high education, pre-professional practices

Contact Address: Bélgica Bermeo Cordova, Universidad Politécnica de Madrid, Campus, Av. Puerta de Hierro, nº 2, 4, 28040 Madrid, Spain, e-mail: bnbermeo@hotmail.com

Examining the Emancipatory Potential of the Counter-Discourse of *La Via Campesina* by Conducting a Foucauldian Discourse Analysis

NIKOLA BLASCHKE, REGINA BIRNER

University of Hohenheim, Inst. of Agric. Sci. in the Tropics (Hans-Ruthenberg-Institute), Germany

In the face of global crisis like the world food price crisis of 2008 or the economic crisis due to the COVID-19 pandemic in 2020 and consequent upheavals for agricultural producers and consumers, there is increased interest in building alternative agricultural systems which are more resilient and sustainable. One of the most radical suggestions is articulated by the international peasant organisation *La Via Campesina* which positions itself opposite the established organisations with their peasant-led approach of food sovereignty which struggles to change the current "agro-industrial" system towards one that is democratically controlled by the people, bringing about fair power relations. A broad academic literature theorizes and supports this counter-discourse and attributes to it a huge emancipatory potential in political and economic terms.
However, the debate about finding meaningful alternatives would benefit much from complicating the claims of *La Via Campesina*, which could for example be attempted from a post-structuralist perspective. Following Michel Foucault, an analysis of the way *La Via Campesina* conceptualize "knowledge" and "power" is conducted and deepened by the comparison with its "counter" organisation, the World Farmers' Organisation. Both organisations' key documents are studied for this purpose.
Overall, the analysis revealed many emancipatory aspects in the discourse of *La Via Campesina*, such as the powerful suggestion to radically redefine the criteria based on which success is measured by putting the needs of the vulnerable into the centre, signifying a change of discourse towards more emancipation. However, following Foucault's scepticism about alternative movements proclaiming the creation of a space which is free of "power", the way *La Via Campesina* conceptualizes change brings to the surface some problematic notions: for example, the movement's internal contradiction between land owners and landless labourers is not made a subject of discussion, but instead the line of conflict is exclusively drawn between "the people" and "evil" neoliberalism. This covering-up of power relations within the movement for the sake of a "cleaner" discourse may carry with it new forms of suppression. This suggests calling for more ambiguity in their discourse to be able to offer meaningful solutions for past, present and future crisis.

Keywords: Counter-discourse, emancipation, food sovereignty, Focauldian discourse analysis, *La Via Campesina*, world farmers organisation

---

Contact Address: Nikola Blaschke, University of Hohenheim, 70595 Stuttgart, Germany, e-mail: nikola_blaschke@yahoo.com

Processing of Neglected Crops for Promotion of Food and Nutrition Security: the Case of Natural Guava Fruits in Kenya

Duke Omayio[1], George Abong'[1], Michael Okoth[1], Charles Gachuiri[2], Agnes Mwang'ombe[3]

[1]*University of Nairobi, Dept. of Food Science, Nutrition and Technology, Kenya*
[2]*University of Nairobi, Dept. of Animal Production, Kenya*
[3]*University of Nairobi, Dept. of Plant Science and Crop Protection, Kenya*

Sub-Saharan Africa is characterised by significantly high rates of micronutrients deficiencies due to limited access and utilisation of fruits as a result of high postharvest losses and minimal adoption of postharvest technologies especially in the rural areas. In Kenya, the guava fruit is a neglected crop and is majorly characterised by naturalized varieties from randomly dispersed seeds despite the favourable climatic conditions for the crop's commercial cultivation. Although the fruit is highly nutritious, its nutritional potential remains extremely underutilised owing to low consumption and limited processing and preservation as natural guava value addition is non-existent. Guava fruits are climacteric and are highly perishable leading to huge losses when in season. Consequently, the annual losses have been estimated to be more than 11,000 tons valued at 1.1 million US dollars contributing to increased poverty and hidden hunger as a result of low income generation due poor marketability and low consumption in households where the fruits grow. The current study has led to the development of natural guava processing and market-testing of nutritious and safe blended nectars with boosted iron, zinc and vitamin A levels. These are affordable guava processing techniques that can be adopted at the household levels in order to produce commercially viable, safe and nutritious natural nectars to tackle malnutrition as well as improve guava farmers' households. Stable products with longer shelf life will increase guava utilisation after the fruits are out of season, minimise post-harvest losses and ensure regular supplies from the production areas to consumers in peri-urban and urban centres while the income generated will improve the food security status of guava farmers' households. However, technical barriers have to be overcome and this will be realised through technology transfer to farmer groups and processors. If adopted, the findings will help in developing policies that will strengthen the guava value chain in Kenya in order to fully exploit the fruits' economic and nutritional potential and promotion of nutritional security and diversity by enhancing the availability of processed local guavas.

Keywords: Fruit, guava, losses, natural, nutrition, processing

Contact Address: Duke Omayio, University of Nairobi, Dept. of Food Science, Nutrition and Technology, 29053, 00625 Nairobi, Kenya, e-mail: dukegekonge@yahoo.com

Food and Nutritional Security in Brazil: Socio-Political Scenario from 2015 to COVID-19

Hayanne Rodrigues Carniel Cavalcante[1], Thiago Henrique Costa Silva[2], Nara Rabia Rodrigues do Nascimento-Silva[3], Luciana Ramos Jordão[2]

[1]*University Center Alves Faria, Law College, Brazil*
[2]*Federal University of Goiás (UFG), Law College, Brazil*
[3]*Federal University of Goiás (UFG), Nutrition, Brazil*

Over the years, Brazil has gained worldwide recognition in the fight against hunger and food insecurity, being a reference for other parts of the world. This fact occurred after the understanding of hunger as a social, political, and economic problem, not just as a natural phenomenon. Food and nutritional security structure were built with policies aimed at reducing poverty and inequities, with a focus on human development, such as "Fome Zero" and "Brasil sem Miséria", resulting in Brazil leaving the Hunger Map in 2014. However, despite all the achievements, recent analyses reveal the increase in poverty numbers in Brazil and great possibility of returning to the Hunger Map. According to the semi-annual report on the Latin American and Caribbean Region, developed by the World Bank in April 2020, since 2015 the reduction of poverty and inequality seems to be stagnant, and this year's projections of the International Poverty Index increased from 4.4 in 2019 to 7.0 in 2020. Thus, this research aims to analyse the current Brazilian scenario, the measures taken and the impact already observed, through a qualitative approach, anchored in historical and dialectical methods, through bibliographic and documentary research. It was noted that in 2019, Provisional Measure 870/2019 was instituted by the current government of President Bolsonaro, which modified the structure of the federal government, reducing and bringing together ministries, and disrupted the Organic Law on Food and Nutritional Security (Losan nº 11.346 / 2006), extinguishing the National Council for Food and Nutrition Security (CONSEA) and, consequently, damaging the food and nutrition security system (SISAN). Currently, Brazil has no new National Plan for Food and Nutritional Security, referring to the years 2020-2024, as this would be the responsibility of CONSEA. Amid this deficit and insecure scenario, due to Covid-19, social problems can be amplieded, increasing the level of food insecurity. Thus, observing and analysing the actions of public authorities and organised civil society are the next tasks of this research.

Keywords: Corona virus, food entitlement, pandemic, public policy

Contact Address: Luciana Ramos Jordão, State University of Goias, Law College, Av. Perimetral Norte N. 4356 C. 88b Cond. Alto da Boa Vista, 74445500 Goiais, Brazil, e-mail: lr.jordao@me.com

Effects of Climate Change on Food Security of Smallholder Farmers in Nepal and on Migration

Giri Prasad Kandel, Miroslava Bavorova

Czech University of Life Sciences Prague, Fac. of Tropical AgriSciences, Dept. of Economics and Development, Czech Republic

Climate change endangers food security mainly of smallholder farmers. Food insecurity is an important push factor of migration and strategy to safeguard livelihood in subsistence farming communities. In Nepal, most of the mountainous households are experiencing effects of climate change such as change in patterns of temperature and precipitation. To cope with climate change scenarios, people are using different adaptation strategies such as use of improved crop varieties, improved irrigation systems and technologies as well as off-farm activities (for instance seasonal migration). This research is intended to understand the link between climate change adaptation of smallholder farmers, food security and migration using a case study in the Baglung district (below 3,500 m - Tarakhola: 2,000-2,400 m) and Mustang district (lower Mustang below 3,800 m - Kunjo: 2,420 m, Marpha: 2,700 m, Muktinath: 3,664 m) in Nepal. The main aim of my study is to explore the theoretical background and interlink the concepts of migration with food security. Further, the use of traditional and modern climate change adaptation strategies will be connected with the food security and livelihood of smallholder farmers as a part of the concept. It is also anticipated to figure out how the topographic and demographic factors influencing the adaptive behaviour and strategies. Mixed method approach with quantitative questionnaire survey and qualitative expert interviews will be used in my field survey to account for various topographic and climate conditions at the end of year 2020. A total of 240 samples respondents (60 from each village) will be selected using simple random sampling technique. And about 15 qualitative expert interviews will be conducted. The quantitative collected data will be analysed along with descriptive statistics (mean, median, mode and etc.) and inferential statistics (multi-factor models). In the end of my study, I will be able interlink different factors such as adoption of climate change, food security, livelihood and migration.

Keywords: Adaptation, food security, Himalayan range, impacts, livelihood, Nepal, perception, smallholder farmers

Contact Address: Giri Prasad Kandel, Czech University of Life Sciences Prague, Fac. of Tropical AgriSciences, Dept. of Economics and Development, Kamycka 1281, 16500 Prague, Czech Republic, e-mail: kandelg@ftz.czu.cz

Willingness to Accept Lock-Down in COVID-19 Pandemic and Effect on Livelihood in Southwest Nigeria

Temitayo Adenike Adeyemo

Nigerian Institute of Social and Economic Research, Dept. of Agriculture and Food Policy, Nigeria

Would an individual be willing to die of hunger or from a disease outbreak in Nigeria? This research question came at the premise of the global health crisis induced by COVID-19 outbreak and its impact on the livelihood of individuals. Livelihoods become vulnerable when unable to cope with shocks and the outbreak of COVID-19 has been a source of such a shock to the global community and especially to developing economies like Nigeria. The main response to the spread has been the lockdown of economies for certain periods. However, the lockdown may impact on the livelihood of citizens and thus reduce its effectiveness. As a result of the restriction in mobility amid COVID-19, this study used an online survey to generate information on individual's Willingness to Accept (WTA) an economic lockdown as a means of preventing the spread of the virus in Nigeria from 75 lower to middle class individuals in Southwest, Nigeria. Respondents with vulnerable livelihoods had lower food and nonfood expenditure patterns. Survey analysis showed that up to 61 % of the respondents were willing to accept a short period of lockdown, at an estimated WTA of not more than 4 days and income loss of N 8,538.20 (22 USD) per day. Also, the findings showed that individuals who were likely to lose income, with large households, and large food expenditure patterns were less willing to accept a lockdown. Respondents were more willing to use a combination of sanitary and social distancing measures rather than a complete lockdown. Although this online survey was relevant for respondents in the lower to middle class group, it was able to show the vulnerability of livelihood to protracted lockdown response to a disease outbreak. This is a retrospective survey and the theme of this study may be more evident if an extended survey involving actors in the informal sector was carried out. An extended study will enable an estimation of the value of economic life across different sectors of the Nigerian economy. This will be useful in developing responses to shocks such as this pandemic. The study therefore recommends that socio-economic and livelihood contexts must be taken into consideration when enforcing disease prevention measures, while social protection must be enhanced to reduce the effect of such measures on livelihood.

Keywords: COVID-19, livelihood, lock-down, willingness to accept

---

Contact Address: Temitayo Adenike Adeyemo, Nigerian Institute of Social and Economic Research, Dept. of Agriculture and Food Policy, P.M.B 5, U.I Post Office, Ojo, Oyo Road, Ibadan, Nigeria, e-mail: adeyemotemitayo@gmail.com

Evaluating the Relationship Between Solid-Food Waste, Environment and Economic Security among Malnutrition in Nigeria

Opeyemi Anthony Amusan

University of Ibadan, Center for Petroleum, Energy Economics and Law / Amiesol Resources Konsult, Nigeria

Solid-food waste generation is estimated at 126.2 million tonnes and 239.8 million tonnes of carbon-dioxide equal by 2020 ending. This huge solid -waste costs Nigeria $750 billion annually while millions of Nigerians are hungry and poor. Nigeria also ranks very-low in nutrition with the highest number of malnourished children under 5 years in sub-Saharan-Africa. 37 %-of-Nigeria-children are stunted, 18 %-wasting and 20 % underweight - these stunning figures rank Nigeria as the second highest globally. While developed countries have been able to manage waste properly for increased environment-and-economic-security, this is farfetched in Nigeria. Since solid-food waste can cause health, environment and socioeconomic problems, there is need to investigate the relationship between solid-food waste, environment and economic security. The main objective of this study therefore was to evaluate the relationship between solid-food-waste, environment and economic security among malnutrition in Nigeria.

Data on waste-management-practices were obtained through structured-questionnaires randomly administered on 210 households in Nigeria. Experts'-workshops-and-interviews were organised for key-officials within relevant-industries to elicit technical-and-economic information. The relationship between: waste, environment and economic security in Nigeria was examined for years 1981-to-2017. While waste-management-practices were evaluated using descriptive-and-inferential-statistics, Autoregressive-distributive-lag (ARDL) was used to determine the relationship between solid-food-waste, environment and economic security. Pollution/Health-risks (69.1 %), limited-resources / funding (44.8 %), lack-of-technical-skill (23.8 %) and inadequate-management-skill (18.1 %) are some identified challenges. 94.3 % and 96.2 % supported polluters'-pay-principe and dissemination of public-information on food-packaging as-well-as waste-reduction-reuse-recycling as part of waste-management practices respectively. 97.1 % of annually-generated-waste are solid-waste, which confirms the Waste-Habit-of-Nigerians as 57 %-organic/food-waste, 27 %-plastics, 5 %-glass, 5 %-metal and 4 %-others.

Contact Address: Opeyemi Anthony Amusan, University of Ibadan, Center for Petroleum, Energy Economics and Law / Amiesol Resources Konsult, P.O. Box 23039 Post-Office Agbowo, 20012 Ibadan, Nigeria, e-mail: amusanope@gmail.com

126.2 million tonnes food-waste equaling 239.8 million tonnes of carbon-dioxide and $750 billion is generated yearly in Nigeria. 95 % are willing-to-pay for waste-management. Hypothesis-test yields a significant result at $p < 0.05$ which shows that waste-management challenges has effect on health issues / pollution in Nigeria. ARDL-model F-statistics of 30.7805 confirms the long-term-relationship between measured variables related to solid-food-waste generation, environment and economic security. ARDL-model also confirms the inverted-correlation between economic-growth and environmental-degradation of Environmental-Kuznet-Curve's hypothesis. At 0.0048 p-value, the estimates enjoy the support of statistical-significance at -5 %.
Undertaking established waste-management significantly limits the impacts on health-environment-socioeconomic-wellbeing. The research shows that improved-funding and dissemination of public-information on food-packaging, as-well-as waste-reduction reuse-recycling enhance social-acceptability of waste-management practices. This research also shows that solid-food-waste has significant impact on environment-and-economic-security.

Keywords: Environmental quality, health, socioeconomic wellbeing, waste management

Collection, Use and Commercialisation of Indigenous Plant Species by Households Living in Barotse Floodplain, Zambia

Moonga Mbozi, Vladimir Verner

Czech University of Life Sciences Prague, Fac. of Tropical AgriSciences, Czech Republic

Indigenous plant species play crucial role in the livelihoods of households in tropical regions. These species are used to meet various needs, such as maintaining food security, generation of additional income, or preservation of traditional knowledge. In these days they have been recognized as key stone of agrobiodiversity and valuable genetic resource that might help to mitigate negative impacts of climate change. Their importance has been increasingly highlighted not just by researchers, but also by policy makers. The main aim of this contribution is to provide an overview of existing literature on economic aspects of indigenous species utilization by households living in Barotse floodplains in western part of Zambia. Special attention is given to external and internal household factors related to collection and use of these species. Furthermore, overview of current research and development projects focusing on indigenous species in target will be provided.

Keywords: Food security, households, Miombo woodlands, sustainable agriculture, Zambia

---

Contact Address: Moonga Mbozi, Czech University of Life Sciences Prague, Fac. of Tropical AgriSciences, 165 00 Prague, Czech Republic, e-mail: xmbom001@studenti.czu.cz

Food Security and Food Quality among Vanilla Farmers in Madagascar: The Role of Contract Farming and Livestock Keeping

Jessica Andriamparany[1], Hendrik Hänke[2], Eva Schlecht[3]

[1]*University of Göttingen, Animal Husbandry in the Tropics and Subtropics, Germany*

[2]*University of Goettingen, Dept. of Agricultural Economics and Rural Development, Germany*

[3]*University of Kassel / University of Goettingen, Animal Husbandry in the Tropics and Subtropics, Germany*

Around 67 % of Madagascar's population is malnourished. Yet little is known about the effects of recently booming vanilla prices on food security in Madagascar's SAVA region, the largest vanilla producing area globally. This study analysed food security and diet composition of local vanilla farming households (HHs, n=140) by means of a 12-month longitudinal food survey. Household Dietary Diversity Score (HDDS), Food Consumption Score (FCS), Food Security Index (FSI) and the contribution of protein from Animal Source Food (ASF) were used to characterise diet composition, food security and nutrient intake. Data was complemented with baseline, agro-economic, longitudinal and field-plot information to determine factors influencing food security and ASF protein contribution using a stepwise generalised linear model.

Many HHs (74 %) were food insecure with insufficient calorie, vitamins A and E intakes, but had an acceptable protein intake. Consumption of rice, the principal source of carbohydrates, was stable across the year. Compared to other regions in Madagascar, local diets were moderately diversified (HDDS = 6.9) with a medium dietary share of ASF protein (about 50 %). HH size ($p < 0.001$) and cash income from rice sales ($p < 0.001$) were the most important factors influencing FSI, while cash crop income ($p < 0.01$) and number of income sources ($p < 0.01$) were more important for the dietary share of ASF protein. Contracts with vanilla exporters and livestock ownership did not improve food security. Although many vanilla actors run social and environmental programs in the SAVA region, more needs to be done to improve diet quality and strengthen farmers' resilience to food insecurity.

Keywords: Animal source food, dietary diversity, food security index, seasonality, vanilla farmers

---

Contact Address: Hendrik Hänke, University of Goettingen, Dept. of Agricultural Economics and Rural Development, Göttingen, Germany, e-mail: hhaenke@uni-goettingen.de

Agroecology: Way to Reduce Social and Environmental Inequalities and to Realize the Right to Food

Thaisa Toscano Tanus
Salgado de Oliveira University, Law Course, Brazil

Agroecology, a term first used in the 1930s, can be considered a transdisciplinary science that combines the three dimensions of sustainability: economic, social and environmental, combining ecology and agriculture. During the following decades, this concept gained an idea of environmental awareness, becoming a project of social and environmental development differentiated from the agricultural models that predominate in some developing countries, such as Brazil. Such models involve environmentally unsustainable agricultural techniques in violation of constitutionally established rights or rights provided for in international treaties. They cause environmental impacts such as soil degradation, air and water contamination, damage to human health through the indiscriminate use of pesticides, among others, which mainly affect the most vulnerable social groups. In this way, they increase the rural exodus and favour the concentration of income and lands in the hands of large producers, aggravating social inequalities and environmental injustice. Within this framework, Agroecology and food production in the 21st century are emerging in the midst of a global conjuncture of social and environmental inequalities and serious climate changes, as issues of the decision-making agenda around the world, especially in developing countries. The negative effects of climate changes mainly affect agriculture, reducing land use, affecting food production yields and consequently limiting food availability. In addition, population growth, pollution and degradation of natural resources make up this state of crisis, making it difficult to realise the social right to food, provided for in the Universal Declaration of Human Rights and constitutions of almost all countries in the world. Thus, agroecological public policies are essential to ensure access to food and nutrition security for the entire world population, in a cooperative effort among nations. Based on an exploratory, documental and legislative research, this study therefore proposes to seek convergent points between Agroecology, from the perspective of an instrument for reducing social and environmental inequalities, and the guarantee of access to food for the entire population, in order to promote environmental justice.

Keywords: Environmental justice, public policies

Contact Address: Thaisa Toscano Tanus, Salgado de Oliveira University, Law Course, Rua 250 nÃ□Â°mero 59 Setor Coimbra, 74535350 GoiÃ□Â□nia, Brazil, e-mail: toscanothaisa@gmail.com

Livelihood Strategies and the Role of Baobab (*Adansonia digitata* L.) Fruits in Poverty Alleviation in the Dry Lands of Sudan

Ismail Adam[1], Yahia Adam[1], Dagmar Mithöfer[2]

[1]*University of Khartoum, Department of Forest Management, Sudan*

[2]*Humboldt Universität zu Berlin, Albrecht Daniel Thaer-Institut für Agrar- und Gartenbauwiss., Germany*

This study aims to (i) assess the contribution of baobab fruits to rural income and poverty alleviation; (ii) analyse the livelihood strategies pursued by rural households, and (iii) identify the factors that influenced households' choice of livelihood strategies in rural Sudan. Data from 374 households surveyed in West Kordofan, North Kordofan, and Blue Nile was collected in 2017 and 2018. We employed a class cluster analysis to determine the optimal number of livelihood clusters and assigned individual households to particular clusters. Regression models were used to examine the factors influencing the livelihood strategy choices. The results showed that baobab fruits contribute from 7 % to 18.5 % of the total annual household income in the study areas. This additional income from baobab fruits contributes to the reduction of both the poverty headcount index and income inequality in the study area. The results also revealed a high diversity index through the livelihood strategies in the study areas. This study provides evidence that households' asset endowments and contextual factors have an important influence on the choice of household livelihood strategy. The results of regression indicated that the gender of the household head, primary and secondary level of education, tropical livestock units, and land size were positively significantly associated with the selection of the livelihood strategies. However, age of household head, household size, and distance to market were negatively correlated with the livelihood strategies selection. This study concluded that baobab fruit play an important role in supporting livelihoods, and therefore provide an important strategy against income inequality and promoting poverty reduction for households. This study highly recommends that policies should focus on enhancing the productivity of agricultural land plots owned by households rather than increasing households' access to common property resources. In addition, effective pro-poor policies should be targeted towards assisting the poor to shift to higher-return activities, such as wage employment and non-farm (business) by investing in education and improving the road infrastructure in rural areas.

Keywords: Diversification, environmental income, livelihood assets, poverty alleviation, Sudan

Contact Address: Ismail Adam, University of Khartoum, Department of Forest Management, Al-Ameen Al-Kareb, 13314 Khartoum North, Sudan, e-mail: somaasamaa2@gmail.com

Climate Change in the Municipality of Morrinhos, Goiás, Brazil: A Survey through Oral Reports from the Elderly Population on Climate and Environmental Changes, Over the Years

Thays Dias Silva[1], Barbara Luiza Rodrigues[1], Igor Gomes de Araújo[1], Rebeca Barbosa Moura[1], Luciana Ramos Jordão[1], Letícia Versiane Arantes Dantas[1], Niury Ohan Pereira Magno[2], Victória Cardoso Carrijo[1]

[1]*State University of Goias, Law College (Southeast Campus: Morrinhos), Brazil*
[2]*Federal University of Goiás (UFG), Law College (Southeast Campus: Morrinhos), Brazil*

Climate change is a matter of global interest currently being discussed by political and environmental leaders, with the aim of proposing a balance between environment and sustainable development. In Brazil, agriculture is one of the most important economic activities in the country. On the other hand, according to the Brazilian Institute of Geography and Statistics (IBGE), the federal agency responsible for geographic, economic and populational statistics, the agricultural activity is the most responsible for deforestation. There are difficulties in achieving a balance between agricultural activities and the environment. The preserved environment is very important to combat climate change. According to the United Nations (UN), forests are great allies of human beings in the fight against climate change, since they absorb about 2 billion tons of CO_2 per year. The work aims to get to know the perception about climate changes in the Municipality of Morrinhos, Goiás, Brazil, and, for that, it conducts research with the elderly population through oral reports. This research is of great importance for society, so that it perceives climate changes, based on the experiences of the elderly and the population's understanding of the direct effects on their daily activities and their quality of life. For Environmental Law, the research is relevant, as it shows how the population is affected by climate change and allows directing this branch of Law towards an increasingly effective protection of the environment, since it is a diffuse right stipulated by the Brazilian Constitution. In order to obtain a result of the research, a semi-structured questionnaire is prepared to collect primary data from the elderly population in the municipality of Morrinhos, by means of a random sample, about climate changes and their causes. The methodology to be used will be empirical research, using data from IBGE to identify the number of elderly people in the city of Morrinhos. The investigation will be carried out via telephone calls, obtaining oral reports on changes in the climate and the environment.

Keywords: Climate Change, environmental law, oral reports, sustainable development

Contact Address: Barbara Luiza Rodrigues, State University of Goiás, Law College (Southeast Campus: Morrinhos), AnÃ¡polis, Brazil, e-mail: barbara.rodrigues@ueg.br

This Is Our Land – Resource Conflict and Arable Crop Production: Evidence from Southwest Nigeria

Tofameh Kindness Isumonah, Oluwafunmiso Adeola Olajide

University of Ibadan, Dept. of Agricultural Economics, Nigeria

It has been said that Africa's soil is the next gold mine. But land resource is becoming a major source of contention between stakeholders; this threatens crop production activities and ultimately the food security situation of the region. But empirical evidence on the drivers of the conflict, the measurement of conflict, the extent and direction of its effect on arable crop production, particularly in developing countries where food production is largely dependent on small holder farmers is required. This study provides some evidence by examining farmer-herdsmen conflict in some rural communities in Ogun state, Nigeria. A multistage sampling technique was used to select 150 farmers from who data were collected through oral interviews with the aid of a structured questionnaire. The data were analysed using descriptive and regression tools. The results show that 60 percent of the respondents had no alternative source of income; the average farm income was about $ 250 per annum and over 70 percent had experienced fights with herdsmen over their land. The drivers of the conflict include land encroachment by herders, killing of stray cattle by farmers, decreasing economic productivity and decreasing consumption rates. Conflict-index was generated from farmer responses; the logistic regression shows that it has a potentially negative effect on food crop production in the area. The study recommends local community action for effective management of the situation through bridge building. Further research is also required to examine possible synergies and models of community action that will diffuse tensions effectively.

Keywords: Land, community, conflict, food security, production, tension

Contact Address: Oluwafunmiso Adeola Olajide, University of Ibadan, Dept. of Agricultural Economics, Ibadan, Nigeria, e-mail: preciousfunso@yahoo.com

The Food Security and Water Insecurity Paradox in Iran; The Case of Feed Production - Feed Import - Irrigation Water Consumption Nexus

Alireza Ahmadi[1], Tinoush Jamali Jaghdani[2]

[1]*Independent Consultant, Germany*

[2]*Leibniz Inst. of Agricultural Development in Transition Economies (IAMO), Agricultural Markets, Germany*

The agricultural sector in Iran is accounted for more than 90 % of water consumption. This sector put huge pressure on endangered water resources which even can cause water insecurity to many water basins in the future. Currently, in Iran around 91 % of the meat demand and 98 % of raw milk are supplied domestically which make the provision of the feed products essential.

In order to enhance the food self-sufficiency of meat and dairy products, the general agriculture's policy of Iran is pushing for the self-sufficiency of the main feed items. While the fodder crops have relatively high irrigation water requirements, the plan of increasing their production will threat the already endangered groundwater resources more. In other word, food self-sufficiency in Iran which is equalised more or less to food security is closely linked with environmental issues and water insecurities. Therefore, decision makers need to have a balanced strategic importing and production plan for feed items in order to prevent water insecurities.

In this study, we apply nexus concept which has been developed to analyse nexus-related interlinkages. We have used the provincial available data on red meat and dairy production on one side and the nexus between fodder crop production, feed import and irrigation water consumption on the other side. We use the available data on water footprint of fodder crops and regional irrigation water data on fodder crops to estimate the water requirements. By using the nexus approach, these data are compared to water resources condition in each province to examine the agronomic water efficiency.

The primary results of this study shows that feed crop production should be reviewed very critically in many provinces as its promotion especially in those provinces depending mainly on groundwater resources can cause danger to inhabitants by increasing their future water insecurities.

Keywords: Feed production, food security, nexus, water security

Contact Address: Tinoush Jamali Jaghdani, Leibniz Inst. of Agricultural Development in Transition Economies (IAMO), Agricultural Markets, Theodor-Lieser-Str. 2, 06120 Halle (Saale), Germany, e-mail: jaghdani@iamo.de

Education - Knowledge - Extension

Crop Index Insurance for Climate Resilience Enabling Higher Agricultural Investments: An Experimental Implementation in Kyrgyzstan

Laura Moritz, Lena Kuhn, Ihtiyor Bobojonov

Leibniz Institute of Agricultural Development in Transition Economies (IAMO), Agricultural Markets, Germany

In times of climate change agricultural producers in the Global South need an incentive to continue investing into their farm businesses. As index insurance is often discussed as an instrument to strengthen farmers' climate resilience, it allows overcoming fundamental investment barriers and producing crops that are desirably needed to feed the population – independent of whether this implies conservative or more agroecological farming. However, index insurance take-up rates are yet low, in particularly among the most vulnerable smallholders. Research argues that smallholders' risk-averse nature, low monetary and land endowments justify this insufficient risk management but these factors alone cannot explain differences in the adoption decision. Our study is the first to also explore the influence of trust, enhanced by understanding, peer effects and subjective as well as objective risk characteristics on the demand for marketable and unsubsidised crop index insurance. To unveil relations we conducted comprehensive and realistic experiments with a sample of 144 representative Kyrgyz smallholders from a predominantly non-irrigated area in 2018. Applying a Heckman two-step approach, our results show significantly positive peer imitation effects for the binary uptake decision as well as for insurance products characterised by more costly premiums favoring higher compensations. This highlights the importance of the collective on individual decision-making. Further, the results suggest that sophisticated product understanding, measured in perfectly comprehending the insurance concept theoretically and having practical non-agricultural insurance experience, and stated trust into the insurer encourage index insurance adoption. While subjective climate risk also has a positive implication, objective climate risk has a negative effect. These opposing relations may result from a yet not well established national insurance market that hinders to trust the insurer to adequately compensate for real and high yield losses but intrinsically longing for insurance to hedge when feeling climate threats. Lastly, we observe that insurance contracts with higher compensations encounter a higher demand. Theoretically, these findings challenge the perception of internalised decision-making processes, which are based on economic factors and invariable individual- and farm-specific characteristics. Practically, it underlines the importance of community-based extension treatments and trust towards the adoption of new farm and risk management solutions.

Keywords: Agricultural index insurance, climate risk, Kyrgyzstan, rural development

Contact Address: Laura Moritz, Leibniz Institute of Agricultural Development in Transition Economies (IAMO), Agricultural Markets, Theodor-Lieser-Straße 2, 06120 Halle (Saale), Germany, e-mail: moritz@iamo.de

Cocoa Farmers' Expectations and Motivation When Converting to Organic Agriculture

LINA TENNHARDT, GIANNA LAZZARINI, CHRISTIAN SCHADER

Research Institute of Organic Agriculture (FiBL), Dept. of Socioeconomics, Switzerland

The global demand for organic cocoa is growing. In order to assure its future supply, more smallholder cocoa growers in the tropics must adopt organic agriculture (OA). However, little is known about farmers' expectations of and their motivation to go through with organic certification as well as prevailing obstacles. Yet, this knowledge is crucial to properly support cocoa farmers in their OA conversion process and facilitate long-term relationships with value chain actors. We carried out semi-structured interviews with 205 cocoa farmers currently converting to OA in Central Uganda to examine their perception of OA. The farmers in our sample first heard of OA from the export company who organised the group certification (42 % of respondents) or the group's lead farmer (29 %). Our results show that farmers perceived OA as a system in which chemicals are banned (63 %). Only 32 % described it as a system that promotes the use of organic inputs. 16 % of farmers were not able to describe OA at all. Perceived benefits of OA were lower production costs (36 %), reduced health risks for farmers (28 %), and improved soil fertility (24 %). The reported disadvantages include difficult pest and disease management (35 %) and ineffective organic inputs (15 %). To address the perceived high pest and disease pressure, 35 % of farmers continued applying synthetic pesticides, while only less than one third were using preventative management practices (32 %) or inputs (23 %). It is very likely that the organic price premium, which farmers will receive once certification is accomplished, will satisfy farmers' expectation of a higher cocoa farm-gate price. However, our results also indicate that price premiums may not be a sufficiently large incentive to deter farmers from incompliance with organic standards. This applies especially if the understanding of OA is limited and the wish for certification did not emerge from within the farmer group, but was proposed from outside. Based on our findings, we suggest certification organisers to emphasise on farmer sensitisation on OA and its principles as well as capacity building that addresses specific farming issues. This could increase farmers' benefits from OA and enable them to practice organic by conviction.

Keywords: Africa, cocoa production, motivation, sustainable agriculture

---

Contact Address: Lina Tennhardt, Research Institute of Organic Agriculture (FiBL), Dept. of Socioeconomics, Ackerstrasse 113, 5070 Frick, Switzerland, e-mail: lina.tennhardt@fibl.org

The Intention of Syrian Youth to Work in Agriculture: Exploring the Drivers

Ibrahim Salman, Miroslava Bavorova, Kindah Ibrahim

Czech University of Life Sciences Prague, Fac. of Tropical AgriSciences - Dept. of Economics and Development, Czech Republic

Young peoples' engagement in agricultural employment increasingly becomes a global challenge. Especially educated youth often choose not to work in this sector and shift their career to other sectors. In Syria, even among agriculturally educated students, this has become a persistent challenge and contributed to a labour shortage in agriculture. This study conducted in 2019 aims at investigating the factors that affect the intention to work in agriculture among tertiary education students in Syria.

A sample of 150 students from the Faculty of Agriculture at Tishreen University (Latakia, Syria) was interviewed face-to-face using a structured questionnaire. We used a binary logistic regression model to examine the factors influencing the students' intention to future work in agriculture (dependent variable, 1= yes/0= no). The results revealed that the students have a higher intention to work in agriculture if they had some farming experience before entering university. Their intention to work in agriculture is 27 times higher than their peers with no previous farming experience ($p < 0.05$), which was the strongest predictor in the model. Another significant factor that increases the probability of intention to work in agriculture was the father's occupation as a farmer (odds ratio is 0.26, $p < 0.05$). Contrarily, positive parents' opinions on agricultural jobs, friends' influence on studying agriculture, and contentment to the rural way of living significantly decrease the intention to work in agriculture with odds ratios of 0.6, 0.7 and 1.14 respectively. The results reveal that there is a potential to attract the young educated Syrians to work in agriculture. However, this needs to make the agricultural sector and rural areas appealing to the young generation. Providing support and investment for practical agriculture training before starting tertiary education could motivate the students to seek agricultural employment. Additionally, investment in improving the rural infrastructure (especially with relation to modern technologies) will undoubtedly attract more youth to rural living, however, this is a very difficult task in particular because of the conflict in Syria.

Keywords: Agricultural labour, intention to work in agriculture, Latakia, youth

---

Contact Address: Ibrahim Salman, Czech University of Life Sciences Prague, Fac. of Tropical Agrisciences, Kamycka 1293, 16500 Prague, Czech Republic, e-mail: ibrahim.90.salman@gmail.com

Analysis of Factors Determining the Productivity of Rice in Terai Belt of Nepal

Sushma Banjara[1], Chandeshwar Prasad Shriwastav[2], Ujjal Tiwari[1]

[1]*Agriculture and Forestry University (AFU), Agricultural Economics and Agribusiness Management, Nepal*

[2]*Agriculture and Forestry University (AFU), Soil Science and Agri-Engineering, Nepal*

Rice is the main staple crop to contribute to ensure food security in Nepal. Rice contributes about one-fifth of the total agriculture GDP and occupies the largest share in terms of area and production in Nepal. However, rice production and productivity are not satisfactory, and the country has to import rice in a large quantity every year. Terai, the southern belt of Nepal is known as the food-belt of the country, where more than eighty percent of farm households are actively engaged in rice production. The study aimed to assess the factors determining rice productivity among rice growers in the Terai belt of Nepal. The study also examined the existing constraints faced by rice growers during rice production and marketing. Farmers having a land size less than two hectares were selected randomly by three-stage sampling procedure from the registered farmer groups in Prime Minister Agriculture Modernisation Project (PMAMP) Rice Zone, Siraha district in 2018/2019. In addition to the secondary information, a pre-tested semi-structured survey schedule was used to collect the primary information from rice growers. The data were analysed with the aid of descriptive and multiple regression statistics. The results of the multiple linear regression model revealed that access to irrigation facility ($p = 0.000$), educational status of farmers ($p = 0.005$), and farmers' participation in the training programs related to rice production ($p = 0.034$) have a significant positive effect on rice productivity. Furthermore, an indexing technique used to rank the constraints faced by the farmers showed that lack of irrigation facility (0.77), incidence of diseases and pest (0.64), unavailability of labour (0.58), low market price of the produce (0.51) and lack of quality seeds (0.44) were the major hindrances faced by farmers during rice production and marketing. Therefore, the findings of the study underscore the need for strengthening training programs and the proper arrangement of irrigation facilities for increasing the productivity of rice. Similarly, provision of quality inputs, technical support for disease and pest management, farm mechanisation, and proper pricing of the produce could convert the constraints into opportunities and encourage rice growers to increase the rice productivity.

Keywords: Farm mechanisation, food security, food-belt, irrigation, multiple linear regression

Contact Address: Sushma Banjara, Agriculture and Forestry University (AFU), Agricultural Economics and Agribusiness Management, Panauti-8 Kharibot, 45209 Panauti, Kavrepalanchok, Nepal, e-mail: sushmabanjara151@gmail.com

Youth in Livestock and the Power of Education, the Case of "Heirs of Tradition"

Natalia Triana-Angel[1], Mauricio Ariza-Aya[2], Jacobo Arango[1], Stefan Burkart[1]

[1]*International Center for Tropical Agriculture (CIAT), Trop. Forages Program, Colombia*
[2]*Alquería, Colombia*

Joint, collaborative efforts between private companies and research centres in the cattle sector are still scarce. Limited resources and divergent agendas have hindered the consolidation of such alliances that can improve our understanding of the challenges and possibilities of the livestock industry, and a deeper comprehension of producer's needs. Here we approach those topics through a close examination of the case study "Heirs of Tradition", an initiative carried out by Alquería's (a major Colombian dairy company) farmer training programme. The objectives of this work are first, to assess the impact achieved through Alquería's educational programme amongst young farmers over the last seven years. Second, to highlight the company's recent alliance with CIAT and the CCAFS/Livestock Cross CGIAR Research Programs (CRP) in the context of the Heirs of Tradition program, and how this partnership strengthens the programs' productive, socioeconomic and environmental components during the 2020–2021 period. We based our analysis on information obtained through workshops, focus groups, in-depth interviews and the review of primary and secondary sources. Preliminary findings first suggest that cooperation between different actors is crucial to address a major threat in livestock production: generational relief and the massive migration of young people from rural to urban areas. Second, education initiatives and technical support can both transform, to a certain extent, the low levels of schoolings amongst rural producers and contribute to closing the gender gap that persists in the Colombian livestock sector. Third, alliances, such as the one between Alquería-CIAT-CCAFS/Livestock CRP, are critical for enriching and bolstering existing programmes by introducing new, urgent areas of expertise. Finally, we seek to shed light on the achievements and the lessons learned from "Heirs of Tradition", highlighting the subjects and issues that could be better addressed, and overall emphasising how the continuity of these ventures favours knowledge transfer, empowers communities and benefits livestock producers across the country.

Keywords: Education, generational relief, impact, knowledge transfer, private-sector-science partnerships, sustainable intensification

Contact Address: Stefan Burkart, International Center for Tropical Agriculture (CIAT), Trop. Forages Program, Km 17 Recta Cali-Palmira, Cali, Colombia, e-mail: s.burkart@cgiar.org

Using a Hot Spot Analysis as the Basis for Target Group Specific Training Materials on Insect Rearing

Isabelle Hirsch, Martin Hamer, Darya Hirsch
Bonn-Rhein-Sieg University of Applied Sciences, Germany

In Europe, the consumption of insects as food and feed is still in its initial stages compared to many Asian, Latin American or African countries, where insects are a fundamental part of human nutrition. Nevertheless, insect rearing for feed is on the rise over the last few years, especially in Europe, with the focus on black soldier fly (*Hermetia illucens*) production. This development raised the question if insect production is and remains sustainable in future. For measuring the sustainability of the European insect value and supply chain, we used the Sustainability Hot Spot Analysis (SHSA), developed by the Wuppertal Inst. for Climate, Environment and Energy GmbH (Germany). Based on literature research, we identified hot spots within the phases of insect rearing and insect processing especially for energy consumption. Based on the knowledge of European insect supply chains, different Burmese experts were interviewed. The interviews aimed at the understanding of insect value chains in Myanmar with special consideration of the social and environmental aspects of the SHSA. The interviewed stakeholders were e.g. from zoology/agriculture department of universities, employees of the chamber of agriculture, farmers and Thai insect rearing experts. The broad expertise of the interview partners highlighted that hot spots in the insect value chain are especially present for the social aspects, like the lack of knowledge and trainings. Similar to the European insect market, the energy consumption during the phases of rearing and processing is a considerable hot spot as the energy supply is a major challenge in our project regions. Locally adapted strategies for insect rearing and processing must be developed. If insect rearing reaches a larger scale the amounts of insect excrements and inedible body parts could become another hot spot and need consideration. These findings are valuable for designing sustainable insect rearing trainings that are suitable for local mini-farmers. Based on the results of the SHSA, a first round of trainings have been realised in Myanmar and Madagascar. The next planned step is to run more interviews and combine the knowledge from European and Burmese insect business in the development of detailed and comprehensive training materials.

Keywords: Economy, farmer trainings, insect farming and processing

---

Contact Address: Isabelle Hirsch, Bonn-Rhein-Sieg University of Applied Sciences, International Centre for Sustainable Development (IZNE), Rathausallee 10, 53757 Sankt Augustin, Germany, e-mail: isabelle.hirsch@h-brs.de

Willingness to Pay for Postharvest Technologies and its Influencing Factors among Smallholder Mango Farmers in Kenya

Esther Mujuka, John Mburu, Chris-Ackello Ogutu, Jane Ambuko
University of Nairobi, Dept. of Agricultural Economics, Kenya

Global food security for the over 9.1 billion people by 2050 remains a key policy challenge. Focus on increasing productivity is unsustainable due to the inelasticity of land and the expected pressure on it. There is consensus across the globe that focus should shift to reduction of postharvest losses, estimated at 30 % globally and at least 40 % in the fruit subsector in Kenya. Reduction of these losses require adoption of postharvest loss reduction technologies which are acceptable to smallholder farmers. Thus, this study sought to assess the acceptability of brick coolers, charcoal coolers and solar driers in Kenya. Multistage sampling technique was used to select 320 respondents in Embu and Machakos Counties from which empirical data was collected. A double hurdle model was used to estimate WTP for these postharvest technologies and the conditioning factors. Marital status, initial bid, agricultural group membership, market access and income from mangoes significantly and positively influenced probability to pay for the postharvest technologies. Probability to pay for postharvest technologies was significantly influenced by gender negatively and positively in Embu and Machakos, respectively. Factors that were found to positively and significantly influence the WTP amount for postharvest technologies were initial bid, agricultural group membership and income from mangoes. On the flipside, experience, credit access, market access, land tenure and age significantly influenced the WTP amount negatively. Further, we found low access to extension and awareness on the postharvest technologies. To intensify adoption of more effective postharvest technologies, there is need for the government to intensify extension programmes to create awareness on the need for the postharvest technologies. The estimated WTP amount for all the postharvest technologies suggested that most of the farmers would prefer the technologies to be offered at lower than the current market prices. Short term price subsidies could spur awareness on postharvest technologies that were found to be low and eventual adoption of the technologies. Farmers who operate under informal land tenure systems were not willing to pay for postharvest technologies. The government needs to strengthen tenure security to avert uncertainty.

Keywords: Double hurdle, postharvest loss, postharvest technology, WTP

Contact Address: Esther Mujuka, University of Nairobi, Dept. of Agricultural Economics, 29053, Nairobi, Kenya, e-mail: esthermujuka@gmail.com

Climate Change Adaptation of Smallholder Tea Farmers in Ilam, Nepal

Steffen Münch, Miroslava Bavorova
Czech University of Life Sciences Prague, Fac. of Tropical AgriSciences - Dept. of Economics and Development, Czech Republic

The government of Nepal intends to increase domestic production and export of tea (*Camellia sinensis*). Climate change effects have a negative impact on Nepalese agriculture, which employs 74 % of Nepal's population. The identification of factors influencing the adaptation behaviour towards climate change among tea farmers in Nepal helps to reveal weaknesses and supports the successful implementation of the tea export strategy. Our research focused on the adaptation behaviour of 91 smallholder tea farmers in Ilam, one of Nepal´s major tea producing districts. Approximately 87 % of the respondents were aware of the term climate change. The average number of applied adaptation strategies by the farmers was three out of six proposed options. Adaptation measures such as crop diversification and soil conservation were used by most farmers. Strategies related to irrigation were implemented by only a few respondents. Agroforestry and the usage of more climate-resilient tea cultivars had a balanced share of whether being deployed as a coping strategy. By using multiple and binary logit regression we identified factors influencing the adaptation towards climate change. Socio-demographic variables did not show any statistically significant effect on the degree of climate change adaptation. However, information sources (other tea farmers, internet and extension services), as well as institutional factors (cooperative membership and credit access) positively influenced the degree of climate change adaptation among the sample. The main constraints to climate change adaptation were a lack of governmental support, insufficient information as well as no access to credits. Improved interaction between the government and the tea farmers is crucial for increased productivity of the Nepalese tea farming sector. In addition, easier credit access and the provision of effective extension services could support the farmers in adapting to climate change. Most participants were concerned about their future as tea farmers. Therefore, the need for further researching this climate-vulnerable industry is becoming inevitable.

Keywords: Coping strategies, barriers to adaptation, climate change impacts, Nepalese tea industry, tea export strategy

---

Contact Address: Steffen Münch, Czech University of Life Sciences Prague, Fac. of Tropical AgriSciences - Dept. of Economics and Development, 160 00 Praha, Czech Republic, e-mail: muench@ftz.czu.cz

The Role of Agricultural Science Knowledge Transfer to Promote Food and Nutrition Security

Patrick Artur Groetz[1], Heinrich Hagel[2], Joern Precht[1]

[1]*Stuttgart Media University, Dept. of Audiovisual Media, Germany*

[2]*University of Hohenheim, Food Security Center, Germany*

Agricultural science strives to improve people's basic livelihoods. In this respect, it has achieved great success to improve food security in developing countries in the last decades. However, the agricultural sector has increasingly got a reputational problem in public opinion. Since large parts of society no longer have direct contact with agriculture, they often draw their knowledge from media coverage. In addition to the media, citizens' initiatives, interest groups and non-governmental organisations (NGOs) have an ever-increasing influence on public discussions and policy-making about agricultural issues. For that reason, recent findings and outputs have to be transferred more effectively in a direct way into public debate. This study aims at investigating whether and to what extent universities, as important part of public agricultural research, could at least partially adopt this type of knowledge transfer for a partly non-academic clientele, or perhaps already do so in order to present a more balanced picture of agricultural realities.

The methods used here are based on a comprehensive literature research. In addition structured telephone interviews were conducted with agricultural scientist of the University of Hohenheim in order to discuss and analyse the topic from the point of view of practicing science based on this case study.

Science communication must be geared to the target group, occasion, medium and format. This requires considerable effort and communication experience and competence. The results obtained paint a clear picture of agricultural science perceptions of this topic. They highlight the challenges, but also opportunities that scientists encounter and see in public relations and the media presentation of agricultural science topics. From this, relevant measures and implications for the transfer of knowledge between actors in agriculture, society and agricultural sciences can be identified, which on the one hand can result in more objective reporting. On the other hand, university science communication helps to raise public awareness that public funds are used efficiently and sensibly in the scientific system.

Keywords: Agricultural sciences, food security, knowledge transfer, media coverage, NGOs, perception of agriculture, public opinion, public relations, science communication

---

Contact Address: Patrick Artur Groetz, Stuttgart Media University, Dept. of Audiovisual Media, Nobelstrasse 10, 70569 Stuttgart-Vaihingen, Germany, e-mail: ahgroetz@hotmail.com

Typology of Farmers' Perceptions Regarding Water Conservation: a Road to Food Security

Maryam Tajeri Moghadam[1], Masoud Yazdanpanah[2], Katharina Löhr[3], Stefan Sieber[3]

[1]*University of Tabriz, Dept. of Extension and Rural Development, Iran*

[2]*Agricultural Sciences and Natural Resources University of Khuzestan, Iran*

[3]*Leibniz Centre for Agric. Landscape Res. (ZALF), Sustainable Land Use in Developing Countries (SusLAND), Germany*

Water is a key resource for food production. Water scarcity, along with rapid population growth, urbanisation, and climate change, has become one of the world's most challenging problems. As a result, its impact on agricultural production and food security has caused widespread concern around the world. In the agricultural sector, as a food producer and the largest consumer of water in the world, it is necessary to manage and protect water. In this regard, farmers' water protection behaviour is an important factor for food security and ending hunger. While farmers' perceptions of water conservation are directly related to their attitudes and behaviours toward water scarcity, understanding their perceptions is important for policymaking. Farmers have heterogeneous attitudes, values, viewpoints, and behaviours, thus showing different perceptions and behaviour on water conservation strategies and policies. The aim of this study is to caracterise and classify farmers' perceptions regarding water protection in northeastern Iran (Neyshabur plain) with farmers selected using a stratified sampling method. The results of the K-Means Cluster Analysis showed that farmers can be divided into four clusters. Comparison of the four clusters reveals that farmers in the first cluster have a higher level of egalitarianism worldview, farmers in the second cluster have a higher hierarchy worldview, farmers in the third cluster have a higher level of individualism and finally, the fourth cluster has a higher level of fatalism. These different worldviews of farmers reflect the heterogeneity of farmers' attitudes and behaviours regarding water conservation in a specific region. Identifying these different perspectives can be useful in planning and policy-making in dealing with water scarcity so that each group of farmers can be offered specific strategies and policies to increase their impact and reduce costs.

Keywords: Farmers' perceptions, food security, typology, water conservation

Contact Address: Masoud Yazdanpanah, Agricultural Sciences and Natural Resources University of Khuzestan, Mollasani, 744581 Ahvaz, Iran, e-mail: masoudyazdan@gmail.com

Potential of the Green Way Application for Data Collection on Crop Economics in Myanmar

Anna Braun

Bern University of Applied Sciences; School of Agricultural, Forest and Food Sciences HAFL, International Agriculture, Switzerland

This study addresses the potential of the Green Way mobile application to record economic data on crop production from farmers in the Gulf of Mottama, Myanmar. The paper aims first at identifying farmers' topics of highest interest and the main sources of information used. Second, the potentials and challenges of the Green Way app for data collection are summarised. The third aim is to study farmers' willingness to record and share economic data. Farmers in the area are interested in cropping techniques, weather conditions, market prices, fertiliser and its application, seeds and its prices as well as pests and diseases. Mass media, especially the TV and the radio, as well as personal communication with other farmers or neighbours and extension services are the most important sources used. The digital knowledge and use of smartphones and mobile applications for agricultural information is still relatively low. At the time of the survey, the Green Way app was known by 33 % and used by 13 % of all farmers. The app could be used as information source for several topics of interest mentioned above.

The farming record feature has been incorporated into the Green Way to simplify the collection of economic data on crop economics. As a challenge, this farming record feature requires precise recording of economic data by the farmers. Farmers receive training on the use of smartphones and advice on farming practices in exchange for their data. However, data privacy is an issue, which should be included in the policy of the app.

Regular recording of crop economics could bring many advantages for farmers, reaching from less barriers for certification to a better overview of production branches and therefore improved farm management. Withal, the recording of data needs to be monitored closely in order to develop the app and the recording in a farmer-friendly and effective way. Such a digital change requires time for adaptation to the new technologies and training in order to educate farmers on the purpose and benefits of data recording.

Keywords: Data collection, mobile application, Myanmar

Contact Address: Anna Braun, Bern University of Applied Sciences; School of Agricultural, Forest and Food Sciences HAFL, International Agriculture, Bruderholzallee 100, 4059 Basel, Switzerland, e-mail: annachristina.braun@students.bfh.ch

Who Is Afraid of Postcolonial Theory? Practical Considerations on Epistemic Freedom for Development-Oriented Research

Annapia Debarry
University of Bonn, Department of Geography, Germany

Postcolonial theories are often regarded as "high theory" and therefore not relevant to practice-oriented research. We owe attention for epistemological questions in development research primarily to postcolonial, decolonial and feminist studies. However, Global South voices are still struggling for epistemic freedom, as Sabelo J. Ndlovu-Gatsheni pointed out for the African case in a book published in 2018, in which he argues that "[...] global coloniality operates as an invisible power matrix that is shaping and sustaining asymmetrical power relations between the Global North and the Global South". Although practical development research in particular has a direct influence on many people in the Global South through recommendations in policy documents and their implementation, power relations in knowledge production are rarely questioned. This research addresses the practical implications of feminist postcolonial approaches to knowledge production and practice-oriented development research by drawing on feminist contributions to postcolonial epistemology, in particular Spivak's essay "Can the Subaltern Speak" (1987) and C.T. Mohanty's contributions (1984, 2003). An ongoing study on socio-cultural effects of rural transformation in Northern Ethiopia that combines feminist theory with practice-oriented approaches to gender-transformative change is taken to reflect processes of knowledge production. In specific, the study uses autoethnographic field notes to analyse how the research has shaped and reproduced power relations in the process of knowledge production. The study shows eg. how identity influences research. Gender, age and race have given the researcher a certain access to the field and transported a corresponding body of knowledge, while other segments were denied her. This contribution exemplifies that awareness of unequal power relations in academic North-South cooperation can lead to a more reflective use of data and thus to profound and holistic knowledge generation. Through the engagement with postcolonial theory, existing power relations can be uncovered and questioned and thus contribute to strengthening epistemic freedom of the Global South and the decolonization of knowledge.

Keywords: Epistemology, Ethiopia, gender-transformative change, North-South cooperation, rural transformations

Contact Address: Annapia Debarry, University of Bonn, Department of Geography, Meckenheimer Allee 176, 53115 Bonn, Germany, e-mail: a.debarry@uni-bonn.de

The Scaling-Up Potential of Agroforestry Systems in Colombia: A Comparative *ex-ante* Assessment Across two Contrasting Regions

Tatiana Rodríguez[1], Katharina Löhr[1], Michelle Bonatti[1], Martha Lilia Del Rio Duque[1], Marcos Lana[2], Stefan Sieber[1]

[1] *Leibniz Centre for Agric. Landscape Res. (ZALF), Sustainable Land Use in Developing Countries (SusLAND), Germany*

[2] *Swedish University of Agricultural Sciences, Crop Production Ecology, Sweden*

Agroforestry systems (AFS) have been recognised as land-use strategies that contribute to biodiversity conservation, climate change adaptation, and mitigation, as well as sustainable rural livelihoods. However, these potential contributions remain unrealised as so far AFS spreading remained limited. This results in increasing demand by development programmes to find effective ways to scale-up AFS so that more people can benefit from them. In this sense, it is crucial to identify beforehand what conditions are needed to effectively disseminate AFS in a particular context. This study assesses the scaling-up potential of AFS for cocoa farming and cattle ranching, in two different regions of Colombia. The methodology is based on questionnaires conducted with 20 AFS experts from different institutions (NGOs, research institutes, local government institutions) promoting these systems in both regions using a snowball sampling. These questionnaires were completed through the ScalA-PB mobile app, an *ex-ante* assessment tool of the potential for scaling-up projects or strategies that support agricultural good practices within post-conflict settings. They include seven scaling-up dimensions: (a) AFS attributes such as their affordability or their complexity to implement; (b) capacities of implementing organisations and (c) their strategies for scaling-up AFS; (d) institutional framework at the national level that includes government support to disseminate AFS; (e) institutional setting at the local level such as the presence of AFS supporting formal or informal rules; (f) economic conditions at the local and regional level; and (g) attitudes of communities towards AFS including their willingness to introduce them. The preliminary results suggest that AFS scaling-up potential is perceived slightly higher for cocoa farming than for cattle ranching in both study regions. For both farming systems, AFS scaling-up is hindered by insufficient access to financial means by farmers to afford the cost of the systems and the lack of stable and differentiated markets that guarantee reasonable prices for products derived from them. The *ex-ante* assessment of AFS scaling-up potential based on expert perceptions indicates which conditions need to be improved when promoting agricultural and policy interventions regarding AFS. So, there is a need in these two regions to develop strategies that facilitate financial means for carrying the cost of AFS implementation and recognise their non-market values.

Keywords: ScalA-PB, scaling-up, sustainable agriculture

Contact Address: Tatiana Rodríguez, Leibniz Centre for Agric. Landscape Res. (ZALF), Sustainable Land Use in Developing Countries (SusLAND), Eberswalder Straße 84, 15374 Müncheberg, Germany, e-mail: rodriguez@zalf.de

Comparison between Socio-Economic Conditions of the Employees of two Cocoa Plantations in Colombia

LAURE LEDOUX, OLIVIA AMOUZOU, RAHUL AVI, JULIE BEAUDOIRE, ANDRÉAS BODOUKIAN, VALENTIN GIRARD, JÉRÉMIE LANNOU, ARNAUD MINGAT, JUSTINE SEUSSE, LUDOVIC ANDRES, DANIEL J.E. KALNIN

ISTOM; Ecole supérieur d'agro développement international, ADI-SUD Innovation and Agro/Food-développement international, France

The objective of this study is to carry out a diagnosis of the living and working conditions of the employees on two cocoa plantations of one company. The aim is to identify different factors that are inducing the exodus of employees from the cocoa production areas by understanding the underlying social mechanisms present at the two locations studied: Necocli and Casanare.
The study was conducted in order to secure the presence of well-trained employees and hence allow for the production of high-quality fine aroma cocoa (*Theobroma cacao*). For the study both qualitative and quantitative data was collected. This data was collected during 3 weeks of fieldwork with 10 interviewers. Data were pocessed using a quantitative approach in order to be able to interpret the results regarding different themes ranging from the study of work contexts, new mechanisms; understand and document the mechanisms in the two sites; explanation of the parameters that make up these mechanisms; proposing and exploring the links between the different results obtained during interviews. This work leads to the understanding of the basic mechanisms, the factors that induce the exodus of young people from the fields.
One of the key outcomes of the study is that it can be seen that the creation of a climate of trust by taking an interest in the life of the employee outside of work and participating in agricultural work is of high importance for the satisfaction of workers with their living conditions outside their working sites and thus their willingness to stay in job and in the region. An important recommendation is to promote a communication and awareness campaign about the profession of cocoa field employees, but especially among young people.

Keywords: Cacao, fieldwork, living conditions, working conditions

Contact Address: Daniel J.E. Kalnin, ISTOM; Ecole supérieur d'agro développement international, ADI-SUD, Innovation and Agro/Food-développement international, 4 rue Joseph Lakanal, 49100 Angers, France, e-mail: d.kalnin@istom.fr

The Role of Extension and Famer Groups in Adopting Agricultural Technologies

Pius Nnahiwe, Jiří Hejkrlík, Miroslava Bavorova

Czech University of Life Sciences Prague, Department of Economics and Development, Czech Republic

Despite the inherent potential and merits of adopting modern agricultural technology, the present-day farmer in sub-Saharan Africa is yet to catch-up with the rest of the world in harnessing this potential. Thus, this study examines factors affecting modern agricultural technology adoption in the coastal regions of Kenya. A multistage comprising of stratified, quota and snowball sampling approach was designed and used for data collection in 2018 with a vision of more representative data from a set of 15,000 local farmers located in three major cashew producing counties in the country. The logit and multiple linear regression models were used to analyse a sample of 372 smallholder cashew farmers in the coastal regions of Kenya. Logit regression analysis was used to establish the relationship between modern agricultural technologies such as fertilizer use and chemical spraying with other variables of interest. Linear regression model was used to investigate appropriate planting density and the consequent effect of adoption on farmers economic performance. The results show that access to extension services and group membership both have significant positive effects on adoption of modern agricultural technologies namely fertilizer usage and appropriate planting density. Regarding the economic performance, only appropriate planting density show a statistical significant positive effect. With pesticides application showing no effect on economic performance and fertilizer usage had a negative effect. Thus, the study recommends promoting the extension of disseminating benefits of adopting appropriate planting density among cashew farmers. Furthermore, proffering a relevant policy implication based on the findings to further strengthen existing farmer groups and encourage formation of new groups with an aim of introducing modern technologies to boost the agricultural sector's performance.

Keywords: Cashew, chemical spraying, fertiliser usage, planting density, sub-Saharan Africa

Contact Address: Pius Nnahiwe, Czech University of Life Sciences Prague, Department of Economics and Development, Kamycka 129 16500 Prague – Suchdol, 16500 Prague, Czech Republic, e-mail: nnahiwe@ftz.czu.cz

Applied Cognitive and Applied Evaluation of Citrus Producers as Related to Agricultural Operations for Citrus Orchards (Al-Gorair Region - Merowe Locality - Northern State - Republic of Sudan)

Amna Hassan Widaa, Ahmed Hamdi Allagabo

Federal Ministry of Agriculture, Agricultural Extension and Technology Transfer, Sudan

Studying in countries such as ours in the third world is very important, especially with the spread of illiteracy and lack of education among large groups of the population, especially among the rural population, who are mostly farmers, In addition to poor agricultural extension coverage. The aim of the study is cognitive and applied evaluation of agricultural practices for citrus producers, to achieve this aim the study was conducted in the al-Gorair region - Northern State - Republic of the Sudan in 50 orchards that grow citrus trees, which contain about 2454 citrus tree in total with an average of 52 tree per orchard. A questionnaire was prepared to measure the farmers' knowledge and application level of agricultural operations in citrus orchards on three axes (axis of citrus orchards establishment, tree pruning operations, other agricultural operations of irrigation, fertilisation, identification of insects, diseases and control) Each of those three axes contained a group of items. The data were analysed and frequencies and percentages were used to describe the knowledge about the elicited issues with regards to the Citrus Orchards in the study area. The study clearly shows that there is a general weakness in knowledge and application of agricultural operations in citrus orchards. The study recommends that the responsible authorities (agricultural extension, Horticultural Sector, Agricultural Research in the Northern State - Republic of Sudan) intensify activities that increase the knowledge of farmers in Al-Gorair region on agricultural operations in citrus orchards to increase knowledge percentage and therefore an application of good methods recommended to increase the citrus productivity which it will increase the farmers income in the region.

Keywords: Agricultural operations, al-Gorair region, citrus orchards

Contact Address: Ahmed Hamdi Allagabo, Federal Ministry of Agriculture, Agricultural Extension and Technology Transfer, 6 Square, 79371 Khartoum North, Sudan, e-mail: ahmedhamdi80@gmail.com

EFForTS-Education – Knowledge Transfer Regarding Research on Tropical Lowland Rainforest Transformation Systems into Indonesian Teacher Education

Finn Kristen Matthiesen, Susanne Bögeholz

Georg-August-University Goettingen, Didactics of Biology, Germany

Land-Use Change (LUC) from tropical rainforest towards domination of crops, e.g. oil palm and rubber, has environmental and socioeconomic effects. The Collaborative Research Centre *EFForTS* (*E*cological and Socioeconomic *F*unctions of Tropical Lowland Rain*for*est *T*ransformation *S*ystems – Sumatra, Indonesia) investigates such LUC issues in Jambi Province.

Science-based knowledge as generated by EFForTS on land-use transition is crucial for informed decision-making concerning more sustainable land-use. LUC is a factually and ethically complex, controversially discussed socioscientific issue (SSI). Up to now, teachers are not sufficiently prepared to teach such SSIs and there is a demand for educational material on SSIs in Indonesian classrooms. For bringing knowledge on highly relevant SSIs into society, teacher education is a motor, since educators act as multipliers and change agents.

Our research questions are: How can EFForTS-related SSIs be reconstructed for SSI teaching and learning in Indonesian teacher education? And, how effective is the resulting training with educational course concepts and the corresponding materials?

We build our research on the SSI teaching and learning model according to Sadler, Foulk and Friedrichsen (2017). Doing so, we aim at fostering competencies regarding decision-making, socioscientific reasoning, including scientific inquiry, perspective-taking and dealing with complexity.

To answer our research questions, we collaboratively design educational resources on predominant LUC issues in Professional Learning Communities (EFForTS scientific researchers, educational researchers, teacher educators) and we qualify (Indonesian) teacher educators to teach such LUC issues. We test, formatively evaluate and refine the educational course concepts and materials for different Indonesian universities on four islands. Furthermore, we summatively evaluate the effects to gain information on the impact of the training. We use triangulation of classroom observations, analyses of documents and artefacts, focus group discussions as well as questionnaires on self-efficacy beliefs to teach LUC issues and competence assessment for decision-making and perspective-taking.

Contact Address: Finn Kristen Matthiesen, Georg-August-University Goettingen, Didactics of Biology, Waldweg 26, 37073 Göttingen, Germany, e-mail: finn.matthiesen@uni-goettingen.de

We shed light on our approach on how to transfer EFForTS knowledge on LUC issues into teacher education by presenting an example of an educational course concept and corresponding materials for the University of Jambi (prestudy). Furthermore, we inform about the following steps in the public relation project of EFForTS on teacher education and its evaluation.

Keywords: Education for sustainable development, Indonesia, science education, socioscientific issue, sustainable land-use, teacher education

Adoption - Adaptation - Agriculture

Determinants of Adaptation Strategies to Climate Change: Implications on Sustainable Agriculture in Ghana

Enoch Bessah[1], Emmanuel Donkor[2]

[1]*Kwame Nkrumah University of Science and Technology, Dept. of Agricultural and Biosystems Engineering, Ghana*

[2]*Rhine-Waal University of Applied Sciences, Fac. of Life Sciences, Germany*

The agricultural sector of Sub-Saharan Africa including Ghana is rain-fed and less capital intensive. Unreliable and unfavourable climatic conditions have exerted negative effects on the sustainable supply of food. Therefore, food insecurity and poverty are prevalent in these countries, particularly Ghana. Smallholder farmers tend to respond to climate change by using different adaptation strategies. However, existing studies focus on analysing the determinants of farmers´ adaptation strategy decision without a considerable evaluation factors influencing the number of adaptation strategies used by farmers. Our paper therefore contributes to narrowing this knowledge gap by rigorously examining the determinants of the number of adaptation strategies adopted by 344 farmers selected from ten districts within the Pra River Basin around across three Ashanti regions of Ghana. We applied the Tobit regression model in the empirical analysis. We find that human capital variables such as primary education, secondary education, tertiary education, household size, and marital status shows significant positive effects on maize farmers' adaptation strategies to climate change whereas labour constraint exerts a negative effect. Institutional variables like extension constraint, credit constraint, and land tenure constraint are negatively related to adaptation strategy decision. Community infrastructure such as electricity, encourages farmers to adapt to climate change. We also find that perception variables – access to weather information, awareness of climate change and perception on amount of rainfall have positive effects on farmers' adaptation decision. To achieve sustainable agriculture and food systems, it is paramount to strengthen the adaptive capacity and resilience of vulnerable farmers by integrating their human capital, institutional variables, infrastructural development and climate perception variables into the national climate policy in Ghana.

Keywords: Adaptation strategies, climate change, sustainable agriculture, Tobit binomial regression

Contact Address: Enoch Bessah, Kwame Nkrumah University of Science and Technology, Dept. of Agricultural and Biosystems Engineering, Private Mail Bag, 00233 Kumasi, Ghana, e-mail: enoch.bessah@gmail.com

Bridging the Information Gap for Increased Livestock Productivity: Evidence from Use of SMS Messaging among Smallholder Farmers in Babati District, Tanzania

Ben Lukuyu[1], Kevin Maina[2], Leonard Marwa[3], Alphonce Haule[4], Julius Githinji[2], Fred Kizito[5]

[1]*International Livestock Research Institute (ILRI), Feeds and Forages Program, Uganda*

[2]*International Livestock Research Institute (ILRI), Kenya*

[3]*Tanzania Livestock Research Institute (TALIRI), Tanzania*

[4]*Babati District Council, Livestock and Fisheries, Tanzania*

[5]*International Insititute of Tropical Agriculture (IITA), Natural Resource Management, Ghana*

Most farming systems in East Africa are mixed with poultry, cattle and small ruminants forming the major livestock component. Population growth has resulted in expansion of cropped areas consequently, resulting in decline of regeneration of soil fertility, crop productivity and feed resources. Inability to feed animals adequately throughout the year becomes the most widespread technical constraint for livestock production. Therefore, there is a need for innovative strategies to alleviate the situation. However, knowledge gaps are still a hindrance to the uptake of livestock technologies. The objective of this article is to assess the effect of ICT based extension services on knowledge, attitudes and practices among smallholder farmers in Tanzania. It focuses on MWANGA platform; a toll short message service (SMS) that connects farmers to vital information on livestock production. Using a random sample of 100 dairy farmers, the study utilises qualitative and quantitative research methods. A baseline was conducted, then short and clear messages on dairy and poultry production were disseminated over a period of 14 weeks. This was followed by an end line survey and focus discussion in Babati District, Tanzania. Quantitative data was analysed using descriptive statistics, whereas qualitative data described the whys.

The findings reveal that the SMS messaging significantly increased farmers' knowledge on the advantages of feed chopping (p-value <0.1). Additionally, from focus group discussions, farmers were able to identify more improved Napier grass varieties. Consequently, increased knowledge resulted in change in practices such as adoption of feed chopping and provision of adequate water for animals. SMS messaging had positive and significant effects on attitudes. At 1 % level of significance, farmers value the importance of keeping improved breeds and believe that better formulated feeds yields better benefits. Moreover, they understand the cost and benefits of different

Contact Address: Kevin Maina, International Livestock Research Institute (ILRI), P. O. Box 30709-00100, Nairobi, Kenya, e-mail: mainakevin.km@gmail.com

technologies & is a key factor in adoption.
MWANGA platform has proved to be a valuable cost-effective extension approach reaching out to as many farmers as possible. The study recommends integration of ICT technologies in extension service provision. Additionally, such platforms should be interactive such that farmers can ask questions or share their experiences on crop and livestock production challenges.

Keywords: Extension, forages, ICT, livestock productivity, technology adoption

Knowledge, Decision Making Capacity and Risk Attitude among Farmers in Vietnam

Huong Dien Pham[1], Duc Loc Nguyen[2], Huynh Nhu Phan[3]
[1]*Banking University of Ho Chi Minh City, Business Administration Dep., Vietnam*
[2]*Southern Center of Agriculture Policy and Strategy, Vietnam*
[3]*Ca Mau Investment Promotion and Enterprise Support Center, Vietnam*

Agriculture can be a risky business as it is based on risks and uncertainty. Without strong technical knowledge, farmers tend to rely heavily on heuristics and subjective judgments to deal with their daily business. It is crucial to understand farmers' practices to provide suitable supports. This study uses data from a long panel household survey to assess farmers' agricultural productivity in Thua Thien Hue (Hue) province of Vietnam combining with data collected from other surveys conducted in 2014 and 2015 focusing on farmers' knowledge, skills and risk attitudes. This study aims to provide an overview of the environment in which farmers do their business in terms of personal constraints. Particularly, we investigate the relations among risk attitudes, farmers' knowledge, management ability and agricultural productivity by using univariate and bivariate analyses. The results indicate a large variation in farmers' knowledge but most of them have low knowledge, both technical and subjective. Farmers' performance has a significant correlation with subjective knowledge but not with technical knowledge. That implies farmers rely their decisions on their subjective experience or self-assessment rather than on scientific evidence. Therefore, it is understandable, most of the farmers reported they received limited support from the extension institutions, meanwhile, those farmers self-assessed to be more risk averse. The farmers have stronger technical knowledge and decision-making ability in livestock production than that in crop production. Furthermore, the more willing to take risks they are, the better they show their knowledge in livestock, but not in crop production. We find that the risk attitude of farmers is linked to technical knowledge and decision-making ability but not to their agricultural performance significantly. This study suggests extension services to fill the gap between the subjective knowledge and technical knowledge by investigating farmers' specific needs to enhance agricultural productivity. In addition, the willingness to take risk is one of the crucial elements needed to consider in agricultural policy making.

Keywords: Agriculture, knowledge, risk attitudes, Vietnam

Contact Address: Huong Dien Pham, Banking University of Ho Chi Minh City, Business Administration Department, 36 Ton That Dam, District 1, Ho Chi Minh City, Vietnam, e-mail: dienhuongpham@gmail.com

Contract Farming Effects on Technical Efficiency of the Export-Oriented Rice Production Sector in Vietnam

Caetano Luiz Beber, Le Ngoc Huong, Ludwig Theuvsen, Verena Otter

Georg-August-University Goettingen, Dept. of Agricultural Economics and Rural Development (DARE), Germany

Measures to increase technical efficiency in emerging and developing economies' agriculture receives great attention by governments, NGOs, private firms and researchers in times of urgent need for poverty reduction and growing resource rivalry in the world. In this regard, there have been a large number of research studies analysing the determinants of technical efficiency (TE) in agricultural production. Some of these studies point on the role of contract farming (CF) participation as an important facilitator of technical efficiency. This study aims at evaluating the influence of CF participation on farming technical efficiency and productivity in a developing country context. A cross sectional sample of 246 Vietnamese export-oriented rice households of the Mekong River Delta is used to provide empirical evidence of the effects of contractual farming (CF) schemes on technological adoption and technical efficiency. Propensity score matching and stochastic frontier model with correction for sample selection are used to control for potential bias arising from observable and unobservable variables respectively. Subsequently we used a meta-frontier framework in order to estimate the effects of contractual schemes on the farms' technology gap ratio, their group technical efficiency and their meta-technology technical efficiency. Our results show that CF has a positive impact in the adoption of better technologies and inputs improving the overall productivity. The group specific technical efficiency is lower for CF farmers in the short run, likely still in the learning process of those new technologies and inputs. This study contributes to the literature of impact evaluation by showing how the integration via contractual schemes can enhance farmers' productivity and consequently their livelihoods.

Keywords: Contract farming, export-oriented rice sector, stochastic frontier analysis, technical efficiency, Vietnam

Contact Address: Caetano Luiz Beber, Georg-August-University Goettingen, Dept. of Agricultural Economics and Rural Development (DARE), Goettingen, Germany, e-mail: cbeber@gwdg.de

Backyard Gardening Innovations: Towards Enhanced Vegetable Consumption for Nutritional Security among Urban and Peri-Urban Dwellers in Central Uganda

Losira Nasirumbi Sanya, Damalie Babirye Magala, Immaculate Ojara Mugisa, Teopista Namirimu, Robert Mutyaba, Monica Muyinda, Beatrice Omounuk Akello

National Agricultural Research Organisation (NARO), Mukono Zonal Agricultural Research and Development Institute (MUZARDI), Uganda

Globally, food and nutrition security remains a challenge especially among the urban poor. With rapid urbanisation amidst high levels of unemployment or underemployment, the urban poor are at a risk of experiencing malnutrition. Studies show that regular consumption of indigenous vegetables can meet nutritional requirements in the human diet thus, strategic interventions for increasing vegetable consumption are critical to address micronutrient deficiency. However, land availability for vegetable production remain a limiting factor among urban and peri-urban (UAP) dwellers in central Uganda. From 2012 – 2017, NARO promoted backyard gardening innovations (BGIs) to enhance production, consumption and marketing of vegetables in UAP areas of central Uganda. This study examined the uptake of BGIs and their subsequent contribution to enhanced vegetable consumption among UAP dwellers in the project sites. A multistage sampling technique was used to select a random sample of 104 beneficiaries. Data were collected using a pre-tested semi-structured questionnaire in a beneficiary survey and sex-disaggregated focus group discussions. Qualitative data was analysed using narrative and thematic techniques while quantitative data was analysed using descriptive and inferential statistics. The results show that 98 % of the respondents adopted at least one of the innovations promoted. Recycled bags (68.0 %) and raised beds (67.7 %) were the most commonly adopted BGIs while food towers (42 %), wooden boxes (35 %) and greenhouses (4.1 %) were the least used. These innovations promoted diversified production with specific techniques being considered suitable for specific or combinations of vegetables. Tomatoes, *Solanum aethopicum* and onions were the commonly grown crops using BGIs. The choice of crops could be a reflection of farmers diverse needs hence targeting crops that meet food, nutrition and income needs. Overall, majority (98 %) of the beneficiaries reported an increase in consumption of vegetables. UAP dwellers need to be empowered to innovatively use locally available materials as a strategy to reduce the cost involved and bolster the benefits. Further research should focus on the cost-benefit analysis of different BGIs and optimal vegetable intercropping combinations based on the existing production systems to enhance their uptake for food, nutrition and income security.

Keywords: Backyard gardens, Central Uganda, innovations, urban and peri-urban, vegetables consumption

Contact Address: Losira Nasirumbi Sanya, National Agricultural Research Organisation (NARO), Directorate of Research Coordination, Plot 11-13 Lugard Avenue, P.O. Box 295, Entebbe - Uganda, +256 Entebbe, Uganda, e-mail: losirasfm@gmail.com

Farmers' Adoption of Conservation Practices: Insights from Diverse Agro - Ecological Regions of Zambia

Samuel Mwanza, William Nkomoki

Czech University of Life Sciences Prague, Fac. of Tropical AgriSciences; Dept. of Economics and Development, Czech Republic

In recent years, high food insecurity, poverty and hunger are critical issues faced in Zambia due to recent declines in food crop yields which among other reasons is attributed to decline in agricultural soil fertility and climate variability. This study investigates the adoption of conservation agricultural practices among the smallholder farmers in distinct Agro ecological regions of Zambia. Using a questionnaire survey on 182 farmers from six districts representing three Agro ecological regions, descriptive analysis using chi square test was employed to assess the association of adopting tillage methods, soil protection practice and crop rotation. Results on minimum tillage indicates region I and III adopting more planting basins with 80 % and 54.1 % respectively, while region IIa adopts ripping more (78.7 %). Retaining crop residues was largely practised by all regions with region IIa leading (67 %) in cover crops. On average 85 % of smallholder farmers practice crop rotation across all the three ecological regions. The multiple linear regression model revealed that credit support, participation in group membership, soil protection reasons, increased yields, and perceived variability in precipitation are some of the factors influencing the adoption of multiple conservation practices. Extension services, farmer cooperatives, and conservation agriculture literature are found to be critical sources of information in promoting conservation practices. Increased yields, soil protection, reduced labour, and mitigation towards variability in precipitation are the main perceived benefits of adopting conservation practices. The barriers constitute lack of conservation tools, widespread of weeds, and pests. Improving accessibility to conservation mechanical services and implements, accessibility to conservation practices information tailored according to agro ecological preferences can increase adoption of conservation agriculture, promote sustainable use of resources and food security in Zambia.

Keywords: Adoption, agroecology, barriers, benefits, conservation practices, information sources, smallholder

Contact Address: William Nkomoki, Czech University of Life Sciences Prague, Fac. of Tropical AgriSciences; Dept. of Economics and Development, Kamycká 129, 16500 Prague, Czech Republic, e-mail: nkomoki@ftz.czu.cz

Silence Speaks Out Loud: Armed Conflict and Bovine Livestock in Colombia, a Historical Perspective

Natalia Triana-Angel, Stefan Burkart
International Center for Tropical Agriculture (CIAT), Trop. Forages Program, Colombia

The convoluted nexus between bovine livestock and the dynamics of armed confrontation in Colombia is a terrain open for exploration. While a vast array of archival sources suggests a historical, problematic connection between livestock production, land dispossession and rising violence in rural settings, academic narratives remain scarce. While livestock activities have been largely understood as vital cultural and economic practices in Colombia and Latin America and much has been written on the history of the country's multi-faceted civil war, both phenomena appear to be disconnected from scholarly interpretations. By building on both print media archives and scant, yet path breaking secondary sources, we emphasise two salient perspectives on the topic. First, one literature segment that understands large-scale bovine livestock as a driving force of forced migration and dispossession affecting small producers and peasant communities across Colombia. A second, more recent trend recognises the critical role of bovine livestock as an opportunity for rural development in impoverished regions, highlighting the importance of livestock systems for different agents, including returning peasants, youth, and even former combatants. Drawing from the experience of Central American communities who, also devastated by recent civil wars, have understood the pivotal function of bovine livestock production as an engine for change and improved livelihoods, we propose a third possible interpretation: one that accounts for the convoluted historical connections between wartime dynamics and cattle and dairy production in Colombia, acknowledges its capacity to empower rural communities in post-conflict contexts and deciphers academic silences as testimonies of its own, violent times. By reconciling divergent postures, our goal is to initiate the conversation around difficult, controversial tropes while seeking to provide methodological and theoretical explanations that can further our understanding on the subject.

Keywords: Empowerment, livelihoods, livestock, post-conflict, rural development

Contact Address: Stefan Burkart, International Center for Tropical Agriculture (CIAT), Trop. Forages Program, Km 17 Recta Cali-Palmira, Cali, Colombia, e-mail: s.burkart@cgiar.org

The Role of Seed Systems in the Adoption of Improved Forages: The Colombian Case

Karen Enciso Valencia, Manuel Diaz, Natalia Triana-Angel, Stefan Burkart

International Center for Tropical Agriculture (CIAT), Tropical Forages, Colombia

Forage improvement processes, which began in Latin America in the 1980's, have resulted in the release of new cultivars and hybrids superior in terms of productivity, sustainability and adaptability. Increasing the adoption rate of these technologies stands as one of the most promising strategies for the sustainable intensification of bovine livestock production in the tropics. In Colombia, 22 cultivars have been released (through formal channels) since then, most of them specifically aimed towards the country's tropical lowlands (0–1200 m a.s.l.). While positive impacts are found and documented within productive systems implementing these technologies, adoption rates remain low. This research expands on the roles and dynamics of both Research & Development (R&D) institutions and seed supply companies as potential explicative factors behind the processes of adoption and diffusion of forage technologies. We used a qualitative approach and developed a meta-analysis that addresses the functioning of seed systems in developing countries. We also conducted focus groups and semi-structured interviews with key agents (research and development centres, seed suppliers, producers and government agencies, among others). Our findings identify a lack of cohesion among R&D institutions and seed supply companies, prompted by their divergent productive goals and means of financing. As a direct consequence, we note several dynamics hindering the adoption of improved forages: a) duplicated efforts and investments resulting in poorly optimised processes; b) "premature" liberation of cultivars from research institutions that lack proper seed availability; c) simultaneous promotion of forage technologies by both actor types as aligned with distortions in the information given to producers; and d) a primacy of interpersonal relations that further complicates adoption and diffusion processes, as it circumscribes technological advances to a non-institutional real*m*. These results illustrate the complex dynamics behind forage technology dissemination, underscoring the critical role of a well-established synergy between institutions that can effectively contribute to overcome bottlenecks lying at the core of technology adoption in the country.

Keywords: Livestock, productivity, R&D, scaling, sustainable intensification, technology adoption

---

Contact Address: Stefan Burkart, International Center for Tropical Agriculture (CIAT), Trop. Forages Program, Km 17 Recta Cali-Palmira, Cali, Colombia, e-mail: s.burkart@cgiar.org

Adaptation to Climate Change by Smallholder Coffee Farmers in the Central Highland of Vietnam

PHUONG ANH NGUYEN, MIROSLAVA BAVOROVA
Czech University of Life Sciences Prague, Fac. of Tropical AgriSciences - Dept. of Economics and Development, Czech Republic

Smallholder farmers are facing many obstacles regarding climate change adaptation. This is also applicable for the Vietnamese coffee growers, whose livelihoods, in many cases, depend solely on coffee production, and face limited access to financial resources. The objective of the planned presentation is to demonstrate the theoretical framework and the factors that are expected to influence the farmers' adaptation behaviour in Vietnam based on the literature review. A descriptive analysis of Vietnam's coffee policy and its credit policy together will be provided. The study aims to explore drivers and barriers to Vietnamese coffee farmers' attention to adaptation strategies by using the Theory of Planned Behaviour (TPB) as the theoretical background. The TPB, specifically, contains three suggested groups of factors to predict people's behaviour: attitude, subjective norms, and perceived behavioural control. Access to credits and finances, among the perceived behavioural control factors, will be an in-depth study.

The planned empirical study will be performed in Dak Lak and Lam Dong province, Vietnam, in 2021. These study areas are selected not only because they are the main coffee-cultivating regions in Vietnam, but also for their current situations in facing climate changes. A mixed-method approach will be used to collect the data, in detail, a semi-structured quantitative questionnaire survey will be conducted (min. 300 respondents) together with 30–50 qualitative interviews with coffee sector actors. Structural Equation Modelling (SEM) will be used to analyse the relationship between behavioural factors and the adaptation intention of the surveyed farmers The research targets, in particular, to extend the knowledge in the field of adaptation behaviour of small farmers.

Keywords: Adaptation strategy, climate change, coffee, credit policy, structural equation model, theory of planned behaviour, Vietnam

Contact Address: Phuong Anh Nguyen, Czech University of Life Sciences Prague, Fac. of Tropical AgriSciences, Kamýcká 129, Prague, Czech Republic, e-mail: nguyenp@ftz.czu.cz

Typology of Farmer Groups in Ukraine on the Example of Agrarian Cooperatives

Alla Treus[1], Oksana Ponomarenko[2]

[1]*Sumy State University, Department of Economics, Entrepreneurship and Business Administration, Ukraine*

[2]*Sumy State University, Department of International Economic Relations, Ukraine*

Cooperation of small farmers is one of the possible ways for agriculture sustainable development. In the paper propose a new typology for agrarian cooperatives that could be used to characterise cooperative movement in Eastern European countries. The paper examined 79 questionnaires of 29 agricultural cooperatives from 8 regions of Ukraine.

Ukrainian legislation separates agrarian cooperatives in two groups: service (or not for profit) cooperatives and production cooperatives. Also, agrarian cooperatives could be divided by a type of activity (milk production, fruits production etc). In Ukraine one more could be, because in this country some cooperatives registered as cooperatives and functioning as farm with one owner. On the other hand, there are agrarian cooperatives that are not registered but farmers informal communities are functioning cooperatives. However, members of such communities different reasons do not want to formalize their business relationships.

Most of cooperatives surveyed (83%) officially registered and functioned as real cooperatives. 10% of agrarian cooperatives were registered as cooperative, but as farmer without any common production or interaction. Farmers gathered colleagues together to get government or donors support, provided only for cooperatives, or farmers who choose such form of registration to get some taxes preferences, etc.

Among cooperatives surveyed, 7% were not registered but really functioned as cooperatives. The most popular reason why farmers do not want to register their communities as cooperatives was long and difficult bureaucratic procedure of registration.

Keywords: Agro-cooperatives typology, cooperation, farmers

Contact Address: Alla Treus, Sumy State University, Department of Economics, Entrepreneurship and Business Administration, 2 Rymskogo-Korsakova St., 40007 Sumy, Ukraine, e-mail: a.treus@econ.sumdu.edu.ua

The Influence of Psychological and Socioeconomic Factors on Farmers Dealing with COVID-19

Masoud Yazdanpanah[1], Maryam Tajeri Moghadam[2], Katharina Löhr[3], Stefan Sieber[3]

[1]*Agricultural Sciences and Natural Resources University of Khuzestan, Iran*

[2]*University of Tabriz, Dept. of Extension and Rural Development, Iran*

[3]*Leibniz Centre for Agric. Landscape Res. (ZALF), Sustainable Land Use in Developing Countries (SusLAND), Germany*

The pandemic of coronavirus is a rapidly growing concern around the world. Through educational programs, policymakers are trying to change farmers' attitudes toward adaptation to the pandemic. The effectiveness of these educational activities depends on designing and conducting research related to factors that change behaviour in order to adapt to corona. While the technological, infrastructure, and educational focus has been usually paid to farmers' behaviour, much less attention has been given to psycho-social factors which may also influence farmers' behaviour. Therefore, the aim of this study is to investigate the psychological and socioeconomic factors influencing the behaviour of adaptation to COVID-19. To achieve this goal, psychosocial structures (perceived happiness, perceived well-being, and perceived stress) and demographic characteristics (age, number of family members, agricultural and non-crop income, reserve rate, land size, and number of livestock) were identified as predictors. The statistical population of this study are farmers in southern Iran. The data was collected by means of a questionnaire based survey that it's validity and reliability were confirmed. The results of the regression show that socioeconomic and psychological variables can explain 35 % of the changes in behaviour related to COVID-19. The results of the study also show that well-being is the most important predictor of adaptive behaviour. The results of the present study can help planners and policymakers to identify ways to improve adaptive behaviour and thereby to increase the effectiveness of educational activities in coping with severe shocks, such as the current corona pandemic. The results of this study also highlight the importance of considering social psychological measures and demographic characteristics in the development and evaluation of educational programs to stimulate adaptive behaviour.

Keywords: Adaptive behaviour, COVID-19, happiness, stress, subjective well-being

---

Contact Address: Masoud Yazdanpanah, Agricultural Sciences and Natural Resources University of Khuzestan, Mollasani, 744581 Ahvaz, Iran, e-mail: masoudyazdan@gmail.com

Potential Impact of New Groundnut Production Technology on Welfare of Smallholder Farmers in Northern Ghana

Bekele Hundie Kotu[1], Nurudeen Abdul Rahman[1], Francis Muthoni[1], Irmgard Hoeschle-Zeledon[2], Fred Kizito[1]

[1]*International Institute of Tropical Agriculture (IITA), Ghana*

[2]*International Institute of Tropical Agriculture (IITA), Nigeria*

This study was conducted to assess the potential welfare impact of applying higher groundnut planting density among smallholder farmers in northern Ghana. We used data from on-farm experiments (N=35), focus group discussions (N=22), and a household survey (N=542 farmers). We followed three steps in our analysis. First, we conducted cost-benefit analysis and risk analysis. Second, we predicted the potential maximum adoption rate and the time to reach the maximum adoption using the Adoption and Diffusion Outcome Prediction Tool. Third, using the results of the first and the second steps, we estimated the potential impact of the technology on poverty at household level using methods such as economic surplus model, econometric model, and the Foster-Greer-Thorbeck method. Results show that gross margin would increase by about 258 % for every 15 cm reduction in inter-row planting from 75 cm (farmers' practice) through to 30 cm. Prediction results show that a maximum of 62 % of groundnut farmers in the study areas are expected to adopt the best plant spacing; this peak rate will take about 10 years to be reached from the initial adoption time. This level of adoption will reduce the incidence of extreme poverty by about 3.6 % if farmers have access to the international groundnut market. The intervention will also reduce poverty gap and poverty severity which means that poor households would be closer to the national poverty line and their inequality would decline as compared to the base case. While the impact on welfare remains positive if farmers rely only on domestic markets, the magnitudes are not as high as when they get access to the international market. This implies that strategies which could improve farmers' access to the international market including trade negotiations with importing countries on tariff and non-tariff barriers and actions targeted to improve the quality of groundnut grain to meet the phytosanitary standards of importing countries will increase the rate of poverty reduction among groundnut growers in northern Ghana.

Keywords: Adoption, Ghana, groundnut, impact, poverty, prediction

Contact Address: Bekele Hundie Kotu, International Institute of Tropical Agriculture (IITA), Sagnarigu, TL6 Tamale, Ghana, e-mail: b.kotu@cgiar.org

Adoption of Rice Parboiling Improved Technologies and Poverty Dynamics in Benin: The Ladder of Life Approach

MAHOUKEDE FLORENT MEDAGBE KINKINGNINHOUN[1], GAUDIOSE MUJAWAMARIYA[2], SALI ATANGA NDINDENG[3]

[1]*University of Abomey-Calavi, Economy and Sociology for Rural Development, Benin*
[2]*Africa Rice Center, Madagascar*
[3]*Africa Rice Center, Ivory Coast*

By improving the quality of rice, reducing postharvest losses, reducing drudgery and improving safety for processors, improved parboiling technology contributes to improved food and nutrition security, income and livelihoods of rural parboiling households and communities. The Ladder of Life tool was used to assess the poverty dynamics by exploring the perceptions and interpretations of different wellbeing groups in communities where improved parboiling technology has been adopted in Benin. Forty-eight focus group discussions (FGDs) and individual interviews were held with 480 male and female rice producers and parboilers in the Central and Northern regions of Benin. Findings show that more female actors produce and parboil rice while male participants are rice producers. The Ladder of Life charts found more male farmers (53 %) than female farmers (50 %) under the poverty line in the investigated communities indicating that slightly more women are out of poverty than men. Also, 15 % of male and female improved their livelihood during the last decade and moved out from poverty. It was commonly recognised by both men and women that improved rice parboiling equipment and practices adoption has significantly contributed to bring women out of poverty during the last decade. Almost all the women who adopted the technology moved up on the ladder of life. In communities where traditional parboiling is still prevalent, the proportion of the poor is largest despite the noted decline for women in the lowest levels from 80 % in the last 10 years to the current 62 % while for men, the poverty situation has stagnated to 71 % for the lowest ladder levels. Men and women who are under the poverty line are aware of improved rice production and parboiling but are not applying them because of the lack of financial support and climate changes. The main factors contributing to moving up or climbing the step of the ladder are access to suitable credit and good and improved technologies, proper planning and management, investment in secondary activities such as animal rearing, agri-food processing, good relationship with people of the upper ladder levels, good networking and hard working.

Keywords: Benin, ladder of life, poverty, rice parboiling

Contact Address: Gaudiose Mujawamariya, Africa Rice Center, Immeuble FOFIFA Ampandrianomby, 1690 Antananarivo, Madagascar, e-mail: G.Mujawamariya@cgiar.org

Markets - Marketing - Value chains

Oral Presentations

Posters

Do Environmentally-Friendly Cocoa Farms Yield Socio-Economic Co-Benefits?

Lina Tennhardt[1], Gianna Lazzarini[1], Eric Lambin[2], Rainer Weisshaidinger[3], Christian Schader[1]

[1] *Research Institute of Organic Agriculture (FiBL), Dept. of Socioeconomics, Switzerland*
[2] *Université Catholique de Louvain, Earth and Life Institute, Belgium*
[3] *Research Institute of Organic Agriculture (FiBL), Sustainabilty Assessment & International Cooperation, Austria*

Chocolate companies are increasingly held accountable for their entire value chains. Thus, they enhance investments in value chain sustainability through third party certification or internal programs. So far, these initiatives have strongly focussed on capacity building for farmers to decrease cocoa farms' environmental impact and increase productivity. The question arises whether this focus on ecological aspects and productivity also has socio-economic co-benefits on farms.

Taking sustainability data from 190 smallholder cocoa farms in Ecuador, we analysed synergies between the environmental and economic dimension as well as social dimension of sustainability. The data was collected using the SMART-Farm Tool, a multi-criteria assessment methodology that assesses farm-level sustainability by operationalizing the FAO Sustainability Assessment of Food and Agriculture Systems (SAFA) Guidelines with 58 sustainability subthemes within the four dimensions of environmental integrity, social well-being, economic resilience, and good governance. We used a Latent Profile Analysis to identify the group of farmers that performed highest (n= 32) and lowest (n=73) across all environmental subthemes, and then compared the groups' performance in the economic and social subthemes using a Wilcoxon test.

Our results indicate significantly higher results ($p < 0.001$) for the high environmental performers compared to the low performers in several subthemes. These include "Internal investment" and "Stability of supply" among the economic, and "Safety and health provisionf" and "Food sovereignty" among the social subthemes. These synergies exist, given more resource-friendly, diverse production practices, which increase a farm's preparedness for the future as well as food self-sufficiency. The lower use of agro-chemicals reduce the dependency of off-farm inputs and decrease farmers' health and safety risks. Further significant differences included "Community investment" and "Public health". In general, however, we detected far more synergies between

Contact Address: Lina Tennhardt, Research Institute of Organic Agriculture (FiBL), Dept. of Socioeconomics, Ackerstrasse 113, 5070 Frick, Switzerland, e-mail: lina.tennhardt@fibl.org

the environmental and economic than between the environmental and social dimensions.

These results are important for chocolate companies, which so far have mostly focused on on-farm activities to improve their value chain sustainability. Achieving improvements in farmers' social well-being might require different interventions that go beyond the farm-level and address systemic issues together with different stakeholders.

Keywords: Cocoa, Ecuador, SMART-farm tool, sustainability assessment, synergies

Uber for Tractors? Opportunities and Challenges of Digital Tools for Tractor Hire in India and Nigeria

Thomas Daum[1], Roberto Villalba[2], Oluwakayode Anidi[3], Sharon Mayienga[3], Saurabh Gupta[4], Regina Birner[1]

[1]*University of Hohenheim, Inst. of Agric. Sci. in the Tropics (Hans-Ruthenberg-Institute), Germany*

[2]*Technical University of Munich, Germany*

[3]*Food and Agriculture Organization of the United Nations (FAO), Italy*

[4]*Indian Institute of Management, India*

Agricultural mechanisation can contribute to agricultural transformation. However, there is a need to find institutional solutions allowing smallholder farmers, who play a key role in agricultural development, to access tractors even though they cannot afford their own. Hire markets hold promise for this, but tractor owners are often reluctant to provide services to smallholder farmers because of high transaction costs. To address this problem, start-ups and tractor manufacturers have developed ICT applications that aim to help smallholder farmers access tractors. This model has been coined Uber for tractors, suggesting strong similarities with the Uber service for ride hailing. Although receiving much advance praise, these models have not been rigorously analyzed. Studying Hello Tractor (Nigeria) and EM3 Agri-Services (India), this paper assesses how such models address the challenges of agricultural markets, which are characterised by spatial dispersion, the concentration of demand around peak seasons, and high transaction costs, among other problems. This paper explores the extent to which such models can help to improve tractor utilisation and access to services by smallholder farmers. The paper acknowledges the potential of ICT-based tractor hire but finds that many of the thornier challenges of agricultural markets – which urban ride-hailing markets do not face – have yet to be addressed. The paper also finds that analog solutions such as booking agents and phone calls still trump digital ones and highlights the need for a supportive environment such as building (ICT) literacy. Last, the paper suggests that the advantages of ICT-based solutions over more traditional ways of organising service markets are more mixed than commonly assumed. In brief, while the Uberisation of mechanisation has appeal, such models are not the silver bullet they are often portrayed to be. More research is needed on how to make such ICT-based efforts work, and it is important not to neglect alternative solutions.

Keywords: Digital agriculture, ICT applications, smallholder farming

Contact Address: Thomas Daum, University of Hohenheim, Inst. of Agric. Sci. in the Tropics (Hans-Ruthenberg-Institute), Stuttgart, Germany, e-mail: thomas.daum@uni-hohenheim.de

The 'Livelihood' Challenge and Sustainable Agriculture: Evidence from Smallholder Cocoa Farming Households in Nigeria

Oluwafunmiso Adeola Olajide[1], Ayodeji Ojo[1], Kehinde Adesina Thomas[2], Molatokunbo Seun Olutayo[3]

[1]*University of Ibadan, Dept. of Agricultural Economics, Nigeria*

[2]*University of Ibadan, Dept. of Agricultural Extension and Rural Development, Nigeria*

[3]*University of Ibadan, Inst. of African Studies, Nigeria*

Global trade includes commodities produced by smallholder farmers from countries, such as Sub-Saharan Africa; as such their ability to participate in agricultural value chains in a sustainable way is important. The complexities underlying farmers' choice within the agri-food systems are often neglected in debates on conventional or agro-ecological agriculture. This study examined such complexities within the Agricultural Policy Research for Africa (APRA) consortium with a focus on Nigeria's Cocoa Value Chain. The trajectories into the sector; current production practices and commercialisation models were examined using a mixed methods approach. A sequential exploratory design was adopted; data were collected through Key Person Interviews, Focus Group Discussions and observation; and surveys. The discussion themes included the development along the value chain; the survey included production, commercialisation, and livelihoods data. The communities were selected through a multistage sampling technique from Osun, Ogun and Ondo states; the respondents were selected through a random sampling procedure from these. The information was analysed using themes; problem tree analyses; descriptive and econometric tools. The results show that trajectories into the cocoa sector include inheritance, labour-employment and marriage; these have several implications for land ownership, control and rights- such as tension between land owners and renters which is boosting illegal mining and logging. Access to land, labour and credit facilities are barriers to participation in the sector; current production practices include a similitude of organic and conventional approaches; farmers have 'mixed' economic trees for multiple streams of income and for social reasons; the commercialisation models show high dominance by produce buyers. The probit analysis shows that planting improved variety could improve the poverty status of farmers; while the tobit analysis shows an evidence of 'over' commercialisation. At the micro level, farmer's decisions on resource allocation are driven by own goals- options that bring immediate 'prosperity' are favoured; the meso level shows existing rural land, labour and credit markets governed by economic and socio-cultural factors; the macro level shows a major failure with respect to rural infrastructure, markets, land and mining policies. Policies that will ease these multi-level complexities need to be implemented if the world would be fed.

Keywords: Agro-ecological, livelihood, multi-level, policies, trajectories, value chain

---

Contact Address: Oluwafunmiso Adeola Olajide, University of Ibadan, Dept. of Agricultural Economics, Ibadan, Nigeria, e-mail: preciousfunso@yahoo.com

Identifying Value Webs of Biomass-Based Resources for Household Food Security: The Case of Local Actors in Ghana

John-Baptist S. N. Naah

University of Bonn, Center for Development Research (ZEF), Germany

The biomass sector is particularly important for supplying essential food and non-food biomass-based resources to many people and countries for a wide range of uses. Yet there is limited deliberation on these important biomass-based resources in Ghana to ensure improved rural household food security. This study was specifically carried out in the Northern and Upper East regions of Ghana to document various kinds of biomass-based resources and their cultural importance to local actors, examine socio-demographic factors influencing local actors' knowledge base on biomass-based resources and identify value webs, challenges, and future actions for sustainable use of biomass-based resources. Individual interviews (using structured questionnaires) were performed to cover 180 local actors in six rural communities. Given the findings of this study, cereals, e.g., maize (*Zea mays*), Guinea corn (*Sorghum bicolor*), rice (*Oryza sativa*), millet (*Pennisetum glaucum*), and legumes, e.g., groundnuts (*Arachis hypogaea*) are cultivated as major crops in the study areas and also considered by local actors as the most culturally important food crop species, while tubers and vegetables are considered minor crops and less culturally important. Ethnicity and residential status of local actors significantly influence local knowledge base on biomass-based resources. Also, the value webs of selected food crop species and their residues are not elaborately developed by local actors and still remained simple and traditional in nature, since no cascading uses of biomass by-products were identified. The local actors mentioned a plethora of challenges, which are largely in the nature of inadequate economic, social, logistics, marketing, soil health and climate-related issues as they negatively affect various stages of their biomass-based production, processing, storage, transportation and trading at the community level. Future actions were thus proffered by local actors, including central government financial support, agriculture-related education, good road networks, ready markets and reduced cost of farm labour. The sustainable utilisation and management of these biomass-based resources and more local value addition are required to help improve local actors' livelihoods, increase family incomes and enhance household food security in poor and vulnerable rural communities.

Keywords: Local actors, smallholder farmers, value webs

Contact Address: John-Baptist S. N. Naah, University of Bonn, Center for Development Research (ZEF), Genscherallee 3, 53113 Bonn, Germany, e-mail: jeanlebaptist@yahoo.co.uk

Fight or Flight: Factors Affecting Local Traders' Decisions to Remain in or Exit the Market

RAKHMA MELATI SUJARWO[1], BERNHARD BRÜMMER[1], THOMAS KOPP[2]
[1]*University of Goettingen, Dept. of Agricultural Economics and Rural Development, Germany*
[2]*University of Siegen, School of Economic Disciplines, Germany*

Intermediaries play an important and often underestimated role in agricultural value chains in developing and emerging economies. In general, a larger number of traders in the market is favourable for competition, as farmers have more choices on whom to sell their products to and can choose their most preferred trader. An assessment of the determinants of market performance in these regions therefore requires a solid understanding of the drivers of small-scale traders' decision-making processes. Especially the decisions on entering, remaining in, and leaving the market alter the fundamentals of the market structure. This study focuses on the factors that influence traders' decisions to remain in or exit the market. To determine the factors that affect the probability of a trader's decision to remain in the market, we employ a binary logistic regression approach, applied to a panel data set, based on three waves of a representative survey with small scale traders in the Jambi province, Indonesia. Numerous studies of similar behaviour among farmers have analysed which factors affect farmers' decisions to remain in or exit the market, which are the most relevant studies in the literature, since, to the best of our knowledge, trader- level studies of agricultural products have not been undertaken. Therefore, this study is intended to fill this gap and to further diversify the existing research on the topic. Results suggest that human capital (education and experience), physical capital (land area, operational vehicle ownership), trading practices (traded product, credit provision), market environment (number of competitors), and socioeconomic factors (trading revenue and trader status) all affect the decision of traders to remain in or leave the market.

Keywords: Binary logistic regression, exit, palm oil, rubber, trader

Contact Address: Rakhma Melati Sujarwo, University of Goettingen, Dept. of Agricultural Economics and Rural Development, Platz der Goettinger Sieben 5, 37073 Goettingen, Germany, e-mail: rakhma.sujarwo@agr.uni-goettingen.de

The Role of Emotion and Rational Self-Interest in Trust Perception: Case of the Dairy Value Chain

Dalel Ayari[1], Lokman Zaibet[2], Ghazi Boulila[1]
[1]*Ecole Superieure des Sciences Economiques et Commerciales, Tunisia*
[2]*Sultan Qaboos University, Natural Resource Economics, Oman*

The problem of coordination and cooperation stands as a cornerstone in value chain management. Although formal institutions guarantee the contractual arrangements, it is recognised the role that human capital can play to ensure the sustainability of these arrangements. Trust among actors in the value chain plays an important role in contracting decision and cooperative membership. There is however a concern to understand trust decisions and the determinants of trust perception. In this paper, trust is decomposed into its components; the emotional and the rational self-interest or calculative trust. The aim is to understand the role of emotional component in trust perception. We assume that non calculative trust and calculative self-interest trust are both present in local rural economies, making this context appropriate to test our hypothesis regarding the importance of emotional trust in networking and therefore contracting in the value chain.

We used a sample of breeders in a local community i) to measure the level of trust amongst breeders and dairy collection centres, ii) to construct the two components of trust, and iii to investigate how these two trust components contribute to generate trust decisions.

The results provide support that emotional trust fosters trust perception; despite opportunistic behaviour and distrust, breeders are able to build trusty links using close relationships. There is evidence that emotional trust is a strong motivating force to guide trust behaviour, whereas forward looking calculus guide trust perception in the case of weak ties. The results suggest the key role of existing social bonding in managing transactions in the local economy and in initiating network cooperation and more formal arrangements. Smallholder dairy farmers could use existing social networks to foster trust and institute sustainable contract as a way to coordinate transactions and improve overall innovation process in the community.

Keywords: Dairy sector, emotional trust, trust decision, Tunisia

Contact Address: Dalel Ayari, Ecole Superieure des Sciences Economiques et Commerciales, Tunis, Tunisia, e-mail: ayaridalel@yahoo.fr

The Impact of COVID-19 Measures on Local Food Systems in Indonesia, Mozambique, South Africa and Zimbawe - a Participatory Digital Co-Research Study

Nicole Paganini[1], Silke Stöber[1], Kustiwa Adinata[2], Nomonde Buthelezi[3], Jennifer Koppelin[1], Fezile Ncube[4], Tandu Ramba[5], Nedim Sulejmanovic[6], Enzo Paolo Nervi Aguirre[7], Kaimuddin Mole[8], Abdulrazack Karriem[9], Daniel Tevera[9], Haidee Swanby[9], Ines Raimundo[10]

[1]*Humboldt Universität zu Berlin, Centre for Rural Development (SLE), Germany*

[2]*Jamtani, Indonesia, Indonesia*

[3]*Urban Research Farmer Cape Town, South Africa*

[4]*Hope Tariro Trust, Zimbabwe*

[5]*Motivator Kondoran, Adaptation and Mitigation of Climate Change, Indonesia*

[6]*Freie Universität Berlin, Germany*

[7]*Berkeley University, United States*

[8]*Universitas Hasanuddin, Indonesia*

[9]*University of the Western Cape, South Africa*

[10]*Universidade Eduardo Mondlane, Mozambique*

The COVID-19 outbreak is spreading rapidly across the globe, forcing national governments to take decisions that have adverse effects on globalised food systems and supply chains. Border closures are disrupting commodity flows and labour force availability. As a result, food producers have difficulties harvesting and selling their produce. Informal economies and social welfare programs have also been affected, with detrimental effects on the most poor and vulnerable.

This study examines the challenges, coping strategies, and innovations of local food producers and city dwellers during the COVID-19 pandemic. The research for this study was farmer- led, meaning that farmers themselves collected the data, co-designed the research, and identified innovative coping techniques to deal with the crisis. Using digital survey tools, the data was collected from producer groups in selected urban and rural centres in South Africa, Mozambique, Zimbabwe, and Indonesia on a weekly basis during the lockdowns in April 2020.

The results are as different as the study regions but they are all the same in one respect: farmers have limited access to their farms and markets, and are forced to develop new approaches to make their ends meet. The most striking outcome of the study is that the farmers do not perceive COVID-19 as the greatest threat to their livelihoods but rather feel threatened by social inequalities and above all, climate change. While the global lockdown experiences highlight the importance of local food systems, there is great need to support agro-ecological and climate change-resilient production systems as well as farmer-led marketing systems.

Keywords: Coping strategies, COVID-19, farmer-led research, participatory research

Contact Address: Nicole Paganini, Humboldt Universität zu Berlin, Centre for Rural Development (SLE), Robert-Koch-Platz 4, 10115 Berlin, Germany, e-mail: paganini@hu-berlin.de

Why Are Good Relationships Along the Value Chain Important? A Cocoa Case Study from Uganda

Kashina Perlinger[1,2], Gianna Lazzarini[2], Christian Schader[2], Urs Niggli[2]

[1] *University of Kassel, Germany*

[2] *Research Institute of Organic Agriculture (FiBL), Dept. of Socioeconomics, Switzerland*

It is expected that global crises, such as the current COVID-19 crisis, will hit developing countries more severely than states with more robust economies. Due to the current disruption of supply chains, many people lose the only income that is essential for their livelihood. Thus, it is all more important to make value chains sustainable and resilient to global crises and unpredictable events. A first step in achieving this can be made through organic certification. As cocoa is increasingly becoming an economically relevant crop for African farmers and the demand for organic cocoa in Western countries is rising, it is important to fully understand these value chains to make them more sustainable. This study aims at analysing the effect of buyer-supplier relationship quality on the understanding of organic farming principles by smallholder cocoa farmers. As the role of the sub-suppliers have been neglected so far in (sustainable) supply chain literature, this research study aims at closing this gap. The overall project sample included 205 smallholder cocoa farmers in Central Uganda, which were in conversion to organic certification at the time of the data collection. Within this sample, 20 semi-structured qualitative interviews were conducted. For capturing the dyadic relationships between sub-suppliers, supplier and buyer, we also interviewed four key actors along the value chain. We found a correlation between a comprehensive understanding of the principles of organic farming and the relationship quality between the sub-suppliers and the indirect buying company. This is due to a high rating of high-quality relationship criteria such as trust, commitment and communication. The indirect buying company achieves this through a strong level of involvement and presence on-site. Nonetheless, the results also show that not all farmers were able to fully convert to organic farming because of challenges such as lacking practical knowledge and a lack of inputs that should have been provided by the direct buying company. Therefore, it is essential to achieve high-quality relationships between all actors and particularly at all levels along the value chain. To achieve this, recommended actions are put forward from the study results.

Keywords: Africa, buyer-supplier relationship, organic cocoa, smallholder farmers, sustainable supply chain

Contact Address: Kashina Perlinger, University of Kassel, 37213 Witzenhausen, Germany, e-mail: perlinger.kashina@gmail.com

The Impacts of Coronavirus on Agricultural Practices and Food Systems in Brazil, Tanzania and Iran

Katharina Löhr[1], Camilo Lozano[1], Michelle Bonatti[1], Ana Paula Turetta[2,1], Charles Peter Mgeni[3,1], Masoud Yazdanpanah[4], Tinoush Jamali Jaghdani[5], Linde Götz[5], Daniel Flemes[6], Stefan Kroll[7], Stefan Sieber[1]

[1]*Leibniz Centre for Agric. Landscape Res. (ZALF), Germany*

[2]*The Brazilian Agricultural Research Corporation (Embrapa), Brazil*

[3]*Sokoine University of Agriculture, Tanzania*

[4]*Agricultural Sciences and Natural Resources University of Khuzestan, Iran*

[5]*Leibniz Inst. of Agricultural Development in Transition Economies (IAMO), Germany*

[6]*German Institute of Global and Area Studies (GIGA), Germany*

[7]*Peace Research Institute Frankfurt, Germany*

Any discussion of nutrition and food security these days requires to also consider the impacts of the current pandemic. Countries are hit by a complex set of burdens by the pandemic. Not only the virus itself but national government policies as well as changes in international trade regulations and policies in consumer countries impact the lives of many people; and particularly those of small-scale farmers. This study investigates the socio-economic and environmental impacts of the current pandemic in Brazil, Tanzania, and Iran with a focus on agricultural practices and food systems. Whilst on a global scale government responses of most countries align, with regards to acknowledging the severity of the coronavirus and imposed measures, the government responses of the chosen countries of study are very different. Though instruments to combat COVID-19 and also the manner of transparency vary across these countries they provide an interesting platform for the study of effects of government responses to COVID-19 in societies. This project analyzes impacts of the novel coronavirus on food systems. The study also wants to explore coping mechanisms of local institutions, the trading and processing industries and small-scale farmers during the corona crises, in order to derive lessons for future pandemic crisis management. Data will be collected along the entire value chain on the levels of governments, private-sector firms, farmer cooperatives and small-scale farmers. A web-based survey will be coupled with in depth interviews to assess impacts and coping strategies.

Keywords: Agricultural practices, Brazil, COVID, food systems, impact assessment, Iran, Tanzania

Contact Address: Katharina Löhr, Leibniz Centre for Agric. Landscape Res. (ZALF), Sustainable Land Use in Developing Countries (SusLAND), Müncheberg, Germany, e-mail: katharina.loehr@zalf.de

Implication of Market Intermediation on Efficiency of Camel Milk Trade in Kenya

Simon Gicheha, Ernst-August Nuppenau

Justus-Liebig University Giessen, Inst. of Agricultural Policy and Market Research, Germany

Notwithstanding investments and interventions by the government and NGOs over the last ten years, the performance of the camel milk value chain in Kenya has remained poor, characterised by poor milk quality, losses and low-marketed volumes as well as producer prices. The market also features a large number of small (individual) traders and a considerable number of large (cooperative) market intermediaries operating along different market channels. Economic theories of increasing returns to scale suggest that presence of increasing returns gives an opportunity for large firms to take advantage of higher productivity to crowd out small firms. The structure of this market is therefore inconsistent with these theories where system-wide efficiency is expected to be large in a market with a few large firms buying directly from producers. Using data from 135 camel milk producers and 193 camel milk traders, this paper investigates whether the variation in scale and size distribution of intermediaries in camel milk market in Kenya can be explained by unexploited scope in efficiency improvement. The importance of market intermediaries in linking farmers to the market is evident as majority of the producers sell their milk at the production point. Analysis of volumes, costs and margins identifies transport, handling and monitoring costs as sources of increasing returns. Increasing returns to distance, transaction size, marketing task and working capital cannot be rejected. Large traders largely utilize public transport buses than motorcycles and cover significantly more distance compared to individual traders. Milk transport by bus is more cost efficient for large volumes over long distances as small loads are more penalized in motorcycle transport than buses. We observe a significant reduction in the density of trade in the dry season and motorcycle transport rival buses among large traders. Obtaining larger volumes in this season demands fetching the milk deeper in the rangelands where transport is only feasible by motorcycles. We therefore conclude that this observed behaviour among large traders is rational given the circumstances. Under similar circumstances, small traders can only capture efficiency gains by aggregating to prevent small volumes from being penalized. This implies that policies should be geared towards institutional innovations that promote cooperation among agents thus capturing efficiency gains.

Keywords: Costs, efficiency, handling, intermediation, margin, market, returns, transaction, transport

Contact Address: Simon Gicheha, Justus-Liebig University Giessen, Inst. of Agricultural Policy and Market Research, Senkenbergstrasse 3, 35390 Giessen, Germany, e-mail: simon.kariuki-gicheha@agrar.uni-giessen.de

Mapping Knowledge Management in a Banana Agro Processing Industry in Uganda: the Case of the Banana Industrial Research and Development Centre

JOSEPH BAHATI

Banana Industrial Research and Development Centre, Uganda

This research conducts a systematic focus on knowledge mapping of the Banana Industrial Research and Development Centre (BIRDC) with the aim of identifying the different critical actors, how they collaborate and share information and how to improve efficiency of their operations from the banana farms through the agro-processing. The main goal for engaging in this research was to identify and map the different knowledge centres, their knowledge needs and interlink the different actors along the production chain. Specifically, the objectives were to identify and categorise the knowledge levels of the different actors in the banana production chain, identify how the stages in the banana production chain are interlinked, investigate the extent to which BIRDC has implemented knowledge management practices such as knowledge creation, sharing and retention through the assessment of existing KM enablers, develop a model that interlinks the different actors to ease information access and to make recommendations to the BIRDC management on KM and mapping. The research design used in this study was a non-probability purposive case study in the pilot banana-growing area located in the five districts of the Greater Bushenyi. Quantitative data were collected from randomly selected managers and technicians at BIRDC and randomly selected banana farmers located in the different sub-counties of the Greater Bushenyi districts. In addition, a GPS was used to collect shape-file information from the research sites and ArcGIS was used to map the study area using some of the quantitative data that was collected. The findings of this study indicated that KM and mapping concepts were not well understood at BIRDC, but after the research, the BIRDC staff viewed the subject as a necessary contribution that would impact on their performance. The findings showed that collaboration of staff within the BIRDC organisation and between BIRDC and the banana-farming communities in creating a meaningful and relevant knowledge environment was essential for BIRDC competitiveness in promoting the commercialisation which is the cornerstone of the organisation. The conclusions and recommendations help develop appropriate BIRDC's wide policies and practices for proper and well-organised methods of integrating and harmonising work processes, collaboration and information sharing.

Keywords: Knowledge management, knowledge mapping, organisational performance, precision decision-making, product traceability

Contact Address: Joseph Bahati, Banana Industrial Research and Development Centre, 26a Lumumba Avenue, 35747 Kampala, Uganda, e-mail: joeb2007b@gmail.com

Value Addition and Off-Season Market Participation among Retailers in the Grasshopper Value Chain in Uganda

EMMANUEL DONKOR[1], DAGMAR MITHÖFER[1], ROBERT MBECHE[2], JOHN N KINYURU[3], DOROTHY NAKIMBUGWE[4]

[1]*Humboldt-Universität zu Berlin, Albrecht Daniel Thaer-Institut für Agrar- und Gartenbauwiss., Germany*

[2]*Jomo Kenyatta University of Agriculture and Technology, Dept. of Agricultural and Resource Economics, Kenya*

[3]*Jomo Kenyatta University of Agriculture and Technology, Dept. of Food Science, Nutrition and Technology, Kenya*

[4]*Makerere University, Dept. of Food Technology and Nutrition, Uganda*

Grasshoppers are a nutritious delicacy for most people in Uganda. They provide essential nutrients such as protein, minerals and vitamins. They are also a source of livelihoods for many people who trade on them including women. However, seasonality and high perishability of grasshoppers coupled with low value addition tend to limit their availability throughout the year. Despite their importance, empirical studies on grasshoppers' value addition and retailers' off-season market participation are limited. Our paper therefore analyses the effect of value addition on retailers' participation in off-season grasshopper markets in Uganda. Five-hundred (500) grasshopper retailers were randomly sampled from Kampala and Masaka Districts of Uganda in December 2019 during the peak season. Data were analysed using multivariate Probit and binary Probit models, where a control function approach was applied to account for endogeneity of value addition. Our results show high participation of retailers in activities that increase product market value, but low participation in those that increase both the product market value and shelf-life. After controlling for endogeneity of value addition, we find a strong positive effect of value addition on off-season market participation. Our results also show mixed effects of retailers' characteristics, institutional variables, transactional variables, storage constraints and managerial constraints on traders' value addition decision and participation in off-season market in the grasshopper value chain. We conclude that to ensure continuous availability of grasshoppers throughout the year, agri-food policy should promote the adoption of grasshopper value addition to increase shelf life stability. Agri-food policy should also prioritise integrating human capacity building and providing affordable credit for retailers in the grasshopper business.

Keywords: Food security, grasshoppers, multivariate probit, retailers, Uganda, upgrading strategies, value addition, value chain

---

Contact Address: Emmanuel Donkor, Humboldt-Universität zu Berlin, Albrecht Daniel Thaer-Institut für Agrar- und Gartenbauwiss., Berlin, Germany, e-mail: edonkor.knust@gmail.com

Where Do Inhabitants of the Parish Kanyanya in Kampala Shop?

Andrea Fongar[1], Youri Dijkxhoorn[2], Beatrice Ekesa[1], Vincent Linderhof[2]

[1]*The Alliance of Bioversity International and CIAT, Healthy Diets from Sustainable Food Systems Initiative, Uganda*

[2]*Wageningen University and Research, Wageningen Economic Research, The Netherlands*

As Africa is urbanizing, the urban population in Uganda increased from 7.4 million in 2014 to 9.4 million in 2017. Thus, the dietary aspiration of the African population is changing, and urban food system have to evolve rapidly to address the new demand. This requires a deep understanding of the current food system. We contribute to the literature by identifying the currently used retail outlets of inhabitants of a low- and middle-income area in Kampala, Uganda.

Kanyana parish (Kawempe division, Kampala, Uganda) was selected, which has similar demographic characteristics to Kawempe division and Kampala city. Three Focus Group Discussions (FGDs) with parish representatives (disaggregated into women, men and youth) were conducted to gather information around the used food sources in 2020.

Eighty-five inhabitants participated in the FGDs. Preliminary results indicate that food stalls (fixed stall location in residential areas) are the main food outlet used by the community. Over 80 % use the local market to purchase their food items, which takes on average 20 min on foot to reach. Around 30 % mentioned shopping in supermarkets. Fresh vegetables and fruits are mostly bought at the local market (51 %) or the market and food vendors/hawkers (22 %). Modern retails (supermarkets) are only used by a minority to purchase meat, eggs, fish or milk and milk products. Fruit and vegetables are consumed daily by 19 % and 34 %, while meat consumption was stated to be consumed monthly or rarely (32 %).

Traditional retail outlets are still the major food supplier in a low- and middle-income parish of Kampala, Uganda. Supermarkets are mainly used to buy milk products, meat and fish. Further, analyses of the FGD results and research are needed to understand the reasons behind the preferred and used retail outlet.

Keywords: Food outlets, Kampala, traditional and modern retail, urban food system

Contact Address: Andrea Fongar, The Alliance of Bioversity International and CIAT, Healthy Diets from Sustainable Food Systems Initiative, Plot 106 Katalima Road, Naguru, Kampala, Uganda, e-mail: a.fongar@cgiar.org

Agricultural Credit's Role in Overcoming Barriers to Sustainable Intensification of Bovine Livestock Production in Colombia

Manuel Diaz, Angie Hurtado, Karen Enciso Valencia, Natalia Triana-Angel, Stefan Burkart

International Center for Tropical Agriculture, Tropical Forages Program, Colombia

The positive effects of agricultural credit in the modernisation of developing economies are well documented and range from increased productivity and reduced vulnerability to seasonality and significant multidimensional poverty reduction. However, credit placement in the Colombian agricultural sector is still low and only 5 % of agricultural credits have investment motivations. On the other hand, the sector has been experiencing a significant growth and diversification of credit lines, including some focused on rural women and youth, and others on promoting sustainability in livestock production. This study attempts to analyse the role of agricultural credit in the upscaling and consolidation of sustainable intensification options for bovine livestock production systems. For this, the study adopts a qualitative approach, based on semi-structured interviews and focus group discussions with key stakeholders (e.g. representatives of financial institutions, policy makers, producer associations), as well as a meta-analysis of recent impact evaluation studies of agricultural credit. Results indicate the presence of structural and market failures. First, the granting of credits has not always corresponded to the spirit of the promotion programs but to political interests and motivations. Second, current credit requirements cause conflicts with producer realities, such as land titles and access to property, and limit accessibility leading to a promotion and proliferation of informal credit structures, with baneful effects on the rural poor. Third, rural poverty limits the effective demand for credit, reducing the incentives for the expansion of private financial services in rural areas. And fourth, low investment in technology in rural areas, which is related to a) missing recognition of the benefits of an innovation due to deficiencies in extension services; b) information asymmetries on the rates of return of new investments and technologies; and c) high transaction costs. The study proposes elements for the mitigation of these failures, such as a limitation of substitute credits (used for purposes other than agricultural activities) and their reorientation towards sustainable intensification (e.g. with silvo-pastoral systems and appropriate extension services), as well as the promotion of co-operative credits, among others.

Keywords: Bovine livestock, credit, rural extension, silvo-pastoral systems, sustainable intensification

Contact Address: Stefan Burkart, International Center for Tropical Agriculture (CIAT), Trop. Forages Program, Km 17 Recta Cali-Palmira, Cali, Colombia, e-mail: s.burkart@cgiar.org

A Transition Towards Higher Added Value of Natural Resource Based Products: Case Study of Acacia Timber Value Chains in Central Vietnam

THI THAM LA[1], JÜRGEN PRETZSCH[1], DIETRICH DARR[2]

[1]*Technische Universität Dresden, Inst. of International Forestry and Forest Products: Tropical Forestry, Germany*
[2]*Rhine-Waal University of Applied Sciences, Fac. of Life Sciences, Germany*

Land and the biomass production it supports are limited. Therefore, sustainable management of these natural resources is needed that sustains and improves their contribution to the local, national and international economy. In Vietnam, plantation forests serve as an important source of raw material for the wood-based industry. About half of the plantation area is currently managed by small-scale producers, with *Acacia auriculiformis* × *Acacia mangium* hybrids being the most prevalent cultivated species. However, up to 80 % of plantation wood is currently used for low value woodchip production. In response to the increasing domestic timber demand and to stimulate higher value uses of plantation wood, the Vietnamese government has introduced a number of policies, such as land tenure or timber value chain (VC) improvement. These interventions aim at the development of high-quality plantation forests and the transition of the domestic timber processing industry towards higher added value products, such as furniture. However, a comprehensive analysis of this sector is lacking so far.
This paper elucidates the financial and economic performance of woodchip, non-FSC furniture and FSC-certified furniture VC from hybrid timber in Thua Thien Hue province, where the species has been cultivated since approximately 20 years. In-depth interviews with 30 timber producers, eight timber traders, one woodchip and one furniture processing and exporting company have been conducted. The findings were validated through direct observations and in six group discussions and 26 expert interviews.
The results demonstrated that small-scale producers and processing firms achieved the highest added value in the FSC-certified furniture VC. Timber traders generated a relatively modest value of 6.90 USD m^{-3} in the chip and 14.60 USD m^{-3} in the non-FSC furniture VC. The total added value of the woodchip, non-FSC furniture and FSC-certified furniture VCs was 31.60 USD m^{-3}, 562.50 USD m^{-3} and 683.60 USD m^{-3}, respectively. Aspects such as higher benefits for participants, higher investment in added value activities and lower negative impacts on natural resources made the VC of the FSC-certified furniture product relatively more efficient in term of economic performance. In contrast, the performance of the woodchip VC was least.

Keywords: Added value, plantation forestry, small-scale producer, sustainability, value chain upgrading

Contact Address: Thi Tham La, Technische Universität Dresden, Inst. of International Forestry and Forest Products: Tropical Forestry, Postfach 1117, 01735 Tharandt, Germany, e-mail: la_thi.tham@tu-dresden.de

Long-Run Effects of the Trans-Atlantic Cotton Trade on Poverty in the American Deep South

MATTHEW RUDH

University of Goettingen, Development Economics, Germany

The growing worldwide demand for cotton in the 1800's and the necessity of large-scale slave labour in its production were the main drivers of plantation slavery in the southern United States. The global trade of goods requiring widespread extractive labour in their production can shape the institutional quality and demographic structure of the region in which the labour is extracted, and have long-lasting impacts on economic growth and human development today. Through a study of the relevant literature, I discuss the theories linking the trade of agricultural goods requiring large-scale extractive labour in their production with institutional quality, human development, and economic growth. Then, focusing on the period in which slavery expanded in the United States between the early 1800s and the abolition of slavery in 1865, I analyse a novel dataset created from a historic article documenting price and demand for cotton in the United States and Europe in the 1800's, census data on U.S. slave population and crop production throughout the 19th century, and modern socioeconomic data on former cotton-producing regions of the US Seep South, to demonstrate a link between the world trade of cotton in the 1800's, the expansion of slavery, and modern-day poverty among African Americans in the southern United States.
In order to present an argument for a potential relationship between the number of slaves in 1860, driven by world trade of US cotton, and poverty today, I present a basic two-stage least squares in which I use a county's adjacentness to the Mississippi River, the main route of cotton trade between the cotton-producing counties in the Mississippi Delta and the European market, to predict the density of the slave population in 1860. I then use this predicted value to estimate the effect of cotton production and trade in 1860 on poverty rates in 2018, at the county level. I also provide some quantitative evidence to demonstrate the close interrelationship between demand for cotton in Europe and cotton production in the United States, as well as the close interrelationship between cotton production and slavery in general in the United States.

Keywords: Agricultural trade, cotton, cotton belt, growth, institutions, plantation, poverty, rural development, slavery, trade, United States

Contact Address: Matthew Rudh, University of Goettingen, Development Economics, Cramerstrasse 11, 37073 Goettingen, Germany, e-mail: matthew.rudh@stud.uni-goettingen.de

Collective Action Opportunities for Upgrading the Value Chain of Small-Scale Furniture Enterprises in Hawassa, Ethiopia

Alexander David Koch[1], Kendisha Illona Soekardjo Hintz[2], Tsegaye Bekele[3], Jürgen Pretzsch[2], Maxi Domke[2], Dietrich Darr[1]

[1]*Rhine-waal University of Applied Sciences, Fac. of Life Sciences, Germany*

[2]*Technische Universität Dresden, Inst. of International Forestry and Forest Products, Fac. of Environmental Sciences, Germany*

[3]*Hawassa University, Wondo Genet College of Forestry and Natural Resources, Ethiopia*

Ethiopia is challenged with deforestation and coinciding wood supply gaps. The wood demand of the expanding furniture sector is projected to continue widening this gap. This sector mainly consists of small-scale enterprises, which face various challenges and are thus often situated in subsistent survival states. A development opportunity emerges when considering that upgrading such enterprises past subsistence has the potential to further generate employment and disseminate market signals towards wood suppliers. A valuable market for wood propels small-scale farmers to further adopt agroforestry practices.

Collective action is a frequently discussed option for countering enterprise development barriers in emerging economies and was thus explored as a potential upgrading mechanism. This study hereby aimed to develop an institutional model of collective action for small-scale furniture enterprises in Ethiopia.

The study applied a value chain analysis and framed the collective action model as a potential upgrading strategy. An explorative and descriptive research approach was applied, encompassing in-depth interviews with 54 wooden furniture enterprises and 12 key informants. The study was conducted in Hawassa, a central location for furniture production.

The interview results gave insight into the enterprises' value chain structure, costs, challenges, and upgrading desires. The most substantial issues were found to be market failures that limited access to factors of production, such as workspaces, machinery and financial capital. In addition, the government's role in supporting these enterprises was found to be weak, as access to support services was observed to be obstructed by selective bias and low capacities.

To counter these issues, a business association of small-scale furniture enterprises is recommended as a collective action upgrading strategy. By considering the local conditions and success factors identified in literature, the study

Contact Address: Alexander David Koch, Rhine-waal University of Applied Sciences, Fac. of Life Sciences, 10365 Berlin, Germany, e-mail: alexander-david.koch@hsrw.org

developed an association model, describing its purpose, institutional arrangements and the characteristics of participating actors. The aims of the association were recommended to facilitate network ties, lower transaction costs with authorities, support the access to business development services, and advocate for improved market conditions. In the long run, the association will help small-scale enterprises collectively develop into innovative and resource efficient businesses, thereby improving local forest-based markets and incentivizing tree growing initiatives.

Keywords: Business association, forest-based market, small-scale enterprises, value chain analysis, value chain upgrading, wood value chain

Stakeholders and Marketing Analysis of African Nutmeg (*Monodora myristica*) in Cameroon

Mbouwe Irene Franceline[1], Jiofack Tafokou René Bernadin[2], Miroslava Bavorova[1], Mustapha Yakubu Madaki[1], Bulus Barnabas[1]

[1]*Czech University of Life Sciences Prague, Fac. of Tropical AgriSciences; Dept. of Economics and Development, Czech Republic*

[2]*Higher Institute of Environmental Sciences / Global Environment Protects, Plant Biology Department, Cameroon*

Monodora myristica Dunal is an edible fruit belonging to Annonaceae family that plays a crucial role in household economy in Cameroon as the seeds of this fruit are widely consumed for his nutritional value. The characteristics of the stakeholders in the value chain and their benefit derived from it are under investigated. Hence, to better understand the value chain of the seeds, the research investigated the socio-demographic characteristic of producers and traders and the marketing channels used. Purposive sampling method was used to select the region and communities due to the large production of the species in Nde district. A sample of 15 producers and 57 traders were randomly selected and semi-structured questionnaires were administrated to respondents. Descriptive statistics were used to describe the data collected and one-way ANOVA was used for analysing the differences among wholesalers, resellers and retailers actors' groups. Profit margin was calculated to estimate the percentage of benefit perceived by traders. Results revealed that men (93%) were predominantly involved in the production, while majority of the women (73.7%) were involved in the trade of seeds. Producers, wholesalers, resellers, retailers and final consumers were the main actors identified and involved in the value chain of the species. Wholesalers gathered the highest quantities of seeds per production cycle (875 kg) compared to resellers (105 kg) and retailers (32 kg). The profit margin obtained by resellers was higher than other actors and varied between US31.07*toUS* 50.82 per bucket of 15 liter according to the abundance and scarcity period. However, women are marginalised in the production process by men due to land ownership control which is in the hands of the men. Similarly, along the value chain despite majority of women involved in the process, the wholesaler actors discriminate women involvement. Therefore, policies that will support resources allocation and gender equality will help to mitigate the challenges.

Keywords: Central Africa, incomes, livelihood, *Monodora myristica*, NTFPs, value chain

Contact Address: Mbouwe Irene Franceline, Czech University of Life Sciences Prague, Fac. of Tropical AgriSciences; Dept. of Economics and Development, Prague 6 - Suchdol, Czech Republic, e-mail: mbouweirene@gmail.com

Bureaucracy, Access to Official Subsidised Credit and the Parallel Credit System for Agriculture in Morrinhos

Victória Cardoso Carrijo, Luciana Ramos Jordão, Barbara Luiza Rodrigues, Thays Dias Silva, Rebeca Barbosa Moura, Letícia Versiane Arantes Dantas, Igor Gomes de Araújo, Niury Ohan Pereira Magno

State University of Goiás, Law College (Southeast Campus: Morrinhos), Brazil

Agribusiness leads Brazilian midwestern economy, among the three states of the region, Goiás stands out in the cultures of soy, corn and rice. The whole country handles a wide range of pesticides while funding production official credit and inducing production through post green revolution technics. Morrinhos has a great number of farmers who depend on subsidised credit provided by the government through public policies that provide below market interest rates in order to achieve greater involvement in agricultural production. However, inability to meet the required criteria, bureaucratic and time-consuming processes make some producers choose their own means finance production. They finance production through credit offered by companies that have a higher rate than the one offered by the official credit program, but which has a less bureaucratic approach to release the amount. However, most of the time these institutions do not require any kind of social and environmental criteria in order to promote sustainable production, on the contrary, they encourage the use of pesticides and establish contract clauses to enforce the of a specific pesticide as a condition to grant the credit.

Thus, this paper proposes an investigation of the difficulties encountered when applying for official credit that make farmers take part in this paralleled credit system even though it ends up less sustainable and more expensive than official credit. The research aims to examine what the biggest challenges for sustainable agriculture in Morrinhos are, and to provide information that can contribute to access to credit, as well as to achieve sustainable practices. The data will be achieved through interviews with some regional producers, simulation of credit application and contact with the regional rural union in order to denote the best way to promote sustainable agriculture and integrate farmers to official credit system.

Keywords: Agriculture, credit, social justice, sustainable development

Contact Address: Luciana Ramos Jordão, State University of Goiás, Law College (Southeast Campus: Morrinhos), Av. Perimetral Norte, n. 4356, c. 88b, Alto da Boa Vista, 74445500 Goiânia, Brazil, e-mail: luciana.jordao@ueg.br

Linkage Between Large Cardamom Value Chain and Food and Nutritional Security in Bhojpur District of Nepal

Arati Joshi, Dharmendra Kalauni
Agriculture and Forestry University, Faculty of Agriculture, Nepal

Large cardamom is the second most exported agricultural commodity of Nepal with significant economic, nutritional, social, and religious importance. It is commercially grown in more than 43 districts of Nepal and more than 67, 000 households are involved in this industry. Nepal Trade Integration Strategy 2010 and Agriculture Development Strategy 2015–2035 have recognised it as the most potential cash crop for export with comparative advantages. Bhojpur is one of the large cardamom producing zone as identified by PMAMP. The purpose of the study was to find linkages between the existing value chain of large cardamom with food and nutritional security in Bhojpur district. Also, the study focuses to reveal different aspects of production economics and marketing. A survey study was conducted with 150 farmers. Further, 10 local, 10 medium, 10 large traders and 10 exporters were selected from Bhojpur, Khadbari, and Birtamod market to study marketing aspects and export performance of large cardamom. Microsoft-Excel was used for data assembling and SPSS and STATA were used for the data analysis. Large cardamom cultivation was found to contribute 42.9 % in total household income and 77.8 % in total farm income. The contribution of large cardamom was the highest among the total household income from different professions in the study area. Cash generated from the selling large cardamom was used to buys cereals, pulses, and fruits which ultimately plays a critical role in maintaining food and nutritional security to an extent. The value chain of large cardamom help to improve the food and nutritional security situation by generating greater economic returns to the value chain actors, influencing food affordability, and diversifying the consumed food sources by improving the purchasing ability of value chain actors.

Keywords: Bhojpur, dietary diversity, food security, nutrition

---

Contact Address: Arati Joshi, Agriculture and Forestry University, Faculty of Agriculture, 10410 Kanchanpur, Nepal, e-mail: arati.joshi.9638@gmail.com

Cricket Production Survey in Thailand

CHAMA PHANKAEW
Kasetsart University, Department of Entomology, Thailand

This research was carried out during 2017–2019 by using questionnaires to get information from 150 farmers, and reviewing of publications. The objective of the study was the investigation of the current state of the cricket sector in Thailand. There are 3 cricket species: house cricket (*Acheta domestica* L.) (15 %), two-spotted cricket (*Gryllus bimaculatus* De Geer) (80 %), and ground cricket (*Teleogryllus mitratus* (Burmeister, 1838)) (5 %) used by farmers in Thailand. However, the main species in the export market is the house cricket, and the local market is two-spotted cricket while the ground cricket is the minor species but the highest price. The cricket farm are about 20,000 farms mainly in the northeast provinces (Kalasin, Khon Kaen, Maha Sarakham, Nakhon Phanom, Nakhon Ratchasima, Roi Et, and Ubon Ratchathani) follow by central provinces (Lopburi, Nakhon Sawan Suphan Buri), north provinces (Chiang Mai, Nan, Phitsanulok and Sukhothai), east provinces (Chanthaburi and Sa Kaeo) and south province (Surat Thani). The farm value of the cricket yield is about 37.36 million € in 2019. The price of cricket powder export is about 40–50 € per kilogram. The key factors of cricket production are feed, water, temperature (35–38°C), relative humidity (60 %), density of cricket population (7,255 adults m^{-3}), and sanitary farm and good health of the farmer. The largest challenges, according to cricket farmers surveyed, are cricket feed, the cricket variety, and product development. The Thai cricket sector will be established as a global standard cricket farm and safe cricket products.

Keywords: Cricket, cricket powder, house cricket, production, two-spotted cricket

Contact Address: Chama Phankaew, Kasetsart University, Department of Entomology, 50 Ngam Wong Wan Rd Lat Yao, 10900 Chatuchak, Thailand, e-mail: fagrcmp@ku.ac.th

Digitizing: A Doorway to the Sustainable Meat Supply Chain in Iran

Afsaneh Ehsani, Tinoush Jamali Jaghdani
Leibniz Institute of Agricultural Development in Transition Economies (IAMO), Agricultural Markets, Marketing and World Agricultural Trade, Germany

In early 2019, a phenomenon called 'meat crisis' in Iran, has led to a significant increase in red meat price over a short period of time. The problems of this chain are so fundamental that despite the increase in the number of domestic livestock, a 188 % increase in meat imports could not solve the crisis. Given the importance of red meat and the profound link between the price of food and food security, we conducted an empirical study in five counties of the eastern province of Khorasan Razavi in Iran, where has the highest production of red meat in the country. The aim of our study is to identify the influential factors in this crisis and to find out the new available digital technologies which can improve the performance and sustainability of the red meat supply chain (RMSC) to reduce the risk of such crises re-emerging. Applying the global value chain governance (GVC) approach allowed us to map the chain and investigate actors, activities, connections, policies and transformations in the local RMSC. Identifying the gaps in RMSC digital maturity and to find the possible application of digital technology was achieved through using the Case study method. The results show that some of the major challenges in Iran's RMSC are due to inappropriate distribution of livestock inputs, lack of production capacity due to the traditional methods, lack of tractability of data from feeding to distributor and the captive governance structure of this chain. We propose that safe tracking and monitoring of the chain can be provided by the blockchain technology along with Radio Frequency Identification (RFID). In other words, the application of these technologies verifies the validity, fulfil the efficiency of information and reduce the opportunistic decisions which can improve decision-making and performance through this supply chain. These technologies increase the responsiveness to the market demands via integrating and maximising the capacity of different sectors to increase productivity and product safety and quality.

Keywords: Blockchain, digital technologies, meat crisis, radio frequency identification, red meat supply chain

Contact Address: Afsaneh Ehsani, Leibniz Institute of Agricultural Development in Transition Economies (IAMO), Agricultural Markets, Marketing and World Agricultural Trade, Theodor-Lieser-Straße 2, 06120 Halle (Saale), Germany, e-mail: ehsani@iamo.de

Market Potential of Hornet (*Vespa* sp.) in Eastern Shan State, Myanmar

Myint Thu Thu Aung, Jochen Dürr
University of Bonn, Center for Development Research (ZEF), Germany

Eastern Shan State, a border gate to Thailand, Laos, and China, is known for its diversity in ethnic groups such as Shan, Burma, Kachin, Lisu, Ann, Akha, Wa, Danu, and Larhu. A big variety of edible insects can be found in that region such as bamboo worms, silkworms, cicadas, hornets, and crickets, which are mostly eaten by these ethnic groups. Unlike other edible insects, hornet harvesting is very dangerous and can even cause death. Therefore, it is very rare in the markets of other regions. The aim of this study was to examine the present use and the market potential of hornets by conducting a face-to-face interviews with local people including hunters, wholesalers, processors, retailers, and consumers. Results show that hornets are the most expensive edible insect, but depending on the size of the hornet, prices vary. It is collected from the deep forest by Larhu people who are skillful in hunting and know very well the nature of hornet. Collecting is seasonal, starting in August and ending in November. Hornets are sold on the market in various forms such as fresh alive, fried, or mixed with alcohol. Besides human consumption, hornets are also utilised as medicine. People believe that the mixture of alcohol with hornet can protect and cure strokes and inflammatory diseases. Depending on the form of the product, the collected amount of the hornets, market distance, location of the hunter's village, habits and economic condition of the villagers, and freshness of the hornets, six types of marketing chains can be distinguished. As hornet is rare and expensive, most of the locals are able to buy them only once per season. According to the wholesalers, almost 75 % of the hornets from Kyaingtone central market is transported to Mongla and Tachileik, borders areas of China, Laos, and Thailand, where 50 % of consumers are foreigners. Thus, hornet is a prospective product for earning foreign income, however, it is difficult to export hornet larvae alive legally. Therefore, commercial processing technologies in order to add value and export to neighbouring countries should be developed.

Keywords: Edible insects, entomophagy, ethnic groups

Contact Address: Myint Thu Thu Aung, University of Bonn, Center for Development Research, Genscherallee 3, 53113 Bonn, Germany, e-mail: myintthuthuaung@gmail.com

Value Chain Analysis of Pineapple (*Ananas comosus*) Production and Marketing from Agroforestry System, Southern Ethiopia

TARIKU OLANA JAWO

Czech University of Life Sciences Prague, Department of Crop Sciences and Agroforestry, Czech Republic

This study was conducted in Aleta Chuko district of Sidama Zone, Southern Ethiopia. Southern Ethiopia is known with a favourable environment for pineapple, coffee and many other horticultural crop productions. However, the practice has been facing several productions and marketing constraints. The main aim of the present study was, therefore, identifying the value chain actors and their roles, mapping the value chain of pineapple production and marketing, and examining the determinants of market supply. Using a purposive sampling technique, 105 households were selected from three representative peasant associations for an in-depth survey. Both primary and secondary data was collected using a combination of Participatory Rural Appraisal tools (semi-structured interviews, key informant interviews, group discussion and direct observation). Descriptive statistics and econometric analysis using Ordinary Least Square (OLS) were used to analyse the collected data. The result revealed that both primary and secondary actors were involved in pineapple production and marketing. The percentage of market margin for the producer, assembler, whole seller, retailer and processor actors were 9.41, 11.86, 18.33, 26.96 and 33.43 %, respectively. Producer (34.20 %), assembler (3.30 %), whole seller (17.39 %), retailer (26.78 %) and processor (18.33 %) were sharing a percentage of profit margins in pineapple production. The result of the OLS regression analysis model indicated that the market supply of pineapple was affected by wealthy status and duration of storage ($p < 0.05$) and price ($p < 0.01$), positively. Enhancing the local actors' capacity through training, providing price and market information, credit and other processing facilities and institutional support could result in increased production and a steady supply of the pineapple products.

Keywords: Actors, pineapple producers, profit margin, value chain

Contact Address: Tariku Olana Jawo, Czech University of Life Sciences Prague, Department of Crop Sciences and Agroforestry, Kamycka 129 Praha 6, 165 00 Prague, Czech Republic, e-mail: jawo@ftz.czu.cz

Institutions

BMZ/GIZ session: BMZ strategy, funding instruments and supported research projects in response to global crises

The BMZ Strategy for International Agricultural Research

Sebastian Lesch

Federal Ministry for Economic Cooperation and Development (BMZ), Head, Division 121: Agricultural Policy, Agriculture, Innovation, Germany

For many years, BMZ has been supporting activities in international agricultural research with about 20 million euros annually and an additional 15 million euros for the year '20 as well as by sending integrated experts to different international research institutions since 2013. Agricultural research remains irreplaceable to tackle our current but also future challenges. We need to provide healthy food while ensuring the sustainable use of natural resources and develop rural areas remain socially, economically, and ecologically vital. Especially within the CGIAR-framework we are supporting the reform to "One CGIAR" to reach less fragmentation and faster problem- and needs-based research solutions.

The COVID19-pandemic has shown us how vulnerable we are to global crises and we have seen that poorer countries are disproportionally hard hit. BMZ is therefore funding a COVID19 support programme of 1 billion euros through budget restructuring and an additional 3 billion euros for ongoing crisis support. Of these funds 200 million euro of the support programme and 600 million euro of the top-up fund will be invested to stabilise food supply and food systems in developing countries. An effort where also agricultural research can contribute significantly. For example, in a partnership with the AfricaRice Center, a CGIAR-centre, we are providing seeds for 300.000 producers in 6 West African countries. We also support farmers in micro-mechanising their operations by establishing service centres that provide trainings and capital for motorized brush cutters in order to cope with decreased labour availability due to the crisis and ensure effective harvesting with lower losses.

BMZ aims to become more agile and effective which is why we have initiated the BMZ-2030 process. During this process, we want to reshape the way we work by providing more close and targeted development aid to especially reform oriented and global partners. One of the key areas of BMZ engagement will be "One World No Hunger" and international agricultural research will remain a focus on which BMZ will be active within the "Agriculture" field of action.

Contact Address: Sebastian Lesch, Federal Ministry for Economic Cooperation and Development (BMZ), Head, Division 121: Agricultural Policy, Agriculture, Innovation, Berlin, Germany, e-mail: sebastian.lesch@bmz.bund.de

Findings of the CIM – BEAF Side Event

Steffen Entenmann, Ulrich Lepel

Gesellschaft für Internationale Zusammenarbeit (GIZ) Gmbh, Advisory Service on Agricultural Research for Development (BEAF), Germany

As a result of the COVID-19 Pandemic, Agriculture and Agricultural Research are likely to undergo major transformations. How the pandemic has affected Agricultural Research for Development until now, and which strategies could be adopted by the research community to efficiently respond to future challenges were discussed in the annual Meeting of CIM-Integrated Experts and Returning Experts (CIM-BEAF Meeting) between 07.-08.09.2020. The experts are placed by BMZ/GIZ at International Agricultural Research Centers, including the Centers of the CGIAR, the World Vegetable Center and The International Centre of Insect Physiology and Ecology (ICIPE). They work on a broad range of topics, including gender, digitalisation, communication and scaling of agricultural innovations.

Not surprisingly, the experts experience severe restrictions for their fieldwork as a result of the pandemic. Many workflows, however, could still be maintained to some degree due to the increased use of digital tools and information and communications technology (ICT). Such virtual communication formats and digital tools were also considered critical for coping – at least to some extent - with the implications of the COVID-19 crisis on agricultural research in the future. In the session, also other approaches and ideas will be presented that could facilitate Research for Development and thus help global food systems to "sustainably recover" from the crises.

Contact Address: Steffen Entenmann, Gesellschaft für Internationale Zusammenarbeit (GIZ) Gmbh, Advisory Service on Agricultural Research for Development (BEAF), Friedrich-Ebert-Allee 32 + 36, 53113 Bonn, Germany, e-mail: steffen.entenmann@giz.de

ATSAF and GIZ Supporting Early Career Scientists - Junior Scientist Program (JSP) and Upcoming PhD Program ACINAR

Folkard Asch

University of Hohenheim, Inst. of Agric. Sci. in the Tropics (Hans-Ruthenberg-Institute), Germany

With around 350 members ATSAF – Arbeitsgemeinschaft für Tropische und Subtropische Agrarforschung e.V. (Council for Tropical and Subtropical Research) connects key players in German development-oriented agricultural research. This includes professors of German universities with a focus on development-oriented agricultural research and a matching teaching concept, that supports the training of young researchers in this field. Contacts to and stays at International Agricultural Research Centres are invaluable in bringing young scientists early on to practice. The "ATSAF - CGIAR++ Junior Scientist Program' ' financed by BMZ via giz BEAF, supports students in completing their Master's degree and enable them to gain practical experience in project-based scientific work at an International Agricultural Research Centre. For this purpose, ATSAF is, in collaboration with giz BEAF, financially supporting Master students enrolled at German universities in conducting their research to create a Master thesis at one of the CGIAR or IARC centres.
As a next career step ATSAF offers a structured PhD programme at the Academy for International Agricultural research (ACINAR). German University Professors and their CGIAR partners supervise PhD students in a CGIAR Research Program. The student receives a 4 year stipend and a research grant enabling him or her to join an international project and conduct development oriented agricultural research in a multidisciplinary team. The academy provides additional training in three obligatory modules and other training components. This programme is financed by BMZ GIZ for initially five years. Application will be possible from October 2020. More information for both programs at www.atsaf.org

Contact Address: Folkard Asch, University of Hohenheim, Inst. of Agric. Sci. in the Tropics (Hans-Ruthenberg-Institute), Garbenstr. 13, 70599 Stuttgart, Germany, e-mail: fa@uni-hohenheim.de

New Heat Resilient Rice Varieties Addressing Climate Change in South and South-East Asia (IRRI)

NESE SREENIVASULU

International Rice Research Institute (IRRI), Applied Functional Genomics and Phenomics, Philippines

Changing climates with rising temperatures pose a major threat to sustaining rice production in Asia and Africa, and to global food security. High night temperatures (HNT) cause rice grain yields in Asia to decline by 10% for every 1°C (>23 °C) rise in minimum temperature in rice-growing seasons, and these losses are currently widespread. Under HNT, increases in rates of dark respiration create an imbalance in source-sink relationship with reduced carbohydrate supply to developing seeds, thereby reducing grain yield and quality. Through our extensive screening of MAGICheat population, promising lines have been identified from multiple locations and seasons experiments and they are valuable materials for beneficiary, like NARES breeders with at least 1.5 to 2 t ha^{-1}. The identified QTLs are good sources for breeding heat-tolerant lines for HNT hotspot regions. Within a subset of lines, increased Rn was related to lowering spikelet number per panicle and thus reduced yield. HNT enhanced the night-time consumption of non-structural carbohydrates (NSC) in stem tissue, but not in leaves, and stem night-time NSC consumption was negatively correlated with yield. Between heading and harvest, the main form of stem NSC remobilisation was starch, not soluble sugar. HNT weakened the relationship between NSC remobilisation and harvest index at both the phenotypic and genetic level. Through genome wide association studies (GWAS), an invertase inhibitor, MADS box transcription factors and a UDP-glycosyltransferase that were identified as candidate genes orchestrating stem NSC remobilisation prominent under control were lost under HNT. Through identification of physiological and genetic components related to rice HNT response, this study can help specify breeding targets to sustain yield stability under climate change. These outputs will be beneficial to breeders, and farmers in the regions of rice productivity suffering with heat stress. The tolerant lines with clear yield advantage under HNT were characterised for grain quality preferences and identified lines matching local preferences of Myanmar. Based on conservative estimates, the potential production loss avoided with the adoption of HNT tolerant rice varieties in Bangladesh and Myanmar is 1.2M and 1.7 M tons of paddy, respectively (total of 2.9 M tons for the two countries). This is equivalent to 296M USD for Bangladesh and 348M USD for Myanmar.

Contact Address: Nese Sreenivasulu, International Rice Research Institute (IRRI), Applied Functional Genomics and Phenomics, College, 4030 Los Banos, Laguna, Philippines, e-mail: n.sreenivasulu@irri.org

Gender Responsive Circular Economy Innovations for Food and Energy Security of Refugee and Host Communities in East Africa

Solomie Gebrezgabher[1], Mary Njenga[2], Ruth Mendum[3]

[1]*International Water Management Institute (IWMI), Ghana*
[2]*World Agroforestry Centre (ICRAF), Kenya*
[3]*Penn State University, Office of International Programs, College of Agricultural Sciences, United States*

In the East African sub-region with about 3.2 million refugees and nearly 5.8 million internally displaced people, competition over resources such as firewood, fertile land and water is becoming a common source of social tension between refugee and their surrounding host communities. Located mostly in dry landscapes with poor soils, the increasing population pressure accelerates the loss of vegetative cover, erosion and land degradation, making small-scale agricultural production challenging. Lack of access to adequate cooking energy forces refugees to trade and barter food aid so as to acquire firewood from host communities. Women suffer in particular under the dwindling accessibility of firewood. To complement the insufficient aid received, few refugees and host communities are attempting to improve their livelihoods through growing crops and trees but are faced with the challenges of low rainfall and poor soils.

In response to these challenges, a project that applies gender responsive circular economy concept was developed and implemented by the International Water Management Institute (WMI), Agroforestry Centre (ICRAF), International Center for Tropical Agriculture (CIAT) and Penn State University. The project adapts gender responsive solutions that recover energy, soil nutrients and grey water while enhancing the functions of agro-ecological systems and building business models. Gender integration is a key component of the project as refugee communities are disproportionately women and children from many cultural backgrounds. Furthermore, the project involves both refugee and host communities in its implementation of activities as the influx of displaced persons and the engagement of the donor community, has changed the living conditions for host communities in ways that require careful consultation with local leadership to avoid unintended and unwanted consequences.

Keywords: Circular economy, East Africa, gender, refugees, resource recovery and reuse

Contact Address: Solomie Gebrezgabher, International Water Management Institute (IWMI), Accra, Ghana, e-mail: s.gebrezgabher@cgiar.org

BMEL session: Building Evidence to foster Sustainable Food and Forestry Systems

Building Evidence to Foster Sustainable Food and Forestry Systems: Sharing Experiences from Research Projects Funded by the Federal Ministry of Food and Agriculture

Silvia Dietz

Federal Ministry of Food and Agriculture (BMEL), Division 123, Research and Innovation, Coordination of Research Area, Germany

The BMEL is committed to contribute to the design of sustainable and resilient agricultural, forestry, aquatic and food systems worldwide. The BMEL promotes context-oriented and locally adapted approaches and supports closing existing knowledge gaps.

Contributing to the improvement of global food security, the BMEL established the funding instrument "Research Cooperation for Global Food Security and Nutrition". Funded projects within this scheme strengthen the German contribution to agricultural and nutritional research by improving food systems in partner countries and building long-term partnerships between agricultural and nutritional research institutions in Germany, Africa, South and Southeast Asia.

In addition, BMEL supports jointly funded cross-continental projects to build a long-term EU-Africa research partnership on Food and Nutrition Security and Sustainable Agriculture (FNSSA).

To promote reforestation and sustainable utilisation of forestry resources worldwide, BMEL also dedicates funds for international research projects. These projects closely collaborate with local institutions to create situated knowledge and build required capacities to identify sustainable solutions for forest management.

The Federal Office for Agriculture and Food (BLE) acts as project executing agency.

The BMEL-Session will provide insights into the diverse contexts and approaches of selected research projects:

- "Processing of edible insects for improved nutrition (ProciNut)", Center for Development Research, University of Bonn
- "Small Fish and Food Security: Towards innovative integration of fish in African food systems to improve nutrition (SmallFishFood)", German Federal Institute for Risk Assessment (BfR)
- "Multifunctionality of mixed agroforestry systems in Kalimantan – Improving data basis and solutions for sustainable biodiversity conserva-

Contact Address: Katharine Tröger, Federal Office for Agriculture and Food (BLE), International Cooperation and Global Food Security (334), Deichmanns Aue 29, 53179 Bonn, Germany, e-mail: Katharine.Troeger@ble.de

tion and rural development (AgroforstKalimantan)", Borneo Orangutan Survival Deutschland e.V., Fairventures Worldwide

The session aims to illustrate the contribution of applied research projects towards strengthening the resilience of agricultural, forestry, aquatic and food systems with respect to multiple risks. In addition to the presentations on three selected projects, the session will provide the opportunity for an enriching exchange.

Keywords: Food and nutrition security, global South, international research cooperation, international sustainable forest management

Practice-oriented further education and support for the transfer of business ideas in the Food Value Chains in Africa

---

Oral Presentations

Posters

Analyses of Value Chains in Food Production in Africa to Create Start up Ideas

Kateryna Tuzhyk, Vadym Petrenko, Bernd Müller
Weihenstephan-Triesdorf University of Applied Sciences, Germany

A deep understanding of the value chain's in agriculture and food production is the basis for entrepreneurial activities and so for income increase and improvement of rural livelihoods in Africa. Success or failure of development in the agricultural sector depends often on the development of strong and effective value chains for the agricultural products. It is not enough to produce agricultural goods but to develop the whole value chain from the input supply over production and processing up to the retail and consumption in order to increase income of farm families.

In the past, the scientific focus was often given to single elements of the value chains such as soil, pesticide use or fertilising. Understanding the holistic concept of value chains allows us to identify weak points of a value chain and to overcome the relevant bottlenecks. This leads to optimised resource utilisation and thus income increase. Furthermore, it opens opportunities to identify business start-ups that can be further developed.

The presentation deals with the example of such start up ideas created from participants of the Postgraduate Course Food Chains in Agriculture at the Weihenstephan-Triesdorf University of Applied Sciences. The elements of a value chain and their interaction under the holistic concept are shown. The process of elaborating a study project from idea over deep problem analysis and suggestion of a cost efficient solution up to implementation in practice is depicted. A focus is also given to the economic analysis and interpretation of the meaning of the single steps. Pros and cons of the approach as well as challenges for the realisation of the start-ups are analysed and discussed in more detail.

The realised start-up gives the opportunity to increase income, create employment in rural areas of African countries, and so improve the living standards of people.

Keywords: Food, start-up, value chains

Contact Address: Kateryna Tuzhyk, Weihenstephan-Triesdorf University of Applied Sciences, Applied agricultural management, Steingruberstraße 1a, 91746 Weidenbach, Germany, e-mail: kateryna.tuzhyk@hswt.de

Improving Cereals and Legumes Productivity

Frédéric le Roi Abalo[1], Yazachew Genet Ejigu[2], Mezgebu Aynalem[3], Tsion Fikre Mulabachew[4], Nguemo Jennifer Achimba[5]

[1]*Animal Production and Fishery Division at National Institute of Agricultural Extension (ICAT), Togo*

[2]*Debre Zeit Acgricultural Research Centre, Ethiopian Institute of Agricultural Research, Ethiopia*

[3]*Debre Markos University, Ethiopia*

[4]*Debre Zeit Acgricultural Research Centre, Ethiopian Institute of Agricultural Research, Ethiopia*

[5]*GIZ AFC, Nigeria*

Although roughly 60 % to 70 % of the continent's population is engaged in agriculture, its production cannot feed all Africans. The population of Africa will double in the next centuries to reach 2.5 billion by 2050. Hence, in 2013 alone, Africa imported 56.5 million tons of wheat, maize, and soybean at the cost of 18.8 billion USD. Crops production in Africa play a vital role in their contribution to food security. However, yields are inferior compared to those in other parts of the world. For instance, yield gaps for maize in the different production situations for four selected locations reveal 76 % in Ethiopia, 34 % in China and 48 % in USA. Low productivity in Africa is also related to poor soil fertility and scarce moisture, as well as a variety of insect pests, diseases, and weeds. While moisture scarcity is responsible for up to 60 % of yield losses in some African staple cereals, insect pests inflict annually substantial crop losses. Weather extremes and scare moisture require more resilient farming systems, and promoting diverse types of cereal and legume crops will help to buffer farming systems. However, improved crop varieties alone will not boost crop productivity unless supplemented with optimum soil, water, and plant management practices as well as the promotion of policies pertaining to inputs, credit, extension, and marketing. Furthermore, biochar and African Dark Earths have been found to improve soil properties and to enhance productivity besides to traditional fertilisers, although their availability and affordability to African farmers remains to be explored. The concept of Integrated Soil Fertility Management (ISFM) has been successfully implemented in some African countries in the Great Lake Region. Some other innovative technologies favourably accepted by farmers are the "Push-pull System" and the System of rice Intensification (SRI) were successfully developed. Our presentation gives some information on the situation of cereals and legumes productivity in Africa in general as well as some strategies developed by stakeholders on field in order to positively impact on productivity and ultimately improves the livelihood of farmers.

Keywords: Agriculture, cereals, legumes

Contact Address: Frédéric le Roi Abalo, Animal Production and Fishery Division at National Institute of Agricultural Extension (ICAT), Lomé, Togo, e-mail: frediabalo@gmail.com

Contribution of Dairy for Food and Nutrition Security

Cleopatra Nawa Kawanga[1], Lovemore Mtsitsi[2], Obright Hamungalu[3]

[1]*Indaba Agricultural Research Institute (IAPRI), Agriculture research and Development, Zambia*
[2]*Civil Society Agriculture Network (CISANET), Malawi*
[3]*District Livestock Production Officer, Zambia*

The global burden to malnutrition is unacceptably high, one in five children is stunted, and 15.9 m children under the age of five are stunted and wasted. Malnutrition is found worldwide and is linked, either directly or indirectly, to major causes of death and disability. More than one third of all child deaths are attributable to under nutrition. Many low- and middle-income countries, particularly in Africa, have not achieved significant reductions in underweight, stunting or vitamin and mineral malnutrition. The global nutrition targets for 2025 are (1) 40 % reduction in stunting in children who are under five, (2) 50 % reduction of anemia in women of reproductive age (15–49 years), and (3) 30 % reduction in low birth weight. The prevalence of anemia in women who are between 15–49 years of age was 32.8 % in 2014. Diets low in nutritious foods is a leading cause of healthy life years lost. Food security: is the situation in which all people at all times have physical, social and economic access to sufficient safe, nutritious food to meet their dietary needs and food preferences for an active and healthy life. The four pillars of food security are availability, access, utilisation and stability. Nutrition security: a situation in which food security is combined with a clean environment, adequate health services, and appropriate care and feeding practices, to ensure a healthy life for all household members. All dairy products belong to the milk pyramid. It is our most nearly perfect food as no other single food can substitute for milk in diet and give a person the same nutrients that you get from a glass of milk. It provides for a steady supply of essential micro and macronutrients (carbohydrates, proteins, fat, vitamins and minerals). Dairy is critical to the nutrition, food security, livelihoods and resilience of hundreds of millions of people throughout the world. It contributes to food and nutrition security through: directly (by improving household diet through increasing access to animal source foods) and indirectly (by improving income and ability to purchase more diverse foods).

Keywords: Agriculture, dairy, nutrition security

Contact Address: Cleopatra Nawa Kawanga, Indaba Agricultural Research Institute (IAPRI), Agriculture research and Development, 26A Middleway Road, Kabulonga, 10101 Lusaka, Zambia, e-mail: cleo2015nawa@gmail.com

Contribution of African Women Entrepreneur in the Development of the Agriculture and Food Industry

Yawa Minnie Elodie Folly[1], Emmanuela Soro[2], Halatou Dem[3], Yousra Soua[4], Mariem Guizeni[5]

[1]*Adili African Vegetables Processing Unit, Entrepreneur, Togo*

[2]*Labo Consulting Afrique, Ivory Coast*

[3]*Tatam Cereales, Mali*

[4]*own start-up Development in Biological Control Sector, Tunisia*

[5]*own start-up Development Potato Processing, Tunisia*

For centuries, agriculture has driven economic growth in countries across the globe, and African nations are following the same path out of poverty. Agribusiness plays a vital role in economic development, contributing a major portion of GDP, employment, and foreign exchange earnings in Africa, where agriculture accounts for 25 percent of the continent's GDP, and 70 percent of employment. Women constitute nearly half of the workforce in this sector in sub-Saharan Africa. They perform most of the production and local marketing of food. However, when it comes to developing research programs, making decisions or leading agricultural development and food industry development work, women are clearly under-represented. It must therefore be noted that only half of the team is actually present in the sector due to difficulties in accessing funding, limited training, and gender disparities in intellectual property, the low percentage of women in science and technology across the continent. Women have so much to bring to the table. By unlocking their potential and utilising women's talents to the full, we will be better placed to leverage science, technology and innovation to solve the chronic and pressing problems facing the continent, particularly with respect to food security.

Keywords: Agribusiness, agriculture, food security, women

Contact Address: Yawa Minnie Elodie Folly, Adili African Vegetables Processing Unit, Entrepreneur, Lomé, Togo, e-mail: minnielodie@hotmail.fr

Promoting Women and Youth in Agriculture for Sustainable Livelihoods

Gift Chitambala[1], Rashid Mpingagjira[2], Joshua Apochi Apochi[3], Ramla Baaba Keelson[4], Salem Hadhami[5]

[1]*Westwood Agri Limited, Zambia*
[2]*Women Economic Empowerment Project, Malawi*
[3]*Financial Cycle GIZ-AFC-AGFIN, Nigeria*
[4]*JKBA Farms, Ghana*
[5]*Plantix (Cooperation with GIZ and start-up PEAT), Tunisia*

Young people and women are increasingly linked to targeted agriculture and food security interventions. In Africa, the argument is that the combination of agricultural value chains, technology and entrepreneurship will unlock a sweet spot for youth and women employment.

Therefore, the main question which begs answers is as follows: which sector provides the most potential to reduce poverty among the youth and women? The agricultural sector seems to provide the answer to this question, despite the increasing evidence suggesting that the sector is not attractive to young people and women.

With the looming employment challenge in Africa, exciting agribusiness ideas as well as policy directions on how the sector can be made attractive to the youth and women remain crucial. Although there is overwhelming empirical evidence on youth employment in agriculture across the globe in general, evidence in Africa seems to be sparse especially with youth and women engaged in agricultural related livelihoods. Generally, while there is an increasing number of studies and interest on youth and employment studies in Africa, very little is known about sustainable livelihoods of youth and women employment in agriculture. Further, sustainable project ideas coming from the youth themselves as well as aspirations existing in agriculture are not well established.

It is against this background that, Salem, Ramla, Josh, Rashid and Gift from Tunisia, Ghana, Nigeria, Malawi and Zambia respectively wish to present their project ideas around their common title which focuses on promoting the youth and women involvement in the agricultural sector for sustainable livelihoods. The following are the project ideas;

- AMAL women cooperative for goat farming (Salem - Tunisia)
- Changing the perception of youth about agriculture through educational strategies and vegetable production (Ramla - Ghana)

Contact Address: Gift Chitambala, Westwood Agri Limited, Lusaka, Zambia, e-mail: gift.chitambala@hotmail.com

• Improving agricultural extension by formation of private extension service of sustainability of agricultural innovations (Josh -Nigeria)
• Improving the role of women in groundnuts value chain (Rashid - Malawi)
• Potential of Agricultural Smart Hub value chains to improve youth unemployment (Gift - Zambia)
These project ideas will therefore centre on promoting women and youth in Agricultural activities for sustainable livelihoods.

Keywords: Agriculture, education, women, youth

Technical-Economic Reference of an Agroecological Model at Togo

Aménenyon Akakpo

NGO LA Colombe, Togo

In order to improve the socioeconomic conditions of vulnerable people, the NGO LA COLOMBE (TOGO) initiated the "agroecological project for the benefit of young girls, young men and women" based on an innovative agricultural model called: "the mixed and tiered agroecological model". It is a model of innovative agricultural practices that responds to the problems of inaccessibility of agricultural land to vulnerable groups, especially women and young people. It is an answer to the questions of agroecological transition at a time when resources are becoming increasingly scarce and conventional practices continue to amplify the degradation of the environment and the living environment of living beings. It constitutes an integrated set of fish farming, breeding and organic market gardening. The system induces a stable and sustainable ecosystem dynamic and allows good sustainable productivity of its ecosystem. This technical-economic reference framework for the model will enable partners to get to know the model well and to make good financing decisions and to be able to assess the impacts on beneficiaries and on the environment. It emerges from this document that the investment costs are relatively high enough and are not within the reach of the targets (the vulnerable layers) of the project. The activity of the project is very profitable from an economic, environmental and social point of view. The net operating results are encouraging.

Keywords: Agriculture, agroecological model, improving socio-economic conditions

Contact Address: Aménenyon Akakpo, NGO LA Colombe, Lomé, Togo, e-mail: akakpoamenyon@gmail.com

Upscaling of Baobab Value Chain in Malawi – A Production and Processing Using Modern Technologies

Hagience K. Phiri
Lilongwe University of Agriculture and Natural Resources, Agribusiness Incubation Center, Malawi

Agriculture has been recognised as the main contributing sector to the GDP amongst most of the African economies. To enhance Agricultural growth, an in-depth understanding of value chains in agriculture and food production is a yardstick for Agribusiness entrepreneurial activities and so for increase in income and improvement of rural livelihoods. This presentation focuses on value chain in the fruit sector with special emphasis on baobab fruit tree in Malawi.
Baobab is one of the fruit trees that grow naturally along the shores of Lake Malawi and Shire River. It is used for food, medicine and income. There is growing demand for baobab products both domestically and internationally but the country's output does not meet the market demand. Low production of baobab is attributed to low levels of technology, poor infrastructure, lack of market information, limited processing knowledge, lack of capital and lack of government support. To address the demand gap, there is need for intervention to upscale production and processing through the adoption of modern technologies through establishment of innovative baobab value chain social enterprise called BaoFoods initiative in Malawi.
The overall objective of this study project idea is to upscale baobab value chain through production and processing using modern technologies. To realise this project idea in practice, the project will follow the following implementation steps; 1) idea validation 2) planning process 3) partnerships 4) location for the baobab value chain social enterprise (BaoFoods Initiative) 5) idea realisation/launch in practice.
The realisation of this study project idea, will give an opportunity to increase income, create employment among the youth and women along the lake shore areas, and hence improve the quality of life of the people.

Keywords: Baobab value chain, social enterprise, upscale

Contact Address: Hagience K. Phiri, Lilongwe University of Agriculture and Natural Resources, Agribusiness Incubation Center, P.O. Box 219, Lilongwe, Malawi, e-mail: hagiencephiri@gmail.com

Improving Honey Production and Marketing: an Application of Outgrower Beekeeping Model in Rural Communities

Solomon Akamiti
Soteria Farms Limited, Ghana

Global honey consumption is projected to exceed $2.8billion by 2024. Despite the increasing demand for honey and its related products, Africa contributes only 7.8 % of global honey market. Ghana's vegetation is favourable for all year beekeeping and honey production but the sector is challenged by the lack of appropriate inputs and equipment, bush fires, lack of financial support, lack of training for beekeepers, adulteration, fear of bees, among others. Despite the numerous challenges faced by the sub-sector, the Federation of Ghana Beekeepers Association (FGBA) estimates Ghana's honey production potential to be at 500,000 tons per annum and worth about $1.5 billion. Despite its great potentials, the beekeeping sub-sector in Ghana has over the years failed to attract adequate public sector support and has largely remained a hobby among a few interested farmers. This can be largely attributed to the general lack of information on the sub-sector which poses a challenge in demonstrating the true opportunities in the industry. This project was aims at proposing and establishing an outgrower beekeeping business model that can be implement in some selected rural communities in Ghana by 2022. The proposed outgrower beekeeping model focuses on promoting community-based beekeeping system that supports women empowerment and youth development. The model seeks to create community-based honey production centres that will be coordinate the collection and processing of honey for onward market distribution. The project concluded that beekeeping appears to be a profitable and sustainable business enterprise which has great protentional. The proposed outgrower beekeeping model suggests to be the best way through which Ghana can increase its honey production whilst improving the livelihoods of farmers and youth in rural communities. This project does not only come with great financial rewards to stakeholders but also social and environmental benefits.

Keywords: Beekeeping, community, development, Ghana, honey, model, outgrower, youth

Contact Address: Solomon Akamiti, Soteria Farms Limited, Accra, Ghana, e-mail: solo9mon@yahoo.com

Improving Soybean Cake Trading through Avian Production

Frederic Le Roi Abalo

Animal Production and and Fishery Division At National Institute of Agricultural Extension (ICAT), Togo

Togo is not a big producer of soy bean in terms of quantity but is the first exporter of organic soybean in West Africa with 22 000 tons of product exported in 2019. The exported organic soybean is mainly for organic animal production sector in Europe and North America. People started now to collect and export conventional soybean. In 2015 the sector brought in about 4 billion XOF. The production is estimated at 78 000 tons in 2019. The sector employed almost 300,000 people. Although exportations provide currencies, it does neither add value nor create lot of employment. Soy bean processing in soybean cake and roasted soybean represent a real potential for Togolese entrepreneurs with a big market related to poultry sector. The poultry sector constitute a big market for soybean processing sector and its by-products. In fact the number of laying hens increased from 229,720 in 2005 to 385,743 laying hens in 2010 (a 40 % increase). We are importing a huge amount of poultry meat every year (almost 18 000 tons). Our local producers are not able to compete with European ones because of lack of competitiveness due to high feed cost. Soybean cake (a by-product) is cheaper and not competing with other food sector. It will surely help reduce feed prices and hence reduce poultry meat and egg production cost for better profitability. Although that important role soybean cake utilisation in feed could play in our poultry sector and the relatively small amount produced locally, actors are facing some issues to sell this by-product. Marketing issues are somehow due to lack of nutritional value analysis facilities, poor packaging without labelling, lack of awareness on poultry sector side, lack of organisation, etc. Our study project aims to improve soybean cake marketing for more income and employment both at processors and farmers side. Our action will be in business idea aspect to collect, analyse and package in an appropriate way soybean cake produced, provide services in profiling the product and extensively feeds, provide services in feed formulation and do some facilitation in actors organisation as well as workshops, etc.

Keywords: Poultry production, soybeans, Togo

Contact Address: Frederic Le Roi Abalo, Animal Production and and Fishery Division At National Institute of Agricultural Extension (ICAT), Lomé, Togo, e-mail: frediabalo@gmail.com

Enhancing Wheat Production and Productivity through an Integrated Agricultural Technologies

Mezgebu Aynalem

Debre Markos University, Ethiopia

Ethiopia is one of developing countries in the world with subsistence agrarian economy. Considerably large number of population is leaving under poverty line. Now a day, among the world's agenda, namely climate change, food security and hunger becoming the leading one. In the horn of Africa particularly in Ethiopia, the societies particularly those areas with shortage of land, poor infrastructures and seasonal migrated people with poor feeding habit as well as unproductive young labour wasted throughout the year attract a great attention worldwide.

Wheat is a cereal crop, which is produced in most parts of the country, Ethiopia. Ethiopia is the second next to Egypt in Africa in wheat production which is 4.54 million tons in 2016. The production of wheat in the country is increasing, with fluctuation, from 2 314 489 in 2008 to 4 537 852 tons in 2016 with the average growth rate of 0.094 %, so this increment is not this much.

In Ethiopia, wheat grain is used in the preparation of a range of traditional food products such as the traditional staple pancake ("*injera*"), fermented bread ("*dabo*"), non-fermented bread ("*hambasha/kitta*"), porridge ("*genfo*"), local fermented beer ("*tella*"), distilled local spirit("*areki*"), etc. This shows that wheat is an important market oriented commodity and a major source of income for many wheat growers in Ethiopia; it is crucial for improving their way of life through and food security of developing countries.

Despite of its all importance, there are many problems that hinder the production of wheat in the country. From these challenges lack of market information, low quality of inputs used, weak market linkage, weak use of technology, low bargaining power of producers and post harvest loss etc .

The main goal of this project is to enhance wheat production & productivity through integrated agricultural technologies in Debre Elias District, Ethiopia. It required a total of € 76 620.00.

Keywords: Ethiopia, integrated agricultural technologies, wheat

Contact Address: Mezgebu Aynalem, Debre Markos University, Debre Markos, Ethiopia, e-mail: mezgebu12aynalem@gmail.com

Income and Employment Opportunities through Cashew Apples Processing in Ghana

George Opare Asare
GM Farm and Consultancy, Ghana

Most of the world's cashew harvest comes from Africa. In, 2018, Africa produced 57 % of the world's volume of cashew which translates to over 15 million MT of cashew apples which is mainly not utilised. The income levels of most cashew farmers tend to be low. Notwithstanding, the potential for creating jobs, achieving higher incomes, improving food and nutrition from growing and processing cashew by-products has not yet been fully harnessed in Africa. Cashew apples, have high nutritional value (five times the vitamin C in oranges) which can be processed into products such as juices, candies, marmalade, jams, jellies etc. but Ghanaian farmers do not have any commercial use for cashew apples. It is anticipated that the crop will increase from its current 90,000 to 400,000 metric tonnes in 10 years meaning more cashew apples will go to waste. The cashew value chain in Ghana is mainly about producing raw cashew nuts (RCN) and exporting to overseas. Only a small percentage of the RCN is processed and marketed within Ghana. Cashew apples is virtually non-existence in the cashew value chain in Ghana. The upgrading of cashew value chain in Ghana with the introduction of cashew apple products will increase the incomes of cashew farmers and women entrepreneurs who would be trained and help to setup up cottage industries for cashew processing, create employment and also consumers will enjoy the high nutritious content of cashew apples. Processing the cashew apples can provide a platform for the economic empowerment of women especially in rural areas. A social enterprise would be established to train women on technical and entrepreneurial skills in processing cashew apples into jams, juice, marmalade and candies. The women would be supported with capital (equipment for processing) through donor funding. The social enterprise will also establish a profit-making processing unit for cashew apples.

Keywords: Cashew apples, Ghana, women entrepreneurs

Contact Address: George Opare Asare, GM Farm and Consultancy, Accra, Ghana, e-mail: goasare@gmail.com

Green People's Energy for Africa: Domestic Biogas to Improve Rural Livelihood in Ethiopia

Hiwot Abayneh Ayele
Hawassa University, Agribusiness and Value Chain Management, Ethiopia

A sustainable rural energy supply is one of the ways to improve rural livelihood as well as to reduce environmental problems. Ethiopia experiences energy and environmental crisis due to sustained reliance on woody biomass to satisfy its energy requirements. The situation gets worse in rural areas where access to modern energy is very low. The country in general has one of the lowest rates of access to modern energy sources, whereby the energy supply is primarily based on biomass. To address such energy related problems, Hawassa University signed prject agreement entitled Green People's Energy: Domestic Biogas to improve rural livelihoods in Ethiopia. The small project aims to improve the rural livelihood through construction of biogas plant in Damot Woyde and Dugna Fango districts in Southern Ethiopia. Target smallholder farmers will have access to construct the technology at their backyard which will be used for lighting and cooking. Moreover, the by-product (slurry) can be used as organic fertiliser. The benefit of the technology in general includes improved health, reduced workload for women and children, improved education, prevent deforestation, improved soil fertility, job creation for local masons and promote rural youth entreprenership. Currently, the project is being implemented by Hawassa University in collaboration with Water, Mines and Energy Bureau and GIZ/EnDev. The project also envisages to bring several professionals together from the University, RBPCU, GIZ/EnDev, TVET and other stakeholder. This will be instrumental to enhance synergy and avoid duplication of efforts and inefficient utilisation of resources. The project follows a participatory monitoring and evaluation approach to ensure the effectiveness, ownership and sustainability of project interventions. Besides this, periodic reports and success stories will be produced to share implementation progresses, challenges, knowledge and best practices.

Keywords: Domestic biogas, energy supply, Ethiopia, rural livelihood

Contact Address: Hiwot Abayneh Ayele, Hawassa University, Agribusiness and Value Chain Management, Hawassa, Ethiopia, e-mail: biscut98@gmail.com

Implementation of A Processing Unit of Cashew into Spreading Cream

Emmanuela Soro
National Polytechnical Institute Houphouet Boigny, Ivory Coast

Côte d'Ivoire is a country whose economy is mainly based on agriculture. Indeed, the agricultural sector represents 28 % of the GDP, 40 % of the country's exports, employs 46 % of the active population and provides a living for two thirds of the population.

However, it must be noted that most agricultural products are sold or exported as raw materials. They are not processed or are only slightly processed. Most of the time, our country don't add value to its agricultural products. In fact, agricultural policy is more focused on the export of raw materials than on processing, which leads to a leakage of value.

This low level of processing and conservation issues also makes the country dependent on exports, and most of the time producers sell products are bought at discount because they fear losses if the products remain in their hands for a long time.

Cashew value chain is not spared. Although Côte d'Ivoire is the world's leading cashew producer with about 900,000 tons, only about 9 % is processed. In 2020, the sector has suffered a loss of about 300 million dollars.

In addition to its excellent organoleptic characteristics, the cashew nut has interesting nutritional virtues and contains active ingredients that give it health benefits. As a result, we believe that the potential of this high potential crop is not sufficiently exploited in Côte d'Ivoire.

One approach to consider is to improve both qualitatively and quantitatively the level of processing. The project we propose is the establishment of a unit for processing cashew into a spreading cream that will be sold both on the local and international markets.

Keywords: Cashew, Côte d'Ivoire, processing

Contact Address: Emmanuela Soro, National Polytechnical Institute Houphouet Boigny, Abidjan, Ivory Coast, e-mail: emmanuelasoro@gmail.com

Establishing New Start-up "*oleacare*" Producing Parasitoid for Olive Fruit Fly Pest Management in Tunisia

Yousra Soua
Own Start-up Development in Biological Control Sector, Tunisia

Tunisia is the most important southern Mediterranean country in the field of olive production and export after the European Union. Actually, for Tunisia, olive growing is the main agricultural activity and its socio-economic role is of paramount importance. Despite this fairly comfortable place on the international market, unit prices of olive oil are largely unfavourable to Tunisia, compared to its main competitors. In fact, residues of pesticides in Tunisian olive-oil are the main obstacle to the creation of value in the export sector.
Tunisian olive producers are using extensively pesticides to control the most serious olive fruit fly pest (*Bactrocera oleae*). The oviposition activity of the females as well as larval feeding are responsible for the destruction of olive fruits. As a result, infested olives completely lose their market value for table consumption and oil production.
In order to enhance the Tunisian organic olive-oil to its full potential on the international market, the rationalisation of the use of pesticides as well as their substitution by eco-friendly method of olive fruit fly control remain OBLIGATORY!
In this context, the aim of the start-up "*oleacare*" is to offer alternative solution to pesticides and to contribute to the development of organic olive-oil production sector in Tunisia. The start-up idea is based on the principle of Biological control, and its main activity of is the mass-rearing of *Eupelmus urozonus* a parasitoid that can accomplish goals of a biological control programme with an acceptable cost / benefit ratio.
The start-up idea is supported by the favourable government policies encouraging investment in innovative technologies for alternatives to pesticides, as well as its commitment to promote biological control for olive tree pest management.

Keywords: Biological control, olive fruit fly (*Bactrocera oleae*), parasitoid (*Eupelmus urozonus*), tunisian organic olive-oil

Contact Address: Yousra Soua, Own Start-up Development in Biological Control Sector, 4180 Djerba, Tunisia, e-mail: souayosra@gmail.com

Build a Sustainable Inclusive Business Model for Rural Communities in Mali through Biogas Technology

Famoussa Dembele
Weihenstephan-Triesdorf University of Applied Sciences, Alumni, Ghana

Population growth and urbanisation, along with changes in lifestyle and consumption, lead to large quantities of solid and liquid organic waste from agricultural, agro-industrial and urban activities. In the absence of an adequate waste management system (value chain), these can cause harm to human health and the environment. Biogas technologies are unique among renewable energy forms in that they can address several challenges in Mali in an integrated manner. They enhance the connections and potential synergies between sectors. However its adoption in the rural and per-urban areas has been very low or inexistent. Most of the biogas systems in Mali have been installed through projects and as at now about 80 % of them are not functioning, the reason for that could be:
• Poor technical service: system not well design to suit the beneficiary need;
• Under feeding of input: restriction of beneficiaries to use only one type feedstock like cow dung while there are other organic waste available on site (food waste, vegetable waste ...etc.).
Looking at the failure of the first pilot projects, many people in rural and per-urban areas of Mali (Southern and Central parts) have adequate and required materials (organic wastes and water) and would like to install bio-digesters on their farms or compound, but they are sceptical about the benefit of the technology because:
• Poor advertisement of the benefit of Biogas: people are not aware of the benefit and importance of having a biogas system. They do not know the business aspect of it which can help them have more revenue streams on your farm (selling of biogas, bio-fertiliser) which save money.
• Require an upfront investment to build the system: most people don't think as an Agripreneur and for that they are not able to create other businesses around the biogas technology as mentioned earlier. The system durability (fixed dome bio-digester) is at least 15 years, with a payback period of 3–5 years when used efficiently.

Keywords: Biogas, Mali, rural communities

Contact Address: Famoussa Dembele, Weihenstephan-Triesdorf University of Applied Sciences, Alumni, Kumasi, Ghana, e-mail: famoussadembele@hotmail.com

Improving Agricultural Extension Service by Formation of Private Extension Group for Sustainability of Innovations

Joshua Apochi Apochi
Financial Cycle GIZ-AFC-AGFIN, Nigeria

Agricultural extension and advisory services plays a major role in Agricultural development. In Nigeria, agricultural extension services have been primarily provided by the public sector, most notably the Agricultural Development Programmes (ADPs) funded through the World Bank in the 1970s. However, the withdrawer of the funding and the subsequent difficulty of government to solely finance the sector has been having a noticeable effect on the inability of the public extension bodies to perform efficiently due to insufficient funding at the state level, inadequate availability of inputs for demo plots, poor logistic support and inadequate staffing. A reasonable number of public extension staff retires annually without replacement or recruitment in the last two decades, leading to the close down of some Agricultural Development Programmes (ADP) offices and some dilapidated buildings. This study project is targeted at creating job opportunities for Agricultural graduate by formation of private extension group / advisors for sustainability of innovations to fill the gap provided by public extension service providers. The private extension agents after undergoing rigorous capacity building, through Training of Trainers (ToT) will work with private agro processing companies in mobilising farmers, support small scale farmers by providing timely advice and link to market, seed and fertiliser companies NGOs in carrying out training and monitoring demo plots. The targeted agents are mainly youths who are already familiar with extension principles and adult learning techniques, therefore adoption and sustainability of innovations will be possible as knowledge transfer will be easy as the agents will closely monitor the farms.

Keywords: Nigeria, private extension services, sustainability

Contact Address: Joshua Apochi Apochi, Financial Cycle GIZ-AFC-AGFIN, 12 Jato Aka Street Itf Road Wurukum, 970211 Makurdi, Nigeria, e-mail: apochijosh@gmail.com

Creation of Agricultural Smart Hub Value Chains to Improve Youth Unemployement in Zambia

Gift Chitambala

Westwood Agri Limited, Zambia

Zambia is a country rich with vast natural resources together with the youthful population that is ready to contribute to national economic development. The country has enough resources to sufficiently produce enough food for national consumption but unfortunately 30 percent of this national output is lost to post harvest losses due to poor developed local value chains. Subsequently, the country spends significant part of the foreign reserves to import finished foods for local consumption at the expense of depreciating local currency.

This project idea will therefore aim at creating agricultural smart hub value chains to improve youth's unemployment in Zambia as shown on the poster. Agricultural smart hub project will create agribusiness hubs in all ten provinces which will be a coworking processing space utilising the locally available raw materials using the machinery in the hub to turn them into finishing products. The agribusiness hubs will utilise the comparative advantage of each of the ten provinces to determine the type of machinery to install in the hub and create a sustainable supply of required raw materials for the hub. For elastration purposes the hub in Copperbelt province will have machineries for processing of Avocado oil/Paste, Fruit juices, Tomato paste and other products due to already locally available fruit trees and horticultural farmers in the province. The processing side of the hub will then process the raw materials into finished products at the same time coordinating with the in-house digitalized IT section to brand and market the finished products.

The Agricultural smart hub project will be a coworking Centre / hub that will setup a network of experts, customers, and agro suppliers to work together with a group of autonomous youth comprising of the graduates and rural youth to collaborate in three economic sectors namely agricultural production, agro processing & technology (ICT) to solve the food challenges. This will be done in order to co-create finished goods and services that will shift Zambia towards the culture of producing and processing its own food by making use of the available local agricultural raw materials and labour through shared value and collaboration.

Keywords: Agricultural smart-hub, value chains, youth employment, Zambia

Contact Address: Gift Chitambala, Westwood Agri Limited, Lusaka, Zambia, e-mail: gift.chitambala@hotmail.com

The Contribution of a Small Scale African Leafy-Vegetables Processing Unit for a More Resilient Future of Women Gardeners in Togo

Yawa Minnie Elodie Folly

Adili African Vegetables Processing Unit, Entrepreneur, Togo

African leafy vegetables (ALVs) are outstanding sources of micronutrients in the world. This makes them helpful in alleviating micronutrient deficiencies common in sub Saharan Africa. The increasing consumption of ALV across Africa is because people have been recently aware health protecting attributed to the compounds they have (vitamin A, calcium, manganese, magnesium and iron).

African leafy vegetables are underutilised in Togo because the gardeners are deficient in the strategies on processing excess produce. Of a production of 40,000 tons of vegetables, almost half has been lost in 2019 (Institut National de la Statistique et des Études Économiques et Démographiques, Togo).

With restricted movement in most west African countries and inside of the country because of COVID-19, many markets for fresh food are closed. This impacts more gardeners, which has difficulties to sell their products and are even force to throw the ALVs due to their perishable nature. Looking at women gardeners more closely become a necessity, due to their principal role in the production of these leaves and the poverty situation generated by the crisis.

Within the frame of a contribution that seeks to ensure a stable revenue to women gardeners and reduce their post-harvest losses, our project aims at developing feasible small-scale ALVs production with improved product quality and expanding marketing possibilities together with a women's cooperative in Togo. From June to September, we were able to work just with two women products and process, in a simply furnished room, their two first harvests (approximately 100 kg of amaranths and 100 kg of jute leaves) for sales as precooked frozen vegetables.

The next planned step is to put the focus on selected cooperative as well as the invest in a processing unit, the technical and financial support, and conduct additional marketing analysis.

Keywords: African leafy-vegetable, food processing, gardeners, resilience, women

Contact Address: Yawa Minnie Elodie Folly, Adili African Vegetables Processing Unit, Entrepreneur, Lomé, Togo, e-mail: minnielodie@hotmail.fr

Modern Rice Processing Techniques Using Solar Dryer and New Packaging in Togo

Ruth Atchoglo

University, Agronomy, Togo

Rice is among those grains that are most consumed in Togo and can be ranked as the third staple food after corn and sorghum. Consumer preferences and demand indices are increasing in both rural and urban areas. As a commodity food, rice is now part of the daily menu of Togolese. Togo imports nearly 50 percent of domestic consumption (150 000 tons for 2017–2018) in order to satisfy rice demand of the population, despite this potential of the country to produce more rice than it needs. In terms of value, the cost of imported rice in 2013 was 20 045 520 Euros. This high dependence on imported rice is due to the low production and the high post-harvest losses of rice. Togo loses up to 40 % of harvested rice, as paddy is spread out to dry with the sun on the soil, on roads or on unprotected spaces. If the post-harvest can be reduced up to almost zero; Togo has to import only 10 % of the rice instead of 50 %. The main objective of this project is to improve the performance of the rice sector in Mission Tové by using innovative processing technology, the Solar Bubble Dryer™ to increase the quantity and quality of the rice produced in Togo. A new created enterprise will work in partnership with farmers, machine suppliers, technical and financial partners. The rice will be dried in good conditions, milled and then packed into 1 kg, 2 kg and 5 kg bags. This rice will be directly reachable in the local market (supermarkets, restaurants, baby food companies, tourist places or similar) through good marketing at reasonable price. By signing a contract between rice producers and the enterprise, the farmers will be encouraged to produce more and to have a sure market for their product. Then, their financial conditions will be improved.

Keywords: Bubble dryer, post-harvest loss, rice, Togo

Contact Address: Ruth Atchoglo, University, Agronomy, Bè-Kpota Au Chalet, Lomé, Togo, e-mail: ruthatcho@hotmail.fr

Index of Authors

www.ingramcontent.com/pod-product-compliance
Ingram Content Group UK Ltd.
Pitfield, Milton Keynes, MK11 3LW, UK
UKHW022002190726
13853UKWH00004B/1674